Monographs on Pathology of Laboratory Animals

Sponsored by the
International Life Sciences Institute

The following volumes have appeared so far

Endocrine System
1983. 346 figures. XV, 366 pages. ISBN 3-540-11677-X

Respiratory System
1985. 279 figures. XV, 240 pages. ISBN 3-540-13521-9

Digestive System
1985. 352 figures. XVIII, 386 pages. ISBN 3-540-15815-4

Urinary System
1986. 362 figures. XVIII, 405 pages. ISBN 3-540-16591-6

Genital System
1987. 340 figures. XVII, 304 pages. ISBN 3-540-17604-7

Nervous System
1988. 242 figures. XVI, 233 pages. ISBN 3-540-19416-9

Integument and Mammary Glands
1989. 468 figures. XI, 347 pages. ISBN 3-540-51025-7

Hemopoietic System
1990. 351 figures. XVIII, 336 pages. ISBN 3-540-52212-3

Cardiovascular and Musculoskeletal Systems
1991. 390 figures. XVII, 312 pages. ISBN 3-540-53876-3

Eye and Ear
1991. 141 figures. XIII, 170 pages. ISBN 3-540-54044-X

The following volume is in preparation

Nonhuman Primates II

T. C. Jones U. Mohr R. D. Hunt (Eds.)

Nonhuman Primates I

With 235 Figures and 10 Tables

Springer-Verlag
Berlin Heidelberg New York
London Paris Tokyo
Hong Kong Barcelona
Budapest

Thomas Carlyle Jones, D. V. M., D. Sc.
Professor of Comparative Pathology
Emeritus, Harvard Medical School
ILSI Research Foundation
1126 Sixteenth Street, N. W., Washington, DC 20036, USA

Ulrich Mohr, M. D.
Professor of Experimental Pathology
Medizinische Hochschule Hannover
Institut für Experimentelle Pathologie
Konstanty-Gutschow-Strasse 8
30625 Hannover, Federal Republic of Germany

Ronald Duncan Hunt, D. V. M.
Professor of Comparative Pathology
Harvard Medical School
New England Regional Primate Research Center
One Pine Hill Drive, Southborough, MA 01772, USA

Distribution rights for North America, Canada, and Mexico by
ILSI International Life Sciences Institute
1126 Sixteenth Street NW, Washington, DC 20036, USA

ISBN-13:978-3-642-84908-4 e-ISBN-13:978-3-642-84906-0
DOI: 10.1007/978-3-642-84906-0

Library of Congress Cataloging-in-Publication Data
Nonhuman primates I/T. C. Jones, U. Mohr, R. D. Hunt, eds. p. cm. – (Monographs on pathology of laboratory animals) Includes bibliographical references and index.
ISBN-13:978-3-642-84908-4
1. Primates-Diseases. 2. Primates as laboratory animals. I. Jones, Thomas Carlyle. II. Mohr, U. (Ulrich) III. Hunt, Ronald Duncan. IV. Series. RA 1199.5.P74N66 1993 619'.98-dc20 92-45285

Softcover reprint of the hardcover 1st edition 1993

Reproduction of the Illustrations: Gustav Dreher, Württembergische Graphische Kunstanstalt GmbH, Stuttgart
Typesetting: Best-set Typesetter Ltd., Hong Kong

25/3130-543210 – Printed on acid-free paper

Foreword

The International Life Sciences Institute (ILSI) was established in 1978 to stimulate and support scientific research and educational programs related to nutrition, toxicology, and food safety, and to encourage cooperation in these programs among scientists in universities, industry, and government agencies to assist in the resolution of health and safety issues.

To supplement and enhance these efforts, ILSI has made a major commitment to supporting programs to harmonize toxicologic testing, to advance a more uniform interpretation of bioassay results worldwide, to promote a common understanding of lesion classifications, and to encourage wide discussion of these topics among scientists. The *Monographs on the Pathology of Laboratory Animals* are designed to facilitate communication among those involved in the safety testing of foods, drugs, and chemicals. The complete set will cover all organ systems and is intended for use by pathologists, toxicologists, and others concerned with evaluating toxicity and carcinogenicity studies. The international nature of the project – as reflected in the composition of the editorial board and the diversity of the authors and editors – strengthens our expectations that understanding and cooperation will be improved worldwide through the series.

Alex Malaspina
President
International Life Sciences Institute

Preface

This book, on Nonhuman Primates, is the eleventh volume of a set prepared under the sponsorship of the International Life Sciences Institute (ILSI). One aim of this set on the pathology of laboratory animals is to provide information which will be useful to pathologists, especially those involved in studies on the safety of foods, drugs, chemicals, and other substances in the environment. It is expected that this and future volumes will contribute to better communication, on an international basis, among people in government, industry, and academia who are involved in protection of the public health.

The arrangement of this volume is based, in part, upon the philosophy that the first step toward understanding a pathologic lesion is its precise and unambiguous identification. Therefore, the microscopic and ultrastructural features of a lesion that are particularly useful to the pathologist for definitive diagnosis are considered foremost. Diagnostic terms preferred by the author and editors are used as the subject heading for each pathologic lesion. Synonyms are listed although most are not preferred and some may have been used erroneously in prior publications. The problems arising in differential diagnosis of similar lesions are considered in detail. The biologic significance of each pathologic lesion is considered under such headings as etiology, natural history, pathogenesis, and frequency of occurrence under natural or experimental conditions. Comparison of information available on similar lesions in man and other species is valuable as a means to gain broader understanding of the process involved. Knowledge of this nature is needed to form a scientific basis for safety evaluations and experimental pathology. References to pertinent literature are provided in close juxtaposition to the text in order to support conclusions in the text and lead toward additional information. Illustrations are an especially important means of nonverbal communication, especially among pathologists, and therefore constitute important features of each volume.

The subject under each heading is covered in concise terms and is expected to stand alone but, in some instances, it is important to refer to other parts of the volume. A comprehensive index is provided to enhance the use of each volume as a reference.

Some omissions are inevitable and we solicit comments from our colleagues to identify parts which need strengthening or correcting. We have endeavored to include important lesions which a pathologist might encounter in studies involving the rat, mouse, or hamster. Newly recognized lesions or better understanding of old ones may make revised editions necessary in the future.

The editors wish to express their deep gratitude to all of the individuals who have helped with this enterprise. We are indebted to each author and member of the Editorial Board whose name appears elsewhere in the volume. We are especially grateful to the Officers and Board of Trustees of the International Life Sciences Institute for their support and understanding. Several people have worked directly on important details in this venture. These include Mrs. Nina Murray, Executive Secretary; Mrs. Ann Balliett and Ms. Cora Moore, Editorial Assistants;

Mrs. June Armstrong, Medical Illustrator; Mrs. Sydney Fingold, Librarian; Ms. Lori MacInnes, Secretary; Ms. Sharon K. Coleman, ILSI Coordinator for External Affairs; Ms. Roberta Gutman, ILSI Publication Manager, and Mrs. Sharon Weiss, Associate Director, ILSI Research Foundation, were helpful on many occasions.

We are particularly grateful to Dr. Dietrich Götze and his staff at Springer-Verlag for the quality of the published product.

March 1993

THE EDITORS
T. C. Jones
U. Mohr
R. D. Hunt

Contents

Contributors

Daniel C. Anderson, D.V.M.
Associate Research Professor
Yerkes Regional Primate Research Center
Emory University
Atlanta, Georgia, USA

Gary C. Armitage, D.D.S., M.S.
Professor and Chairman
Division of Periodontology, School of Dentistry
University of California/San Francisco
San Francisco, California, USA

Gary B. Baskin, D.V.M.
Head, Department of Pathology
Tulane Regional Primate Research Center
Covington, Louisiana, USA

Pamela E. Binkerd
Research Associate
California Primate Research Center
University of California/Davis
Davis, California, USA

Beverly J. Blake
Research Associate, Technical Assistant to Director
New England Regional Primate Research Center
Southborough, Massachusetts, USA

James L. Blanchard, D.V.M., Ph.D.
Head, Veterinary Sciences Department
Tulane Regional Primate Research Center
Covington, Louisiana, USA

Roderick V. Bronson, D.V.M.
Department of Pathology
Tufts University Schools of Veterinary Medicine and Medicine
Jamaica Plain, Massachusetts, USA

Laura V. Chalifoux
Associate in Comparative Pathology
New England Regional Primate Research Center
Southborough, Massachusetts, USA

John M. Gliatto, V.M.D., Ph.D.
Assistant Professor
Department of Pathology
Tufts University School of Veterinary Medicine
North Grafton, Massachusetts, USA

Andrew G. Hendrickx, Ph. D.
California Regional Primate Research Center
University of California, Davis
Davis, California, USA

Ronald D. Hunt, D. V. M.
Director
New England Regional Primate Research Center
Southborough, Massachusetts, USA

Norval W. King, Jr., D. V. M.
Associate Professor of Comparative Pathology
Associate Director of Collaborative Research
New England Regional Primate Research Center
Southborough, Massachusetts, USA

Sherry A. Klumpp, D. V. M.
Yerkes Regional Primate Research Center
Emory University
Atlanta, Georgia, USA

Andrew A. Lackner, D. V. M., Ph. D.
California Regional Primate Research Center
University of California/Davis
Davis, California, USA

Norman L. Letvin, M. D.
Associate Professor of Medicine
New England Regional Primate Research Center
Harvard Medical School
Southborough, Massachusetts, USA

Linda J. Lowenstine, D. V. M., Ph. D.
Associate Professor
Department of Veterinary Pathology
University of California/Davis
Davis, California, USA

Harold M. McClure, D. V. M.
Associate Director for Scientific Programs
Chief, Division of Pathobiology and Immunobiology
Yerkes Primate Research Center, Emory University
Atlanta, Georgia, USA

Wilbur P. McNulty, M. D. (†)
Scientist
Oregon Regional Primate Research Center
Beaverton, Oregon, USA

Keith A. Reimann, D. V. M.
Assistant Professor of Medicine
New England Regional Primate Research Center
Harvard Medical School
Southborough, Massachusetts, USA

E. Donald Roberts
Pathologist
Tulane Regional Primate Research Center
Covington, Louisiana, USA

Morten Schiødt, DDS, Dr. Odont
Department Oral Medicine & Oral Surgery
Central Hospital
3400 Hillerød, Denmark

Che-Chung Tsai, D. V. M., Ph. D.
Head, Pathology Division
Regional Primate Research Center
Primate Field Station, University of Washington
Medical Lake, Washington, 99022, USA

Philip M. Zack, D. V. M., Ph. D.
LTC USA
Chief, Department Experimental Pathology, Pathology Division
US Army Medical Institute of Infectious Diseases Ft. Detrick
Frederick, Maryland, USA

Bernard C. Zook, D. V. M.
Professor of Pathology
Director Animal Research Facility
Department of Pathology
The George Washington University Medical Center
Washington, District of Columbia, USA

Immunodeficiency: An Overview

Keith A. Reimann and Norman L. Letvin

The term "immunodeficiency" encompasses a wide variety of clinical disorders. In fact, the spectrum of these abnormalities is so broad that employing a single term to describe these diverse diseases can lead to imprecision in analyzing the underlying pathologic processes. Thus, the use of a single term to describe aspects of such diverse syndromes as canine granulocytopathy, severe combined immunodeficiency of Arabian foals, and feline panleukopenia can act to diminish rather than sharpen the diagnostic acuity of a clinician or pathologist in interpreting manifestations of disease.

The immune system is comprised of populations of cells with complimentary and interdependent functions. Abnormalities of individual lineages of these cells can lead to quite discrete and distinct immunologic abnormalities and result in different disease susceptibilities. Diseases which affect progenitor or early lineage cells are more likely to result in broader and more dramatic immunologic impairment than those diseases which only affect the more differentiated cell. Through an understanding of the role and function of the individual cell populations which comprise the immune system, one can often predict the clinical manifestations of disease which will result from the loss or impairment of one of those cell populations. Similarly, recognizing discrete disease manifestations can often facilitate the accurate prediction of the immunologic cells which might be abnormal. Therefore, a precise documentation of disease manifestations is important in determining whether these signs of disease are likely to be due to primary or secondary immunodeficiencies and what those immunodeficiencies might specifically be.

In this brief overview we hope to provide a framework in which to view the primary and secondary causes of immunodeficiencies. A discussion of the manifestations of retrovirus-induced immunodeficiency states in nonhuman primates serves as an example of how one might approach the definition of these problems.

Primary Immunodeficiencies

The primary (inherited) immunodeficiencies can be understood within the framework provided by our understanding of the paths of differentiation and maturation of the pleuripotential stem cell. As shown in Fig. 1, the pleuripotential stem cell gives rise to erythrocytes, platelets, granulocytes, monocytic cells, and lymphocytes of both the T

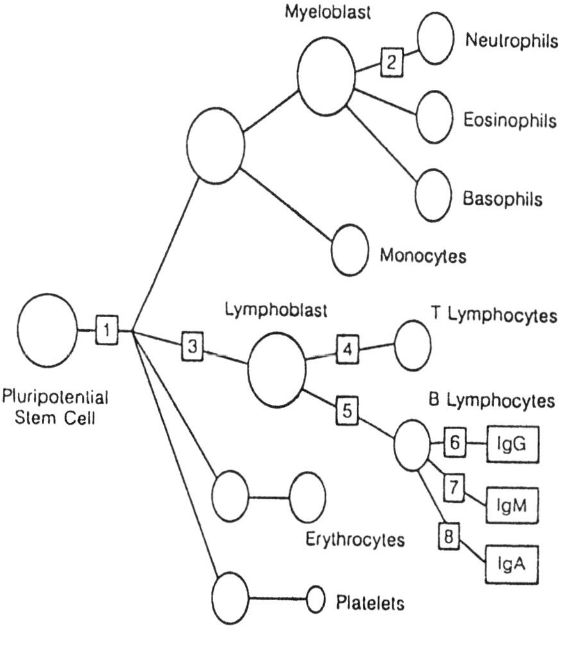

Fig. 1. Pathway of differentiation of hematopoietic cells and the sites of cellular defects in primary immunodeficiency diseases in animals. *1* cyclic hematopoiesis: dogs; *2* granulocytopathy: dogs; *3* severe combined immunodeficiency: horses, dogs; *4* T cell deficiency: cattle, dogs; *5* agammaglobulinemia: horses; *6* IgG deficiency: cattle; *7* IgM deficiency: horses, dogs; *8* IgA deficiency: dogs

and B lineage. Inherited abnormalities in the differentiation and/or function of the progeny of the pleuripotential stem cell are well documented in domesticated animals and humans. A cyclic hematopoietic abnormality has been described in collies (reviewed by Campbell 1985). While all hematopoietic cell lineages are affected, the 10–14 day cyclic nature of this stem cell disorder causes episodic peripheral blood depletion of the shortest lived cell, namely, neutrophils. However, careful measurements have documented cyclic alterations in all other cell lineages as well. In the syndrome of canine granulocytopathy, a defect at a more terminal stage of differentiation occurs (Renshaw and Davis 1979). Neutrophil numbers are not decreased in this disorder; however, these cells lack normal killing function, probably due to an adhesion molecule deficiency. In both cyclic hematopoiesis and granulocytopathy syndrome, the primary clinical manifestation is recurrent bacterial infections.

Combined immunodeficiency diseases involved inherited abnormalities in the maturational lineage of both T and B lymphocytes. This immunodeficiency has been described both in Arabian foals (McGuire and Poppie 1973) and in dogs (Felsberg and Jezck 1982). Affected animals lack the ability to generate humoral and cellular immune responses. As might be predicted, affected foals die from bacterial, viral, and protozoal infections.

More selective defects in T cell maturation or function have been described in cattle (Anderson et al. 1970) and weimaraner dogs (Roth et al. 1980). These animals lack selective cell-mediated immune functions while the ability to generate humoral responses remains intact. Finally, an array of B cell immunoglobulin secretion defects have been defined in a variety of species (Felsberg et al. 1985; Banks and McGuire 1986; Perryman et al. 1977). An awareness of these syndromes in humans and domestic animals will facilitate a recognition of similar syndromes in nonhuman primate species as these animals increasingly come under careful veterinary scrutiny.

Secondary Immunodeficiencies

While primary immunodeficiencies are quite rare, secondary states of immunodeficiency are common in domesticated animals and humans. As outlined in Table 1, immunodeficiencies can arise secondarily as a result of a variety of pathogenic processes. The failure to passively transmit colostrum to offspring is a common cause of neonatal death in calves and foals. Endocrine abnormalities such as growth hormone deficiency can result in cell-mediated immune dysfunction. Commonly employed drugs such as steroids or alkylating agents can induce very specific abnormalities in both granulocytes and lymphocytes. A variety of neoplasms, including lymphomas and multiple myeloma, can result in immunologic dysfunction. Protein malnutrition or single nutrient deficiencies such as zinc or iron deficiency can have significant immunologic manifestations. However, more commonly, infectious agents induce immunodeficiency. These immune deficient states are often quite specific and can indicate to the alert diagnostician the underlying infectious process.

Table 1. Disorders associated with secondary immunodeficiency in animals

– Failure of passive transfer
– Drugs and toxins
– Nutritional imbalances
– Endocrine disorders
– Neoplasia
– Infectious agents

While virus infections can be associated with profound immune suppression, the mechanisms by which viruses induce alterations in immunity are not yet completely understood (Rouse and Horohov 1986). In general, viral infections can cause immunosuppression by at least four distinct mechanisms: (1) direct (cytotoxic) effect of virus replication on cells of the immune system; (2) effect of an immunosuppressive, soluble, viral product; (3) induction of soluble immunosuppressive substances by host cells; and (4) imbalance of immunoregulatory cell subsets. The specific cells affected by these processes will determine the disease which will be manifested.

Among domestic animals, a number of viruses can cause disease syndromes in which immunosuppression, and thus opportunistic infections, are a major component. In canine distemper, the etiologic paramyxovirus is lymphocytotoxic, causing destruction of both mature B and T lymphocytes (Appel 1969). The resulting lymphoid depletion clearly thwarts the host's attempt to contain this infection and is central in permitting the secondary bacterial infections which ensue. Measles virus infection in human and nonhuman primates provides a very similar example. Feline panleukopenia, in contrast, is cytotoxic for all rapidly proliferating cells in the cat (Rohovsky and Fowler 1971). Therefore, the primary manifestations of this disease include diarrhea and bone marrow failure with panleukopenia and death due to myriad bacterial infections. Interestingly, feline panleukopenia virus causes a much milder disease in specific pathogen-free cats.

The feline leukemia virus (FeLV) is a common infection in cats which is associated with a variety of opportunistic infections (Rojko and Olsen 1984). The virus can induce thymic atrophy and lymphoid depletion from lymph nodes and spleen. In addition, infection with this virus is associated with a variety of peripheral blood cytopenias. It is these abnormalities which probably cause the increased incidence of intercurrent infections in cats. There is also considerable evidence that a 15 kDa component of the FeLV envelope glycoprotein can cause in vitro many of the immunologic abnormalities seen in the virus-induced disease. Finally, it is becoming clear that a number of viral infections, including FeLV, can result in immune abnormalities secondary to release of cytokines.

Retrovirus Infections in Nonhuman Primates

The recently discovered retroviruses of nonhuman primates and the diseases which they induce serve as an instructive example of the distinct immunologic abnormalities and resulting secondary infections caused by primary viral infections. Three different families of retroviruses which infect Old World monkeys, simian T-cell lymphotropic virus type I (STLV-I), type D retroviruses, and the simian immunodeficiency viruses (SIV), all have unique effects on the immune system of their host.

STLV-I has nearly complete nucleotide sequence identity to the exogenous retrovirus human T-lymphotropic virus type I (HTLV-I) (Hayami 1986). While HTLV-I infects large numbers of individuals in geographically restricted human populations, clinical manifestations of infection are seen in only a very small percentage of these individuals (Gallo 1984). It has been estimated that less than 3% of HTLV-I-infected humans develop any manifestations of disease. The most common HTLV-I-associated diseases include adult T cell leukemia, tropical spastic paraparesis, and a newly described arthritis. While HTLV-I infects T lymphocytes, the pathogenic potential of this virus appears to be low, and the immunologic defects associated with HTLV-I infection remain poorly defined.

In light of these observations in HTLV-I-infected humans, it is not surprising that STLV-I-associated immunologic dysfunction or disease has not been seen in naturally or experimentally infected macaques. Like HTLV-I, this virus can immortalize macaque T lymphocytes in vitro but has not been shown to induce a T cell leukemia in vivo. Moreover, central nervous system demyelination or arthropathy has not been associated with the infection of a macaque species with STLV-I. If infected macaques only rarely develop STLV-I-associated disease, perhaps a large enough number of STLV-I-infected monkeys have not been studied to detect overt disease. However, it is also possible that STLV-I is simply not pathogenic in macaques. Thus, even subtle immunologic abnormalities may not be found in these infected species.

In the early 1980s, the search for a virus infection which might underlie the AIDS-like syndrome described in captive colonies of macaque monkeys led to the isolation of type D retroviruses from a number of macaque species (Gardner et al. 1988). Following experimental infection with these isolates, macaques were shown to develop hematologic abnormalities including anemia, neutropenia, and lymphopenia, without a selective depression in circulating CD4+ lymphocytes. Transient abnormal responses of blood lymphocytes to mitogens in blastogenesis assays were described in these animals. While cytomegalovirus infections are occasionally seen in these type D retrovirus-infected monkeys, the infections most commonly associated with this retroviral infection are caused by pyogenic bacteria. The spectrum of clinical diseases and the immune dysfunction are distinct from that seen in AIDS virus infections in macaques and humans (see p. 5, this volume).

The SIVs are a family of nonhuman primate lentiviruses with significant nucleotide sequence homology to HIV isolates (Desrosiers et al. 1989). These viruses, in general, are nonpathogenic in their natural host African species. However, in selected Asian monkey species, certain of these isolates induce an AIDS-like disease (Gardner et al. 1988). The clinical spectrum of disease seen in the SIV-infected Asian monkeys differs from that seen in type D retrovirus-infected animals. Lymphomas occur as a common opportunistic tumor. While pyogenic infections are not a major component of the constellation of opportunistic infections seen in SIV-infected monkeys, infections by a variety of viral agents, including cytomegalovirus, adenovirus, and simian virus-40, commonly occur. Finally, *Pneumocystis carinii* pneumonia is a frequent opportunistic protozoal infection in these monkeys. SIV-induced disease in macaque monkeys, therefore, has striking similarities to human AIDS. Thus, it was predictable that the underlying immunologic dysfunctions in SIV-infected macaques would prove to be similar to those seen in HIV-infected humans. A selective decrease in the number and function of CD4+ lymphocytes exists in these animals. These viruses also display macrophage tropism (Ringler et al. 1989). It is probably a combination of the SIV-induced macrophage and CD4+ lymphocyte defects which accounts for the specific immunologic abnormalities seen in these animals.

Thus, three distinct families of simian retroviruses, STLV-I, the type D retroviruses, and SIVs, are responsible for quite distinct effects on the immune system of macaque monkeys. STLV-I can immortalize T cells in tissue culture but does not induce immunologic dysfunction or disease in vivo. The type D retroviruses induce an immunologic dysfunction whose major manifestation is secondary infection by pyogenic bacteria. The SIVs induce specific macrophage and T lymphocyte dysfunction with resultant opportunistic viral infections and lymphomas. Certainly, retroviruses of nonhuman primates which remain to be isolated and characterized may induce other distinct immunologic abnormalities with resulting distinctive clinical manifestations. The awareness of such differences in immunologic dysfunction will speed the definition of new disease syndromes in nonhuman primates. Recognizing that the term "immunodeficiency" can subsume very different clinical entities will also simplify the task of diagnosing underlying viral infectious processes in nonhuman primates. The immunodeficiency states induced by specific viruses and their secondary invaders are covered in detail in the chapters which follow in this volume.

References

Anderson I et al. (1970) Evidence of a lethal trait, A46, in black pied Danish cattle of Friesian descent. Nord Veterinaermed 22:473

Appel MJG (1969) Pathogenesis of canine distemper. Am J Vet Res 30:1167–1182

Banks KL, McGuire TC, Jarrett TR (1986) Absence of B lymphocytes in a horse with primary agammaglobulinemia. Clin Immunol Immunopathol 5:282–290

Campbell KL (1985) Canine cyclic hematopoiesis. Comp Cont Educ 7:57–61

Desrosiers RC, Daniel MD, Li Y (1989) HIV-related lentiviruses of nonhuman primates. AIDS Res Hum Retroviruses 5:465–473

Felsburg PJ, Jezyk PF (1982) A canine model for combined immunodeficiency. Clin Res 30:347–353

Felsburg PJ, Glickman LT, Jezyk PF (1985) Selective IgA deficiency in the dog. Clin Immunol Immunopathol 36:297–305

Gallo RC (1984) Human T-cell leukemia/lymphoma virus. Cold Spring Harbor Laboratory, Cold Spring Harbor

Gardner MB, Luciw P, Lerche N, Marx P (1988) Nonhuman primate retrovirus isolates and AIDS. Adv Vet Sci Comp Med 32:171–226

Hayami M (1986) Simian T-cell leukemia viruses, STLV (HTLV-I-related simian retroviruses). Cancer Rev 1:35–51

McGuire TC, Poppie MJ (1973) Hypogammaglobulinemia and thymic hypoplasia in horses: a primary combined immunodeficiency disorder. Infect Imm 8:272–277

Perryman LE, McGuire TC, Hilbert BJ (1977) Selective immunoglobulin M deficiency in foals. J Am Vet Med Assoc 170:212–215

Renshaw HW, Davis WC (1979) Canine granulocytopathy syndrome: an inherited disorder of leukocyte function. Am J Pathol 95:731–744

Ringler DJ, Wyand MS, Walsh DG, Mackey JJ, Sehgal PK, Daniel MD, Desroisier RC, King NW (1989) The productive infection of alveolar macrophages by simian immunodeficiency virus. J Med Primatol 18:217–226

Rohovsky MW, Fowler EH (1971) Lesions of experimental feline panleukopenia. J Am Vet Med Assoc 158:872–874

Rojko JL, Olsen RG (1984) The immunobiology of the feline leukemia virus. Vet Immunol Immunopathol 6:107–165

Roth JA, Lomax LG, Altszuler H, Hampshire J, Kaeberle ML, Shelton M, Draper DD, Ledet AE (1980) Thymic abnormalities and growth hormone deficiency in dogs. Am J Vet Res 41:1256–1262

Rouse BT, Horohov DW (1986) Immunosuppression in viral infections. Rev Infect Dis 8:850–873

Simian Immunodeficiency Virus Infections

Norval W. King, Jr.

Synonyms. Simian lentivirus infections; simian AIDS-like virus infections; simian AIDS.

Gross Appearance

African species of nonhuman primates harboring persistent simian immunodeficiency virus (SIV) infections do not develop clinical illness or gross lesions. These species include: African green monkeys infected with SIV_{AGM}, sooty mangabey monkeys infected with $SIV_{SMM(\delta)}$, Sykes' monkeys infected with SIV_{SYK}; mandrills infected with SIV_{MND}, and chimpanzees infected with SIV_{CMZ}. The gross findings in various Asian species of macaque monkeys dying of naturally acquired or experimental SIV_{MAC}, $SIV_{SMM(\delta)}$, SIV_{MNE}, or SIV_{STM} infection are variable and nonspecific.

The carcass may be severely dehydrated and emaciated, if prolonged diarrhea and inanition were features of the animal's clinical illness, or well-muscled, if death resulted from an acute, fulminant, opportunistic infection (King et al. 1983; Letvin et al. 1985). There may be an erythematous maculopapular rash visible on the skin of the less-haired portions of the body, i.e., the face, thorax, abdomen, and medial aspects of the extremities (Ringler et al. 1987a). Thymic atrophy is a consistent finding in SIV-infected animals, but its usefulness as a diagnostic finding is restricted to young experimentally infected animals who would normally still have a discernible thymus. Other lymphoid organs, including lymph nodes and spleen, may be hypertrophied, normal in size, or atrophic depending upon the duration of the disease. The findings in other visceral organs generally reflect the nature and extent of the various opportunistic infections to which these animals usually succumb. The gross lesions associated with specific opportunistic infections are described elsewhere in this monograph.

Microscopic Features

No microscopic lesions have been described in African green monkeys, sooty mangabey monkeys, Sykes' monkeys, mandrills, or chimpanzees with their respective persistent SIV infections. In contrast, macaque monkeys dying of colony acquired or experimentally induced SIV_{MAC}, $SIV_{SMM(\delta)}$, SIV_{MNE}, or SIV_{STM} infection have a wide array of microscopic lesions, many of which are associated with the opportunistic infections to which these immunosuppressed animals are predisposed. Active infection with cytomegalovirus, adenovirus, papovavirus, *Mycobacterium avium-intracellulare*, *Pneumocystis carinii*, *Cryptosporidium* sp., *Cryptococcus neoformans*, and *Candida albicans* have been described in SIV-infected macaques (Letvin et al. 1985; Baskin 1987; Baskin et al. 1988; Blanchard et al. 1987; King et al. 1990). The microscopic lesions associated with these opportunists are described elsewhere in this monograph. The lesions described here include only those known or suspected to be a direct effect of SIV_{MAC} and the closely related agents, $SIV_{SMM(\delta)}$, SIV_{MNE}, and SIV_{STM}.

Lymphoid Organs

Lymph Nodes and Spleen

Lymphoid tissues are the principal target organs for SIV. Six distinct microscopic patterns of tissue reaction have been identified in lymph nodes and spleens of SIV_{MAC}-infected rhesus monkeys (Chalifoux et al. 1986, 1987; Ringler et al. 1989). These are: (1) normal morphology, (2) follicular hyperplasia, (3) follicular involution with normal or expanded T lymphocyte domains, (4) depletion of follicular and T lymphocyte domains, (5) a distinctive granulomatous lymphadenitis/splenitis, and (6) a generalized lymphoproliferative syndrome. These patterns represent either successive or alternative stages in a continuum of reactions of lymphoid tissue to persistent SIV_{MAC} infection. Even within an individual animal there can be considerable variation in the pattern of tissue reaction in different lymph nodes at the same stage of infection. Finally, it should be emphasized that, with the possible exception of the granulomatous lymphadenitis/splenitis seen in some experi-

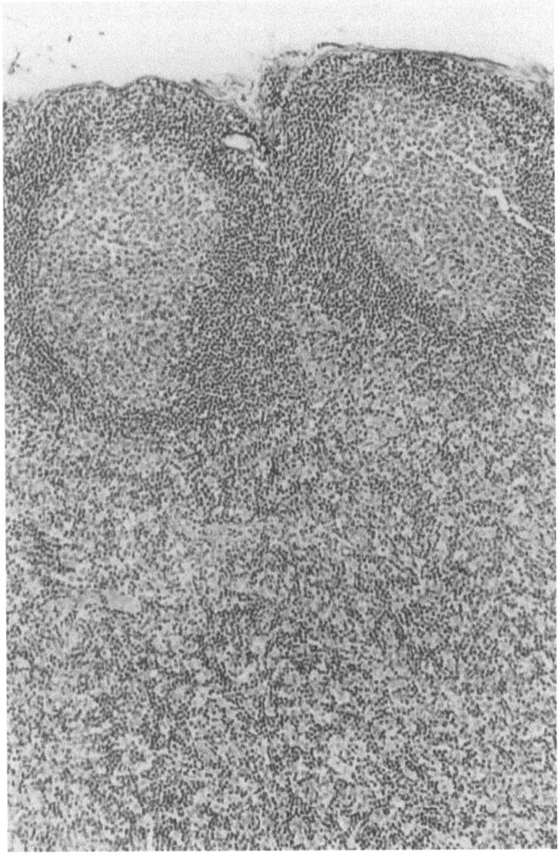

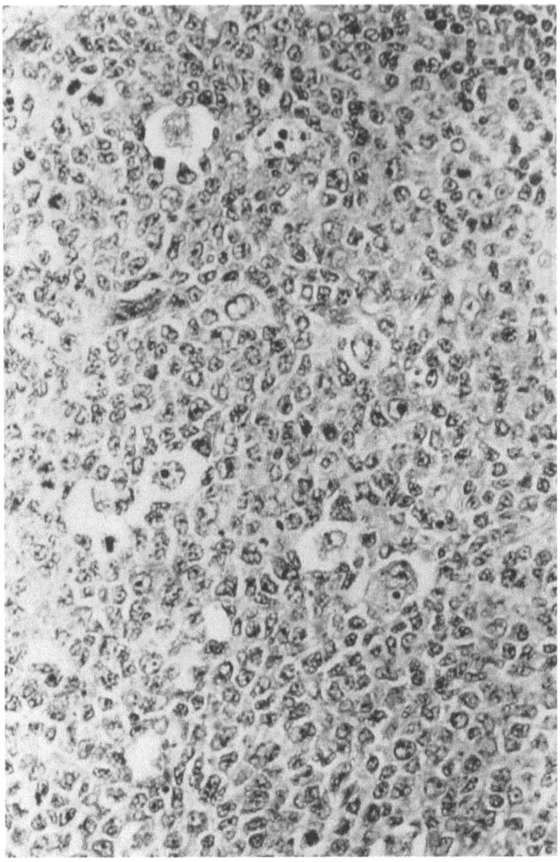

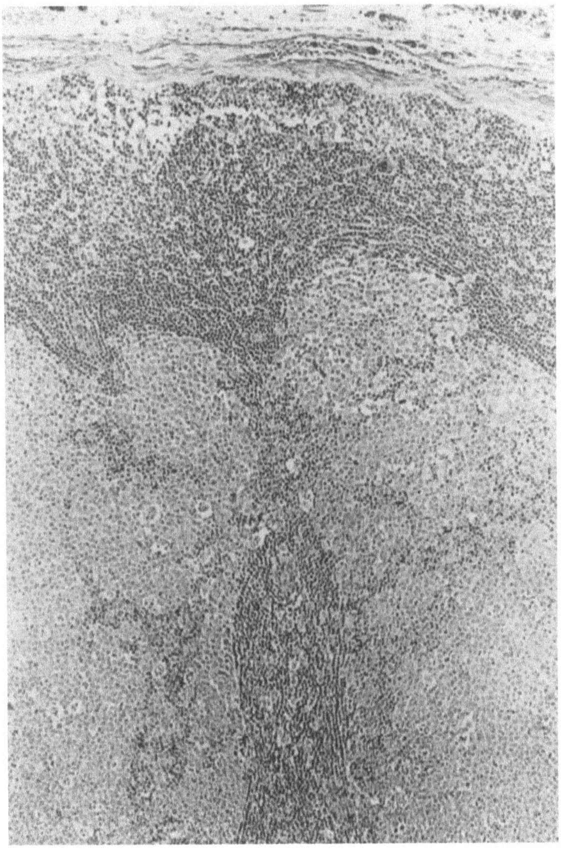

Fig. 2 (*upper left*). Histologically normal lymph node from a rhesus monkey with SIV_{MAC} infection. H and E, ×125

Fig. 3 (*lower left*). Lymph node of an SIV_{MAC}-infected rhesus monkey. Note marked follicular hyperplasia. The germinal centers of lymphoid follicles are large, confluent, and irregularly shaped. H and E, ×125

Fig. 4 (*upper right*). Germinal center of one of the hyperplastic follicles in Fig. 3, in which there are scattered macrophages. H and E, ×400 mentally infected macaques, the microscopic findings on H and E stained tissue sections of lymphoid organs are not pathognomonic for SIV infection.

However, immunocytochemical and in situ hybridization procedures can be used to identify structural proteins and nucleic acids in tissue sections of lesions suspected of being caused by this virus (Ringler et al. 1989; Wyand et al. 1989). The following microscopic patterns have been observed in lymphoid tissues of rhesus monkeys experimentally infected with SIV_{MAC}.

Normal Nodal and Splenic Morphology. During the early stages of SIV_{MAC} infection, peripheral lymph nodes and spleens of rhesus monkeys may be normal microscopically (Fig. 2). Immunohistochemical studies of these tissues, however, reveal the presence of macrophages containing SIV_{MAC} p27 core protein scattered throughout the T lymphocyte-rich areas, mantle zones of follicles, and the medullae and sinuses of nodes and spleens, respectively. At this stage, there may be a few p27+ macrophages in germinal centers and a small number of p27+ T lymphocytes in other parts of the node.

Follicular Hyperplasia. Enlargement of peripheral and visceral lymph nodes occurs around 6–8 weeks postinoculation. Microscopically, proliferation of large follicular center cells results in expanded, irregularly shaped, germinal centers that may become confluent (Fig. 3). Many mitotic figures and macrophages are present within the hyperplastic follicles (Fig. 4). Mantle zones are thinner than normal or absent. The hyperplastic germinal centers contain a network of p27+ follicular dendritic cells (Ringler et al. 1989). Paradoxically, these antigen-presenting follicular dendritic cells are rarely positive for SIV_{MAC} RNA by in situ hybridization and do not contain SIV_{MAC} particles by electron microscopy (Ringler et al. 1989; Wyand et al. 1989). At this stage, the T lymphocyte domains, i.e., the paracortex of lymph nodes and periarteriolar lymphoid sheaths of the spleen, range from normal to slightly reduced in size. The paracortex and periarteriolar lymphoid sheaths generally have reduced numbers of CD4+ T lymphocytes and increased numbers of CD8+ lymphocytes (Chalifoux et al. 1987). Macrophages may also be increased in these areas. The germinal centers and mantle zones of hyperplastic follicles contain scattered macrophages and lymphocytes positive for SIV_{MAC} RNA by in situ hybridization (Wyand et al. 1989).

Follicular Involution with Normal or Expanded Paracortex and Periarteriolar Lymphoid Sheaths. This stage generally succeeds follicular hyperplasia if the SIV_{MAC}-infected animal survives. The germinal centers are reduced to small aggregates of follicular center cells or replaced by a mass of acellular hyaline material (Figs. 5, 6). Mitotic figures and mantle zones are no longer evident. Only a few B lymphocytes remain at the site of involuted follicles (Chalifoux et al. 1987). The reduction in follicles is accompanied by normal or expanded paracortex and periarteriolar lymphoid sheaths. The paracortex of nodes may have increased vascularity. The paracortex and periarteriolar lymphoid sheaths, which normally are composed mainly of CD4+ lymphocytes, at this stage consist mainly of CD8+ lymphocytes and scattered macrophages. The macrophages in the expanded paracortex and periarteriolar lymphoid sheaths are positive for SIV_{MAC} p27 core protein and SIV_{MAC} RNA and contain abundant simian immunodeficiency viral particles within cytoplasmic vacuoles (Ringler et al. 1989; Wyand et al. 1989).

Depletion of Follicles, Paracortex, and Periarteriolar Lymphoid Sheaths. Lymph nodes and spleens with these changes are generally found in animals that have died following prolonged persistent SIV_{MAC} infection. Mesenteric and pancreatic nodes are often more severely depleted of cortical and paracortical lymphocytes than peripheral or other visceral nodes. At this stage, lymph nodes and spleens contain only small nodules of residual follicular lymphocytes. The paracortex is reduced to a thin latticework of reticular cells and a few residual lymphocytes that separate dilated cortical and medullary sinuses. Macrophages containing erythrocytes and hemosiderin are frequently present within the dilated sinuses (Fig. 7). The splenic periarteriolar lymphoid sheaths are also no longer discernible. Nodules of lymphocytes remaining in the cortex consist of B cells, while those left in the depleted paracortex, periarteriolar lymphoid sheaths, and medulla are a mixture of CD4+ and CD8+ lymphocytes (Chalifoux et al. 1987). Only a few cells in these depleted zones contain SIV_{MAC} p27 core protein and only a rare cell contains SIV_{MAC} RNA (Ringler et al. 1989; Wyand et al. 1989).

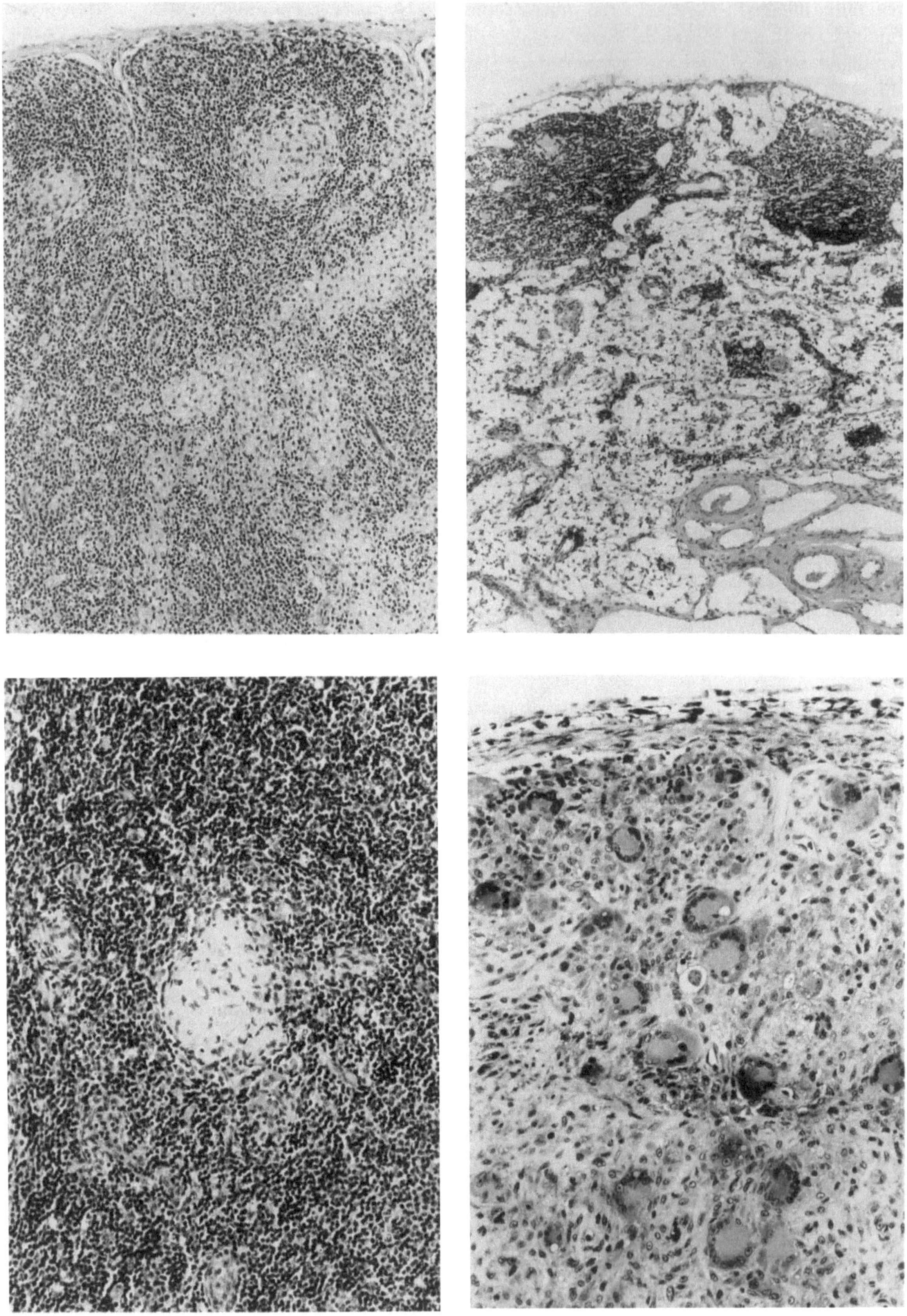

Granulomatous and Giant Cell Lymphadenitis and Splenitis. Some rhesus monkeys infected with SIV_{MAC} develop grossly enlarged peripheral and visceral lymph nodes and splenomegaly that are not due to follicular hyperplasia but, instead, to extensive infiltration by epithelioid histiocytes and varying numbers of multinucleate giant cells. The infiltrating cells obliterate the cortical and medullary sinuses and eventually replace all lymphoid tissue of the cortex, paracortex, and periarteriolar lymphoid sheaths (Fig. 8). Both the epithelioid cells and multinucleate giant cells bear surface antigens indicating a monocyte/macrophage origin. Approximately 10% of the epithelioid cells and essentially all of the multinucleate giant cells contain SIV_{MAC} p27 core protein (Ringler et al. 1989). Small numbers of lymphocytes and epithelioid cells and up to 50% of the multinucleate giant cells may be positive for SIV_{MAC} RNA (Wyand et al. 1989). Some affected nodes may also contain coarse bands of connective tissue that divide the node into lobules. Acid-fast stains of affected nodes and spleens are consistently negative for mycobacteria. This lesion, although distinctive and highly suggestive of SIV_{MAC} infection, has only been observed in a portion of naturally and experimentally infected rhesus monkeys. Interestingly, it occurs most often in macaques infected with the highly virulent PBJ strain of SIV_{SMM} (Fultz et al. 1989; McClure et al. 1989).

Lymphoproliferative Syndrome. SIV_{MAC}-infected animals with this disorder develop nodal and splenic enlargement that also differs from the follicular hyperplasia seen more commonly in animals infected with this virus (King et al. 1983; Chalifoux et al. 1986; Ringler et al. 1989). In this condition, peripheral and visceral nodes become massively enlarged due to a proliferation of solid sheets of large plasmacytoid cells that replace the normal architecture of the node and obliterate the subcapsular and medullary sinuses (Figs. 9, 10). Macrophages are also scattered throughout the sheets of plasmacytoid cells. The proliferating cells in this lesion express both κ and λ light chains confirming their B cell origin. Mitotic figures may be common, but invasion of the capsule of affected nodes has not been observed. Multiple nodular aggregates of similar cells may also be found in other organs and tissues of affected animals, including the salivary gland, lung (Fig. 11), liver, kidney, bone marrow, and thymus. At these sites, the infiltrating cells occasionally assume an ill-defined follicular pattern. The nodules are composed of a large numbers of CD4+ lymphocytes mixed with a smaller population of CD8+ cells and B cells with evidence of plasma cell differentiation (Chalifoux et al. 1986).

◀ **Fig. 5** (*upper left*). Lymph node of an SIV_{MAC}-infected rhesus monkey. Note that the lymphoid follicles have undergone marked involution and hyalinization. At this stage the paracortical lymphoid tissue is hyperplastic. H and E, ×125

Fig. 6 (*lower left*). Hyalinized lymphoid follicle and a portion of the expanded paracortex. Higher magnification of the node shown in Fig. 5. H and E, ×200

Fig. 7 (*upper right*). Lymph node from an SIV_{MAC}-infected rhesus monkey. Note the marked depletion of follicular and paracortical lymphoid tissue. The sinuses of the node are widely dilated. H and E, ×75

Fig. 8 (*lower right*). Lymph node from an SIV_{MAC}-infected rhesus monkey. The cortical lymphoid tissue has been replaced by infiltrates of histiocytes and multinucleate giant cells of macrophage-monocyte origin. H and E, ×200

Thymus

The thymuses of rhesus monkeys dying of SIV_{MAC} infection are invariably atrophic regardless of the age of the animal. There is marked depletion of cortical thymocytes with loss of a distinct corticomedullary junction (Fig. 12). Hassal's corpuscles may be reduced in number and occasionally cystic. Although this is a consistent finding in infected animals, it is not pathognomonic. SIV core protein and RNA are demonstrable in a small to moderate number of lymphocytes and macrophages in both the atrophic cortex and medulla.

Nervous System

Brain and Spinal Cord

Approximately 50% of rhesus monkeys experimentally infected with SIV_{MAC} develop a characteristic meningoencephalomyelitis that resembles the encephalopathy that occurs in a high percentage of human patients with acquired immunodeficiency syndrome (Letvin et al. 1985; Ringler et al. 1987b; Sharer et al. 1987). For reasons that are still not clear, this lesion has

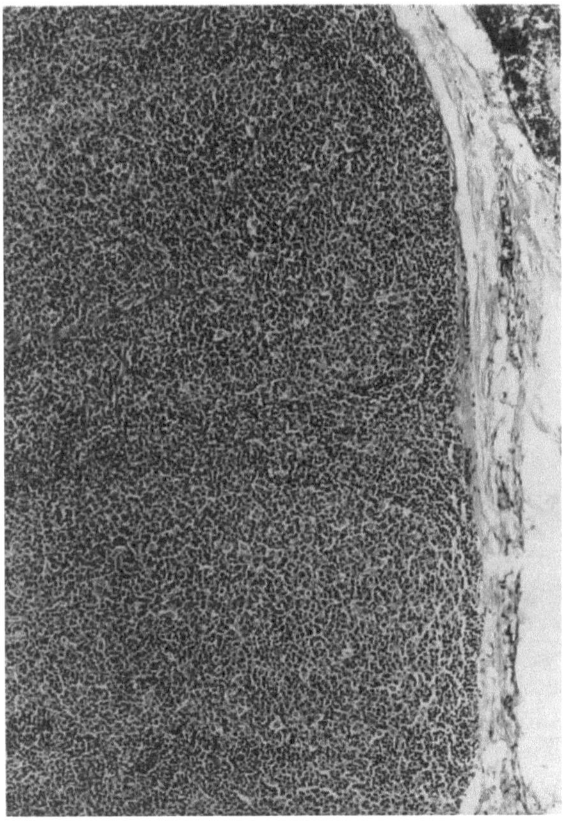

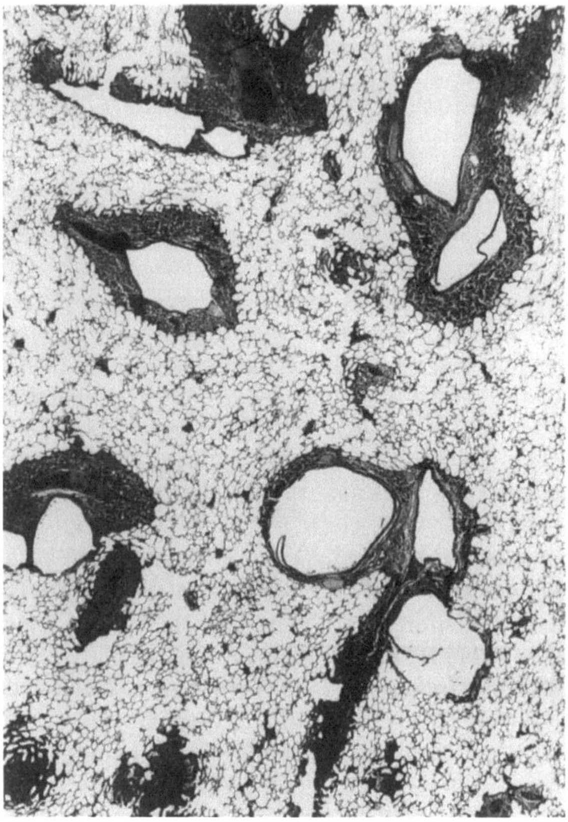

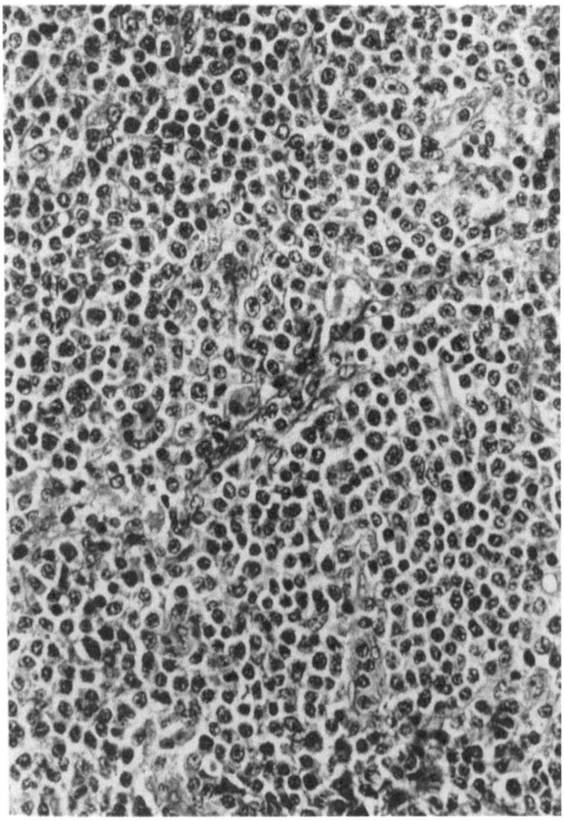

Fig. 9 (*upper left*). Lymph node from an SIV_{MAC}-infected rhesus monkey with lymphoproliferative syndrome. Note that the subcapsular and medullary sinuses and the cortex and paracortex of the node have been replaced by a diffuse proliferation of a uniform population of lymphoid cells. H and E, ×200

Fig. 10 (*lower left*). Higher magnification of a portion of the lymph node depicted in Fig. 9. Note the plasmacytic nature of the proliferating cells. Immunophenotyping studies have revealed that these cells consist of a polyclonal population of B lymphocytes in which there are scattered macrophages. H and E, ×400

Fig. 11 (*upper right*). Lung of the same animal with lymphoproliferative syndrome depicted in Figs. 9 and 10. The prominent perivascular and peribronchiolar cellular infiltrates consist of polyclonal populations of B lymphocytes. H and E, ×40 not been observed in rhesus monkeys or other nonhuman primates with SIV acquired in a colony situation (King et al. 1990). Lesions are found in both the gray and white matter of the brain and spinal cord but are more common in the white matter. The meninges and choroid plexus are less often affected. Typically, there are multifocal perivascular infiltrates of finely vacuolated histiocytes and multinucleate giant cells scattered throughout the cerebrum, cerebellum, brainstem, and spinal cord (Fig. 13). An occasional lesion may also contain small numbers of neutrophils or lymphocytes. There may be a mild loss of myelin around these granulomas. Glial nodules may also be scattered throughout affected brains. The leptomeninges and choroid plexus may also be infiltrated by histiocytes, giant cells, and varying numbers of fibroblasts. By electron microscopy, large numbers of mature SIV_{MAC} particles are present within cytoplasmic vacuoles of the infiltrating histiocytes and giant cells (Letvin et al. 1985; Ringler et al. 1987). In situ hybridization using an SIV_{MAC} envelope probe has identified SIV_{MAC} RNA mainly in histiocytes and giant cells and rarely in cells thought to be glial cells (Wyand et al. 1989).

Respiratory System

Lungs

SIV_{MAC}-Induced Giant Cell Pneumonia. Although the lung is a common site of opportunistic infection by agents such as *Pneumocystis carinii*, *Mycobacterium avium-intracellulare*, and cytomegalovirus in SIV_{MAC}-infected macaques, it may also be the target of a primary pneumonia induced by SIV_{MAC} (Letvin and King 1990). This SIV_{MAC}-induced pneumonia is characterized by extensive infiltration of the pulmonary parenchyma and alveolar spaces by large numbers of foamy macrophages and multinucleate giant cells (Fig. 14). Alveolar walls are thickened by the inflammatory cell infiltrate and there is often a concomitant proliferation of type II pneumocytes.

Pulmonary Lymphoid Hyperplasia. Marked hyperplasia of the peribronchial lymphoid tissue may be seen in some rhesus monkeys infected with SIV_{MAC}, especially during the early stages of infection (Letvin and King 1990). It often occurs when lymphoid organs are also undergoing follicular hyperplasia. This lesion is similar to that seen in children affected with acquired immunodeficiency syndrome (AIDS), but does not progress to a stage at which it would qualify as lymphoid interstitial pneumonia.

Cardiovascular System

Heart

Nonseptic, vegetative, valvular endocarditis has been observed in a small number of rhesus monkeys experimentally infected with SIV_{MAC}, but the role of the virus in its pathogenesis has not been defined.

Blood Vessels

Unexplained arterial lesions have been observed in SIV_{MAC}-infected rhesus monkeys (Letvin and King 1991). This arteriopathy involves large to medium sized vessels, especially the pulmonary arteries. The lesion is characterized by progressive focal to segmental thickening of the intima and media of the artery with fragmentation of the internal elastic lamina. Some vessels become so thickened that their lumens are occluded (Fig. 15). The endothelium of affected arteries varies from being hypertrophied and hyperplastic to focally eroded and covered with mural or occlusive thrombi. Although cytomegalovirus inclusion bodies have been observed on rare occasions in inflamed pulmonary arteries in monkeys infected by SIV_{MAC}, no viral inclusions have been found in this type of proliferative arteriopathy. Moreover, inflammation is not a consistent finding in affected arteries. No viral particles or immune complex deposits have been found in affected vessels by electron microscopy.

Urinary System

Kidney

Glomerular lesions have been described in rhesus monkeys infected with SIV_{MAC} and $SIV_{SMM(\delta)}$ (Alpers et al. 1988; Letvin and King 1991; Baskin et al. 1988) Affected glomeruli are hypercellular due to hypertrophy and hyperplasia of visceral epithelial cells (Fig. 16). Their

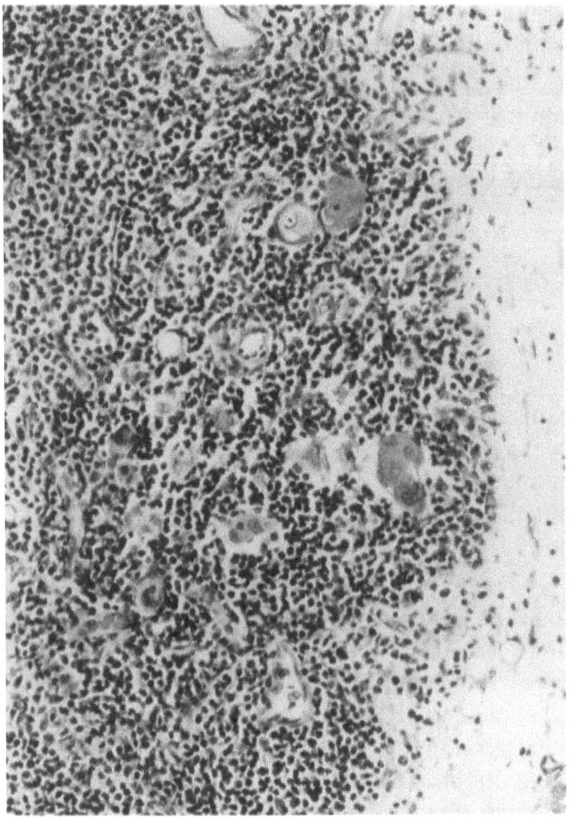

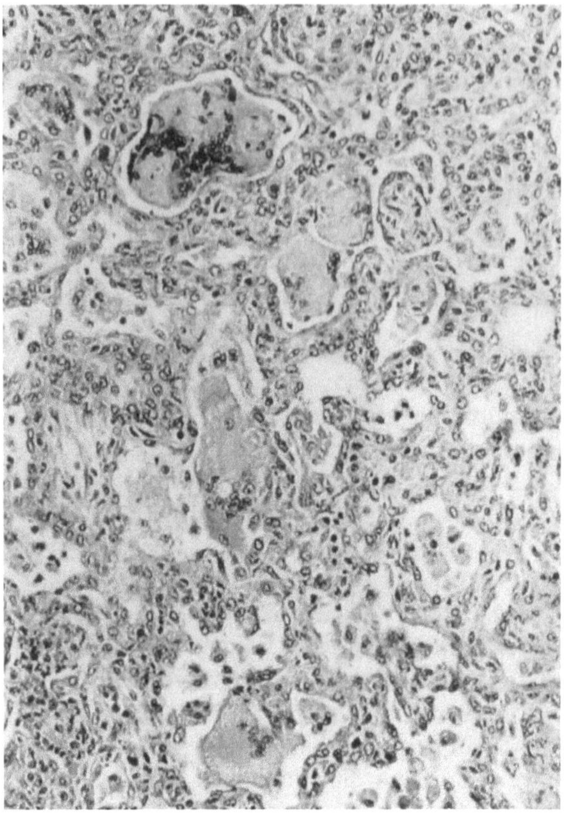

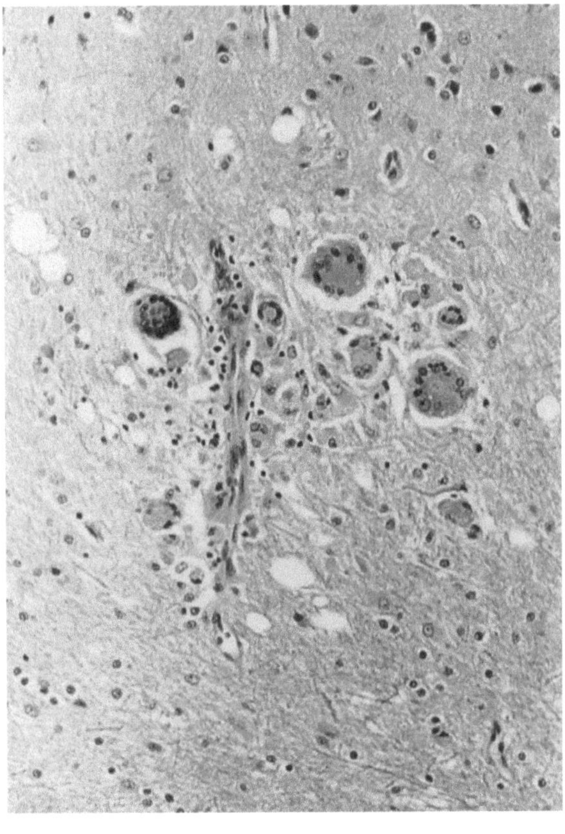

Fig. 12 (*upper left*). Atrophic thymus from a SIV_{MAC}-infected rhesus monkey. There is extensive depletion of cortical thyocytes and scattered residual Hassal's corpuscles. H and E, ×200

Fig. 13 (*lower left*). Cerebral cortex of an SIV_{MAC}-infected rhesus monkey. Note three microgranulomas consisting of perivascular infiltrations of histiocytes and multinucleate giant cells in both the gray and white matter. H and E, ×400

Fig. 14 (*upper right*). Lung of an SIV_{MAC}-infected rhesus monkey with extensive giant cell pneumonia. Note diffuse thickening of alveolar septa caused by mononuclear inflammatory cell infiltration and alveolar spaces contain large numbers of macrophages and multinucleate syncytial giant cells. H and E, ×400

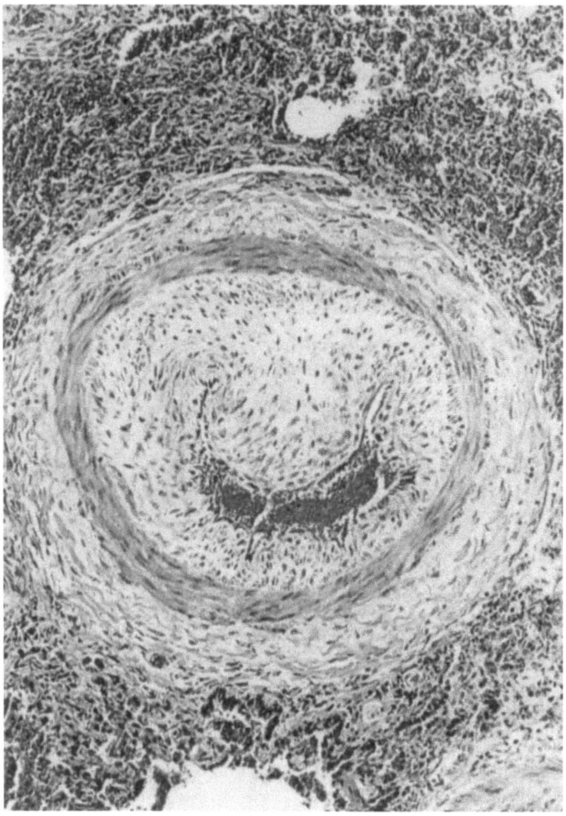

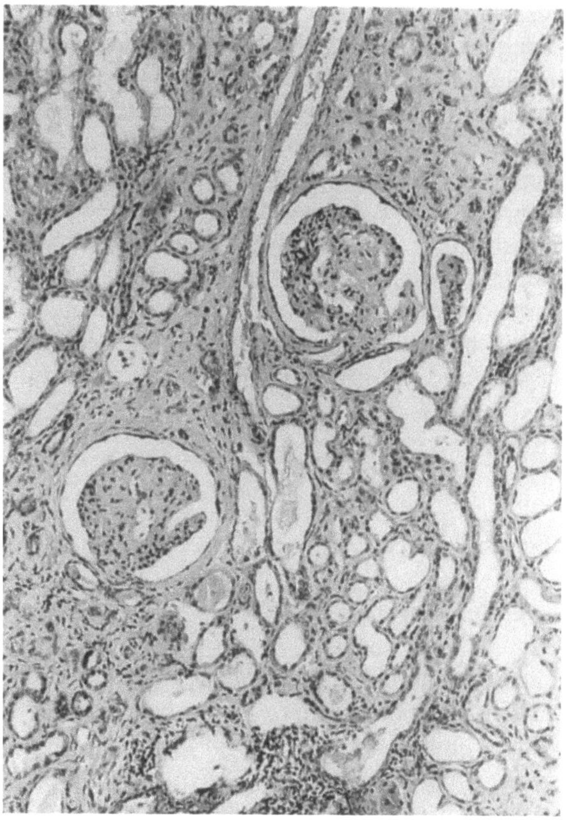

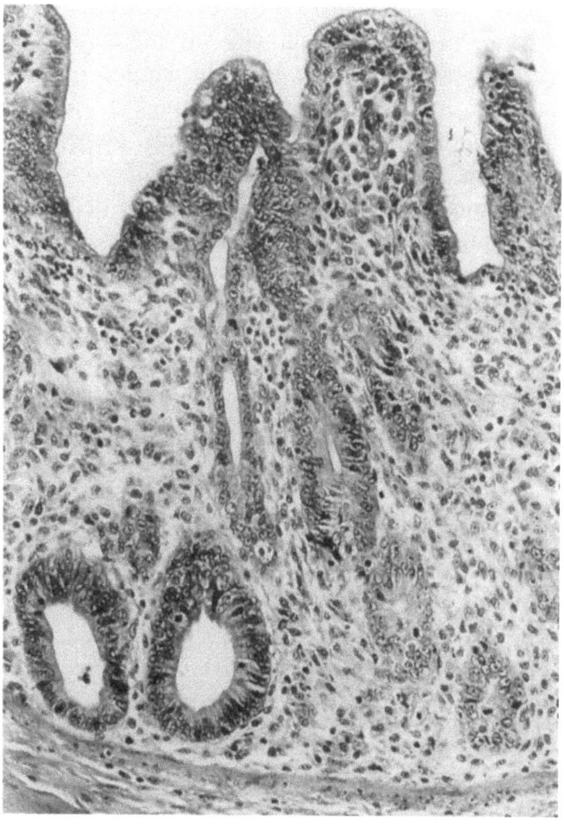

Fig. 15 (*upper left*). Pulmonary arteriole of SIV_{MAC}-infected rhesus. Note marked intimal proliferation with occlusion of the lumen. H and E, ×150

Fig. 16 (*upper right*). Renal cortex of an SIV_{MAC}-infected rhesus monkey in which there is increased cellularity of renal glomeruli, focal glomerular sclerosis, interstitial and periglomerular fibrosis, and associated tubular atrophy. H and E, ×100

Fig. 17 (*lower right*). Ileum of an SIV_{MAC}-infected rhesus monkey in which there is fusion and blunting of villi and moderate hyperplasia of crypts. The lamina propria of the apical portions of villi is distended by infiltrates of lymphocytes, plasma cells, and macrophages. H and E, ×200 basement membranes are thickened and often reduplicated and there is an increase in the mesangial matrix. These changes result in constriction and occlusion of glomerular capillary lumens. The urinary space of severely affected glomeruli may be dilated and protein casts may be seen in distal portions of the nephron. Tubules may become atrophic and the adjacent interstitium may be fibrotic and infiltrated with lymphocytes and plasma cells. Electron microscopy has failed to reveal immune complex deposits or viral particles in affected glomeruli; hence the pathogenesis of these lesions is not known.

Intestinal Tract

Small Intestine

In addition to *Cryptosporidia* and *Mycobacterium avium-intracellulare* (see elsewhere in this volume) infections, which occur commonly in the small intestine of SIV_{MAC}-infected rhesus monkeys, a nonspecific enteropathy is also observed with some frequency (Letvin and King 1990). It is characterized by severe blunting and atrophy of intestinal villi with varying degrees of crypt hyperplasia (Fig. 17). The lamina propria is often infiltrated with lymphocytes, plasma cells, and tingible body macrophages. Depending upon the stage of the disease, the gut-associated lymphoid tissue may be hyperplastic or depleted. This enteric lesion may, in part, be responsible for the protracted diarrhea that occurs in these infected monkeys in which a specific enteric pathogen cannot be identified. In situ hybridization has revealed that SIV_{MAC} RNA is located only in scattered lymphocytes and macrophages within the lamina propria, there being no evidence that SIV_{MAC} infects enterocytes.

Skin

Rhesus monkeys experimentally infected with SIV_{MAC} often develop a recurrent maculopapular rash. Microscopically, the superficial dermal capillaries of affected skin are congested with mild to moderate perivascular infiltrates of lymphocytes (Fig. 18) of the natural killer cell phenotype (Ringler et al. 1987a). Since these cells are often apposed to degenerate Langerhans' cells it has been suggested that the Langerhans' cells could be targets for the cytolytic natural killer cells. However, virus has not been identified in rhesus Langerhans' cells infected with SIV.

Ultrastructure

Electron microscopic examination of lymph nodes, spleen, brain, and other tissues containing histiocyte and giant cell infiltrates from macaques with SIV_{MAC} or $SIV_{SMM(\delta)}$ infection reveals varying numbers of lentivirus particles budding from the plasmalemma of lymphocytes and from cytoplasmic membranes into cytoplasmic vacuoles of histiocytes and multinucleate giant cells (Letvin et al. 1985; Ringler et al. 1989). In the latter cell types, numerous enveloped virions accumulate within membrane-bound cytoplasmic vacuoles. The intracellular assembly of virions is similar to that which occurs with the type C oncornaviruses that form a crescent-shaped nucleocapsid simultaneously with, but just beneath, that portion of the cell membrane that becomes the viral envelope (Fig. 19). Mature virions are 100–120 nm in diameter and have cylindrical or pyramidal-shaped nucleocapsids rather than spherical nucleocapsids as seen in mature C type oncornaviruses.

Differential Diagnosis

Infection with SIV must be differentiated from other conditions that cause immune dysfunction leading to death from opportunistic infections and/or lymphoma. Principal among these is infection with type D retroviruses which occurs with much greater frequency than SIV in many primate colonies. The clinical and pathologic features of type D retrovirus infections in macaque monkeys are described elsewhere in this monograph (see p. 20). Although SIV and type D retrovirus infections can be readily differentiated by serological and virus isolation procedures, serum, blood, or fresh tissues may not be available from animals in which these infections are suspected based upon gross or microscopic findings. In such cases, fixed or frozen tissues can be examined by electron microscopy, immunocytochemistry, and/or in situ hybridization to detect the presence of an active retrovirus infection and to differentiate SIV from type D retrovirus infection (Letvin et al. 1985; Ringler et al. 1989; Wyand et al. 1989).

Biologic Features

Natural History

Naturally occurring SIV infections in native populations of nonhuman primates have only been reported in African species in which they exist commonly as persistent infections unassociated with clinical disease. These infections include: SIV_{AGM} in African green monkeys (*Cercopithecus aethiops*) (Ohta et al. 1988; $SIV_{SMM(\delta)}$ in sooty mangabey monkeys (*Cercocebus atys*) (Fultz et al. 1986; Lowenstine et al. 1986; Murphey-Corb et al. 1986), SIV_{MND} in mandrills (*Mandrillus sphinx*) (Tsujimoto et al. 1988); SIV_{CMZ} in chimpanzees (*Pan troglodytes*) (Peeters et al. 1989), and most recently SIV_{SYK} in Sykes' monkeys (*Cercopithecus mitis*) (Emau et al. 1991). Although several Asian species of the genus *Macaca*, including *M. mulatta* (Letvin et al. 1985), *M. fascicularis* (Daniel et al. 1988; King et al. 1990), *M. arctoides* (Lowenstine et al. 1986), and *M. nemestrina* (Benveniste et al. 1986), are highly susceptible to and die from experimental or colony-acquired SIV infection, this virus has not been found in natural populations of these species (Lowenstine et al. 1986; Ohta et al. 1988). In laboratory-maintained macaques, SIV_{MAC}, $SIV_{SMM(\delta)}$, SIV_{MNE}, and SIV_{STM} infections occur as rare, sporadic, but fatal diseases (Daniel et al. 1988; Benveniste et al. 1986; Lowenstine et al. 1987). SIV_{MAC}, the first of the SIVs to be identified and characterized, was isolated from a rhesus monkey (*Macaca mulatta*) that died of a lymphoma (Daniel et al. 1985). Interestingly, $SIV_{SMM(\delta)}$, SIV_{MNE}, and SIV_{STM} were also originally isolated from rhesus (Murphey-Corb et al. 1986), pigtailed (Benveniste et al. 1986), and stump-tailed (Khan et al. 1991a,b) macaques, respectively, dying of lymphoma. Paradoxically, when these

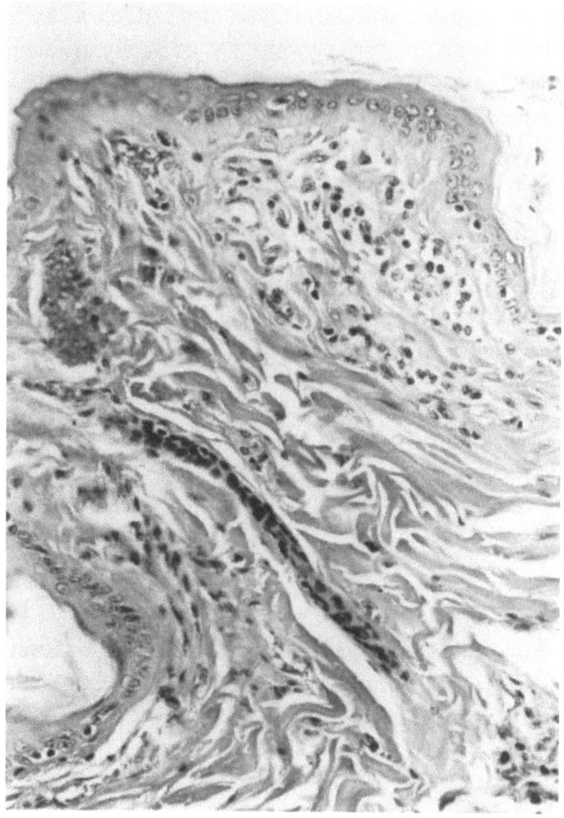

Fig. 18 (*above*). Skin from an SIV_{MAC}-infected rhesus ▶ monkey with an erythematous rash. The superficial dermis contains perivascular infiltrates of lymphocytes. H and E, ×200

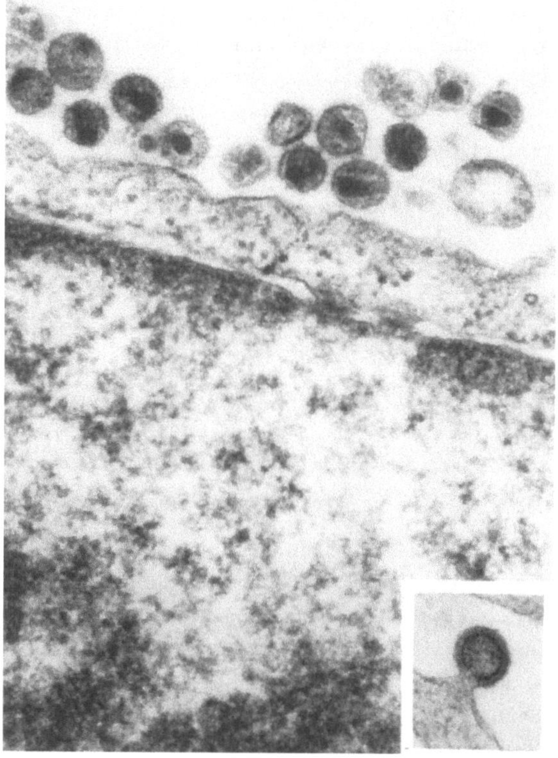

Fig. 19 (*below*). Multiple, mature, extracellular SIV_{MAC} particles. *Inset*, A single immature particle budding from the plasmalemma of a lymphoid cell. Note that the mature particles have a cylindrical or pyramidal nucleocapsid. TEM, ×60000

same species are experimentally infected with these same agents, lymphoma is rarely if ever observed. The explanation for this is not known. However, SIV_{MAC} consistently induces an AIDS-like disease in naturally and experimentally infected rhesus monkeys (Letvin et al. 1985; King et al. 1990). Although SIV_{SMM} (δ) does not cause clinical disease in its natural host (Fultz et al. 1990), the sooty mangabey, when rhesus monkeys are experimentally or incidentally infected with it, a disease is induced that is essentially identical to the one caused by SIV_{MAC} in this species (Murphey-Corb et al. 1986). In fact, the nucleotide sequences of SIV_{MAC} and $SIV_{SMM(\delta)}$ are so similar that it is now believed that SIV_{MAC} probably arose in the recent past from the chance transmission of $SIV_{SMM(\delta)}$ from a subclinically infected sooty mangabey to a rhesus monkey, an aberrant host for which it is pathogenic (Hirsch et al. 1989). Where and when this incident occurred has not been determined. SIV_{MNE} is also a very closely related agent and may have evolved in the same way. SIV_{STM}, by contrast, genetically appears to be an ancient and more divergent form of SIV that is the most closely related to human immunodeficiency virus (HIV)-2 (Khan et al. 1991a,b). Interestingly, SIV_{MAC} and $SIV_{SMM(\delta)}$ and variants thereof also cause a fatal AIDS-like disease in cynomolgus monkeys (*Macaca fascicularis*) (Daniel et al. 1988; King et al. 1990; Kuller et al. 1990) following naturally acquired or experimental infection. African green (*Cercopithecus aethiops*) and cynomolgus (*Macaca fascicularis*) monkeys can be experimentally infected with SIV_{AGM} but do not develop disease (Honjo et al. 1990).

Transmission

The principal mode of transmission of SIV among native populations and captive colonies of nonhuman primates has not been established. Since these viruses are readily transmissible experimentally by infected blood or serum, it is generally assumed that contamination of bite wounds with blood from infected animals is the most likely possibility. Transplacental transmission has also been documented in a rhesus monkey with naturally acquired infection (Daniel et al. 1988), but attempts to demonstrate maternal-fetal transmission of experimental infection at different stages of gestation have thus far failed (Davison-Fairburn et al. 1990). The virus has been experimentally transmitted via the genital mucosal route (Miller et al. 1989). It is also likely that spread within laboratory colonies can occur through careless reuse of hypodermic needles during group venipuncture, drug administration, or tuberculin testing procedures.

Clinical Findings

The clinical signs, symptoms, and course of SIV_{MAC} infection in macaque monkeys vary and depend, of course, upon the nature and organ specificity of the opportunistic infection(s) to which an animal succumbs. During the first 6–8 weeks of infection and continuing for variable periods thereafter, SIV_{MAC}-infected animals often develop a maculopapular rash that is most obvious on the less-haired portions of the body, i.e., face, abdomen, and medial aspects of the extremities. The duration of the rash can be weeks to months. Lymphadenopathy of the axillary and inguinal lymph nodes is also a frequent finding early in the course of infection. It can persist until death in those animals that die early but generally subsides prior to death in those animals with a more prolonged clinical course. Diarrhea leading to dehydration and weight loss is also a frequent finding in SIV_{MAC}-infected macaques.

Hematologic alterations in monkeys infected with the virus are variable and of limited diagnostic value. The most consistent hematologic abnormality is a reduction in the number of circulating CD+ (helper/inducer) lymphocytes in the peripheral blood (Letvin et al. 1985; Letvin and King 1990). Measurement of absolute numbers of circulating CD4+ lymphocytes in SIV_{MAC}-infected rhesus monkeys is an extremely useful prognostic test. Within the first week or so following infection, there may be a transient rise in circulating CD4+ and CD8+ lymphocytes. Thereafter, in animals which survive longer than a year the circulating CD4+ and CD8+ lymphocyte numbers return to normal. Normal values generally persist until just prior to the onset of clinical deterioration, when a progressive decline in the number of circulating CD4+ lymphocytes occurs. Terminally, there may occur a slight increase in CD4+ lymphocytes. In contrast, in those animals which develop rapidly progressive disease a dramatic decline in circulating CD4+ lymphocytes occurs almost immediately following SIV_{MAC} infection and continues until death. While not as dramatic as with CD4+ lymphocytes, circulating CD8+ cells may also decline, especially during the terminal stages of disease.

The clinical course of disease in animals infected with the same isolate of this virus can also be highly variable. Some animals die within 3–4 months after infection, while others survive longer than 2 years. The humoral immune response to SIV_{MAC}, since it reflects CD4+ lymphocyte functional activity, also can be used to predict the clinical course of an infected animal's disease. Those animals that survive the longest develop high titer anti-envelope and anti-gag antibody responses, whereas those animals that develop low titer antibody responses to these antigens usually have a short clinical course (Kannagi et al. 1986).

Pathogenesis

The SIVs have marked tropism for cells that express the CD4 molecule on their surface. These cells include the helper inducer subset of T lymphocytes and antigen-presenting cells of monocyte-macrophage origin. The viruses enter these permissive cells through an interaction between the viral envelope glycoprotein gp 120 and the CD4 molecule which serves as its receptor. Once inside the cell, the single-stranded viral RNA is transcribed via reverse transcriptase into DNA copies of itself which ultimately become integrated into the host cell DNA. Transcription of this proviral DNA results in the production of progeny virus that buds primarily from the surface of infected lymphocytes but mainly into cytoplasmic vacuoles of infected macrophages. In those species in which the virus causes fatal disease, there is a profound depletion of CD4+ lymphocytes leading to severe immune dysfunction and death from opportunistic infection or lymphoma. The mechanism(s) by which CD4+ lymphocytes are depleted in these animals is not known but the possibilities include: (1) accumulation of toxic quantities of viral nucleic acids or structural proteins in the cytoplasm of infected cells; (2) fusion of infected cells expressing the viral envelope glycoprotein, gp120, on their surface with noninfected CD4-bearing cells resulting in syncytia formation and death; (3) lysis of infected CD4+ T lymphocytes bearing viral-encoded antigens on their surface by virus-specific CD8+ (cytolytic) T cells, (4) lysis of infected CD4+ T lymphocytes by an antibody-dependent cellular cytotoxicity reaction and (5) enhancement of apoptosis (programmed cell death) of infected CD4+ lymphocytes. In contrast, infected macrophages, in which viral assembly occurs primarily within cytoplasmic vacuoles rather than on the surface of infected cells, are more resistant to lysis and, in fact, may be responsible for dissemination of virus to non-lymphoid tissue, such as the brain, because of their mobility. Although the immunodeficiency that occurs in macaques infected with the virus is due to the reduction of CD4+ lymphocytes, other lesions found in these animals clearly have another basis.

Etiology

The eight different SIV isolates from various species of nonhuman primates are all classified in the subfamily Lentivirinae of the family Retroviridae. Their genomes have from 35% to 75% sequence homology with the two known human lentiviruses, HIV-1 and HIV-2, the etiologic agents of human AIDS. Due to their genetic relationship to the HIVs, these nonhuman primate viruses have been designated simian immunodeficiency viruses, even though in reality only four of them have been associated with immunodeficiency disorders. The different SIV isolates are currently designated by a three-letter subscripted suffix to indicate the species from which they were originally isolated. They include: SIV_{MAC} from rhesus macaques (*Macaca mulatta*) (Daniel et al. 1985); SIV_{SMM}, also referred to as $SIV_{(\delta)}$, from sooty mangabey monkeys (*Cercocebus atys*) (Fultz et al. 1986; Lowenstine et al. 1986; Murphey-Corb et al. 1986); SIV_{MNE} from pig-tailed macaques (*Macaca nemestrina*) (Benveniste et al. 1986); SIV_{STM} from stump-tailed macaques (*Macaca arctoides*) (Lowenstine et al. 1987; Khan et al. 1991a,b); SIV_{AGM} from African green monkeys (*Cercopithecus aethiops*) (Ohta et al. 1988); SIV_{MND} from mandrills (*Mandrillus sphinx*) (Tsujimoto et al. 1988); SIV_{CMZ} from chimpanzees (*Pan troglodytes*) (Peeters et al. 1989); and SIV_{SYK} from a Sykes' monkey (*Cercopithecus mitis*) (Emau et al. 1991). Genetic sequence analysis of their respective *pol* genes has revealed that the eight different SIVs and two different HIVs fall into four distinct genetic groupings with SIV_{AGM} and

SIV_{SYK} comprising one group, SIV_{MND} a second group, SIV_{CMZ} and HIV-1 a third group, and SIV_{MAC}, $SIV_{SMM(\delta)}$, SIV_{MNE}, SIV_{STM}, and HIV-2 comprising the fourth group (Desrosiers 1990). The SIVs, like all lentiviruses, have additional genes beyond the three basic constitutive genes, *gag* (group specific antigen or core antigen), *pol* (polymerase or reverse transcriptase), and *env* (envelope), that are common to all replicatively competent retroviruses. These additional genes have been termed "accessory" or "novel" genes. Evidence is mounting that they play important roles in the regulation of virus replication and pathogenicity in vivo (Kestler et al. 1991). SIV_{MAC}, $SIV_{SMM(\delta)}$, SIV_{MNE}, and SIV_{STM}, like their human counterpart HIV-2, have six novel genes referred to as: *vif* (virion infectivity factor), *vpx* (viral protein X), *vpr* (viral protein R), *tat* (transactivator), *rev* (regulator of expression of viral proteins), and *nef* (negative factor). The function of some of these viral proteins, i.e., vpx and vpr, has yet to be defined. The recent identification of a pathogenic molecular clone of SIV_{MAC} will greatly facilitate definition of the functions of these novel genes in vivo (Kestler et al. 1990). SIV_{CMZ} and HIV-1 also have six novel genes, but instead of *vpx* they have a gene referred to as *vpu* (virus protein U) (Huet et al. 1990). SIV_{AGM} and SIV_{MND} have only five novel genes. SIV_{AGM} lacks *vpr* and SIV_{MND} lacks *vpx*. The genome of SIV_{SYK} has not yet been fully characterized, but in all likelihood its genomic organization will be similar to that of SIV_{AGM}.

Comparison with Other Species

The striking similarities between SIV-induced disease in macaque monkeys and HIV-induced disease in human beings make the SIV-infected macaque monkey an extremely valuable model for the study of AIDS. Macaques infected with SIV develop many of the same clinical and pathologic abnormalities that occur in AIDS patients and die of the same array of opportunistic infections and lymphomas. An exception is that no lesion completely comparable to Kaposi's sarcoma has been described in SIV-infected monkeys. The clinical course of SIV infection in macaques differs significantly from HIV infection in humans. While disease evolves in HIV-infected humans over a period of years to decades following infection, macaques infected with SIV typically die within 2 years of becoming infected. One of the most intriguing questions in SIV biology is why species such as the African green monkey, sooty mangabey, mandrill, and chimpanzee harbor naturally occurring SIV infections without developing disease, while humans with HIV infection and macaques with SIV_{MAC}, $SIV_{SMM}(\delta)$, SIV_{MNE}, or SIV_{STM} infection uniformly develop fatal disease. A basic understanding of how these species cope with their SIV infections could conceivably enhance our ability to improve and prolong the lives of human beings infected with HIV.

Acknowledgement. This work was supported by Public Health Service Award RR00168 from the National Institute of Health.

References

Alpers CE, Tsai C-C, Thouless M, Kueller L, Morton WR (1988) Focal and segmental glomerulosclerosis (FGS) in primates infected with simian immunodeficiency virus (SIV). Lab Invest 58:3A (abstr)

Baskin GB (1987) Disseminated cytomegalovirus infection in immunodeficient rhesus monkeys. Am J Pathol 129:345–352

Baskin GB, Murphey-Corb M, Watson EA, Martin LN (1988) Necropsy findings in rhesus monkeys experimentally infected with cultured simian immunodeficiency virus (SIV)/delta. Vet Pathol 25:456–467

Benveniste RE, Arthur LO, Tsai C-C, Sowder R, Copeland TD, Henderson LE, Oroszlan S (1986) Isolation of a lentivirus from a macaque with lymphoma: comparison with HTLV-III/LAV and other lentiviruses. J Virol 60:483–490

Blanchard JL, Baskin GB, Murphey-Corb M, Martin LN (1987) Disseminated cryptococcosis in simian immunodeficiency virus/delta-infected rhesus monkeys. Vet Pathol 24:454–456

Chalifoux LV, King NW, Daniel MD, Kannagi M, Desrosiers RC, Sehgal PK, Waldron LM, Hunt RD, Letvin NL (1986) A lymphoproliferative syndrome in an immunodeficient rhesus monkey naturally infected with an HTLV-III-like virus (STLV-III). Lab Invest 55:43–50

Chalifoux LV, Ringler DJ, King NW, Sehgal PK, Desrosiers RC, Daniel MD, Letvin NL (1987) Lymphadenopathy in macaques experimentally infected with the simian immunodeficiency virus (SIV). Am J Pathol 128:104–110

Conner MW, King NW (1989) Acquired immunodeficiency syndromes in non-human primates. In: Harawa SJ, O'Hara CJ (eds) Pathology and pathophysiology of AIDS and HIV-related diseases. Chapman and Hall, London, pp 459–470

Daniel MD, Letvin NL, King NW, Kannagi M, Sehgal PK, Hunt RD, Kanki PJ, Essex M, Desrosiers RC (1985) Isolation of T-cell tropic HTLV III-like retrovirus from macaques. Science 228:1201–1204

Daniel MD, Letvin NL, Sehgal PK, Schmidt DK, Silva DP, Solomon KR, Hodi FS, Ringler DJ, Hunt R, King NW, Desrosiers RC (1988) Prevalence of antibodies to 3 retroviruses in a captive colony of macaque monkeys. Int J Cancer 41:601–608

Davison-Fairburn B, Blanchard J, Hu F-S, Martin L, Harrison R, Ratterree M, Murphey-Corb M (1990) Experimental infection of timed-pregnant rhesus monkeys with simian immunodeficiency virus (SIV) during early, middle, and late gestation. J Med Primatol 19:381–393

Desrosiers RC (1990) HIV-1 origins, a finger on the missing link. Nature 345:288–289

Emau P, McClure HM, Isahakia M, Else JG, Fultz PN (1991) Isolation from African Sykes' monkeys (Cercopithecus mitis) of a lentivirus related to human and simian immunodeficiency viruses. J Virol 65:2134–2140

Fultz PN, McClure HM, Anderson DC, Swenson RB, Anand R, Srinivasan A (1986) Isolation of a T-lymphotropic retrovirus from naturally infected sooty mangabey monkeys (Cercocebus atys). Proc Natl Acad Sci USA 83:5286–5290

Fultz PN, McClure HM, Anderson DC, Switzer WM (1989) Identification and biologic characterization of an acutely lethal variant of simian immunodeficiency virus from sooty mangabeys (SIV/SMM). AIDS Res Hum Retroviruses 5:397–409

Fultz PN, Gordon TP, Anderson DC, McClure HM (1990) Prevalence of natural infection with simian immunodeficiency virus and simian T-cell leukemia virus type 1 in a breeding colony of sooty mangabey monkeys. AIDS 4:619–625

Hirsch VM, Olmsted RA, Murphey-Corb M, Purcell RH, Johnson PR (1989) An African primate lentivirus (SIVsm) closely related to HIV-2. Nature 339:389–392

Honjo S, Narita T, Kobayashi R, Hiyaoka A, Fujimoto K, Takasaka M, Sakakibara I, Mukai R, Ishikawa K, Ohta Y, Hayami M (1990) Experimental infection of African green monkeys and cynomolgus monkeys with SIV_{AGM} strain isolated from a healthy African green monkey. J Med Primatol 19:9–20

Huet T, Cheynier R, Meyerhans A, Roelants G, Wain-Hobson S (1990) Genetic organization of a chimpanzee lentivirus related to HIV-1. Nature 345:356–359

Kannagi M, Kiyotaki M, Desrosiers RC, Reimann KA, King NW, Waldron LM, Letvin NL (1986) Humoral immune responses to T cell tropic retrovirus simian T lymphotropic virus type III in monkeys with experimentally induced acquired immunodeficiency-like syndrome. J Clin Invest 78:1299–1236

Kestler H, Kodama T, Ringler D, Marthas M, Pedersen N, Lackner A, Regier D, Sehgal P, Daniel M, King N, Desrosiers R (1990) Induction of AIDS in rhesus monkeys by molecularly cloned simian immunodeficiency virus. Science 248:1109–1112

Kestler H III, Ringler DJ, Mori K, Panicali DL, Sehgal PK, Daniel MD, Desrosiers RC (1991) Importance of the *nef* gene for maintenance of high virus loads and for the development of AIDS. Cell 65:651–662

Khan AS, Galvin TA, Jennings MB, Gardner MB, Lowenstine LJ (1991a) SIV of stump-tailed macaques (SIV_{STM}) is a divergent Asian isolate. J Med Primatol 20:167–171

Khan AS, Galvin TA, Lowenstine LJ, Jennings MB, Gardner MB, Buckler CE (1991b) A highly divergent simian immunodeficiency virus (SIV_{STM}) recovered from stored stump-tailed macaque tissues. J Virol 65: 7061–7065

King NW (1986) Simian models of acquired immunodeficiency syndrome (AIDS): a review. Vet Pathol 23: 345–353

King NW, Hunt RD, Letvin NL (1983) Histopathologic changes in macaques with an acquired immunodeficiency syndrome (AIDS). Am J Pathol 113:382–388

King NW, Chalifoux LV, Ringler DJ, Wyand MS, Sehgal PK, Daniel MD, Letvin NL, Desrosiers RC, Blake BJ, Hunt RD (1990) Comparative biology of natural and experimental SIV_{MAC} infection in macaque monkeys: a Review. J Med Primatol 19:109–118

Kuller L, Morton WR, Benveniste RE, Tsai C-C, Clark EA, Gale MJ, Hu SL, Thouless ME, Katze MG (1990) Inoculation of Macaca fascicularis with simian immunodeficiency virus, SIV_{MNE}, immunologic, serologic and pathologic changes. J Med Primatol 19:367–380

Letvin NL, King NW (1990) Immunologic and pathologic manifestations of the infection of rhesus monkeys with simian immunodeficiency virus of macaques. J Acquir Immune Defic Syndr 3:1023–1040

Letvin NL, Daniel MD, Sehgal PK, Desrosiers RC, Hunt RD, Waldron LM, MacKey JJ, Schmidt DK, Chalifoux LV, King NW (1985) Induction of AIDS-like disease in macaque monkeys with T-cell tropic retrovirus STLV-III. Science 230:71–73

Lowenstine LJ, Pedersen NC, Higgins J, Pallis KC, Uyeda A, Marx P, Lerche NW, Munn RJ, Gardner MB (1986) Seroepidemiologic survey of captive Old-World primates for antibodies to human and simian retroviruses, and isolation of a lentivirus from sooty mangabeys (Cerocebus atys). Int J Cancer 38:563–574

Lowenstine LJ, Lerche N, Jennings M, Marx P, Gardner M, Pedersen N (1987) An epizootic of simian AIDS caused by SIV in captive macaques in the 1970's. In: Girard M, Vallette L (eds) Retroviruses of human AIDS and related animal viruses. Pasteur Vaccines, Paris, pp 174–176

Lowenstine LJ, Lerche NW, Yee JA, Uyeda A, Jennings MB, Munn RJ, McClure HM, Anderson DC, Fultz PN, Gardiner MB (1992) Evidence for a lentiviral etiology in an epizootic of immune deficiency and lymphoma in stump-tailed macaques (Macaca arctoides). J Med Primatol 21:1–14

McClure HM, Anderson DC, Ansari AA, Fultz PN (1989) Spectrum of disease in macaque monkeys chronically infected with SIV/SMM. Vet Immunol Immunopathol 21:13–24

Miller CJ, Alexander NJ, Sutjipto S, Lackner AA, Getti A, Hendrickx AA, Lowenstine LJ, Jennings M, Marx PA (1989) Genital mucosal transmission of simian immunodeficiency virus: animal model for heterosexual transmission of human immunodeficiency virus. J Virol 63:4277–4284

Murphey-Corb M, Martin LN, Rangan SR, Baskin GB, Gormus BJ, Wolf RH, Andes WA, West M, Montelaro RC (1986) Isolation of an HTLV-III-related retrovirus from macaques with simian AIDS and its possible origin in asymptomatic mangabeys. Nature 321:435–437

Ohta Y, Masuda T, Tsujimoto H, Ishikawa K-I, Kodama T, Morikawa S, Nakai M, Honjo S, Hayami M (1988) Isolation of simian immunodeficiency virus from African green monkeys and seroepidemiologic survey of the virus in various non-human primates. Int J Cancer 41:115–122

Peeters M, Honore C, Huet T, Bedjabaga L, Ossari S, Bussi P, Cooper RW, Delaporte E (1989) Isolation and partial characterization of an HIV-related virus occurring naturally in chimpanzees in Gabon. AIDS 3:625–630

Ringler DJ, Hancock WW, King NW, Letvin NL, Daniel MD, Desrosiers RC, Murphy GF (1987a) Immunophenotypic characterization of the cutaneous exanthem of SIV-infected rhesus monkeys. Am J Pathol 126: 199–207
Ringler DJ, Hunt RD, Desrosiers RC, Daniel MD, Chalifoux LV, King NW (1987b) Simian immunodeficiency virus induced meningoencephalitis: natural history and retrospective study. Ann Neurol [Suppl] 23:S101–S107
Ringler DJ, Wyand MS, Walsh DG, MacKey JJ, Chalifoux LV, Papovic M, Minmassian AA, Sehgal PK, Daniel MD, Desrosiers RC, King NW (1989) Cellular localization of simian immunodeficiency virus in lymphoid tissues: I. Immunohistochemistry and electron microscopy. Am J Pathol 134:385–393
Sharer LR, Baskin GB, Cho E-S, Murphey-Corb M, Blumberg BM, Epstein LG (1987) Comparison of simian immunodeficiency virus and human immunodeficiency virus encephalitides in the immature host. Ann Neurol [Suppl] 23:S108–S112
Tsujimoto H, Cooper RW, Kodama T, Fukasawa M, Miura T, Ohta Y, Ishikawa K-I, Nakai M, Frost E, Roelants GE, Roffi J, Hayami M (1988) Isolation and characterization of simian immunodeficiency virus from mandrills in Africa and its relationship to other human and simian immunodeficiency viruses. J Virol 62: 4044–4050
Wyand MS, Ringler DJ, Naidu YM, Mattmuller M, Chalifoux LV, Sehgal PK, Daniel MD, Desrosiers RC, King NW (1989) Cellular localization of simian immunodeficiency virus in lymphoid tissues: II. In situ hybridization. Am J Pathol 134:385–383

Type D Retrovirus Infection, Macaques

Linda J. Lowenstine

Synonyms. Simian retrovirus infection; simian acquired immunodeficiency syndrome (SAIDS or simian AIDS).

Gross Appearance

With the exception of retroperitoneal fibromatosis, subcutaneous fibrosarcomas, and possibly noma, there is no gross lesion pathognomonic for type D retrovirus infection in macaques. Gross lesions vary considerably depending upon the clinical outcome of infection. Persistently viremic healthy carriers have been recognized who may have no gross lesions (Ahmed et al. 1974; Lerche et al. 1986). Some animals develop only mild illness with transient lymphadenopathy corresponding to a period of viremia from which they recover (Maul et al. 1986). Animals which are fatally infected develop a spectrum of immunologic and hematologic abnormalities that collectively are called simian acquired immunodeficiency syndrome (SAIDS) and may have a disease course ranging from 1 month to several years (Fine et al. 1975; Bryant et al. 1986a; Letvin et al. 1984; Maul et al. 1986). In animals dying acutely, weight loss, splenomegaly, and lymphadenopathy are the most consistent lesions. Pallor due to anemia is common and purpura due to thrombocytopenia may also be seen. The thymus is usually atrophic. Lesions referable to opportunistic infections may also be appreciated grossly, especially oral lesions of ulcerative gingivitis or noma, and interstitial or bronchopneumonias from cytomegalovirus or bacteria (Osborn et al. 1984; Maul et al. 1986; Schiodt et al. 1988).

Animals that die of SAIDS after a longer clinical course generally do not have anemia or thrombocytopenia. Wasting is usually severe and grossly apparent gastrointestinal lesions are common as are pneumonias (Osborn et al. 1984). Contracture and pallor or streaking of the muscles of the hind limbs may be seen in up to 50% of type D virus-infected macaques (Dalakas et al. 1986). Lymph nodes, although initially enlarged, may be smaller than normal at death (Chalifoux et al. 1984; Osborn et al. 1985). Retroperitoneal fibromatosis has been reported in approximately 25% of deaths (17/67) in a colony of Celebes macaques in which type D infection was common (Shigii et al. 1986), and multiple or solitary skin nodules of varying size are occasionally detected (London et al. 1983; Osborn et al. 1984; Tsai et al. 1985b).

Some macaques develop a protracted course in which persistent lymphadenopathy and splenomegaly are important features. This clinical outcome is referred to as SAIDS-related complex and is often accompanied by generalized lymphoproliferative disease (Maul et al. 1986; Heidecker et al. 1987).

The thymus may be grossly normal, atrophic, or occasionally enlarged.

The terms "SAIDS and simian AIDS" have also been used to describe acquired immunodeficiency in macaques induced by the simian immunosuppressive lentiviruses (SIV) described elsewhere in this volume (see previous chapter). There are many similarities in clinical signs and lesions in these two retrovirus infections, but there are some significant differences which may help to distinguish them (Table 2) (Baskin et al. 1988; King et al. 1983; Letvin and King 1990; McClure et al. 1989; Osborn et al. 1984; Giddens et al. 1985; L.J. Lowenstine, unpublished).

Table 2. Comparison of clinical signs and lesions of simian retrovirus (SRV) and simian immunosuppressive virus (SIV) infections in rhesus macaques[a]

Clinical signs or lesion	Type D (SRV)	SIV
Wasting	++	++
Diarrhea	++	++
Generalized lymphadenopathy	++	++
Splenomegaly	++	+
Hematologic abnormalities	++	+
Anemia	++	+
Neutropenia	++	+
Lymphopenia	++	+
Selective decrease in CD4+ subsets of lymphocytes	−	++
Thrombocytopenia	+	+
Altered mitogen blastogenesis		
Phytohemagglutinin	++	−
Conconavalin A	++	−
Pokeweed mitogen	+	++
Pyogenic bacterial infections	++	+
Opportunistic infections	++	++
Viral infections		
Cytomegalovirus	++	++
Adenovirus	+/−	+
Papovavirus (SV40)	+/−	+
Granulomatous bacterial enteritis		
Mycobacterium avium	+/−	++
Rhocococcus equi	+	+/−
Yeast infections		
Candida, etc.	+	+
Pneumocystis carinii	−	++
Protozoa		
Cryptosporidium	+	+
Trichomonads	+	+
Balantidium	+	+/−
Proliferative disorders		
Lymphoproliferative disease	+	+
Lymphoma	−	++
Retroperitoneal fibromatosis	++	−
Subcutaneous fibrosarcomas	++	−
Gingivitis/noma	+	+/−
Encephalitis	+/−	++
Giant cells inflamation in lung lymph nodes or other tissues	−	++
Myositis	++	−
Major vascular or atrial thrombosis	−	+

++ Very common or highly suggestive of this infection; + often reported; +/− reported but rare; − not reported.
[a] From Baskin et al. 1988; King et al. 1983; Letvin and King 1990; McClure et al. 1989; Osborn et al. 1984; Giddens et al. 1985; L.J. Lowenstine, unpublished.

Microscopic Features

As is the case for gross lesions, there is no histologic lesion pathognomonic for type D retrovirus infection in macaques (King et al. 1983; Osborn et al. 1984). In acute infections, pancytopenia occurs in the face of marrow hyperplasia and apparent maturation arrest (Osborn et al. 1984; MacKenzie et al. 1986). Sometimes the hypercellularity of the marrow and the large number of blast cells suggest a myeloproliferative disorder. Changes in lymph nodes during acute or recrudescent viremia are somewhat unique. There is marked expansion of the paracortex by large cells with abundant cytoplasm which when stained with H and E and examined by light microscopy might be mistaken for macrophages (Figs. 20, 21). Immunocytochemistry reveals these cells to be immature T cells of both CD4 and CD8 subsets (Chalifoux et al. 1984; Osborn et al. 1985). Follicles are compressed and flattened beneath the capsule. This atypical interfollicular or paracortical hyperplasia has been noted in human viral infections, especially with the Epstein-Barr virus, but has not been described in any other infection in macaques.

As the infection persists there is strong evidence of marked B cell stimulation with hypertrophic, sometimes misshapen, follicles present in lymph nodes and spleen (Fig. 22) (Meyer et al. 1985; Chalifoux et al. 1984). In cases of persistent lymphadenopathy and splenomegaly, lymphoid follicles appear in a variety of organs and tissues, usually salivary glands, pancreas, kidneys, thymic medulla, and bone marrow. Gut – and bronchus – associated lymphoid tissues increase in volume. Lymphoid follicles may also become evident in pituitary, adrenal, and male and female reproductive tracts. These nodal and extranodal, nodular, lymphocytic infiltrations are often extensive enough to suggest a diagnosis of lymphoma, but they are poly-

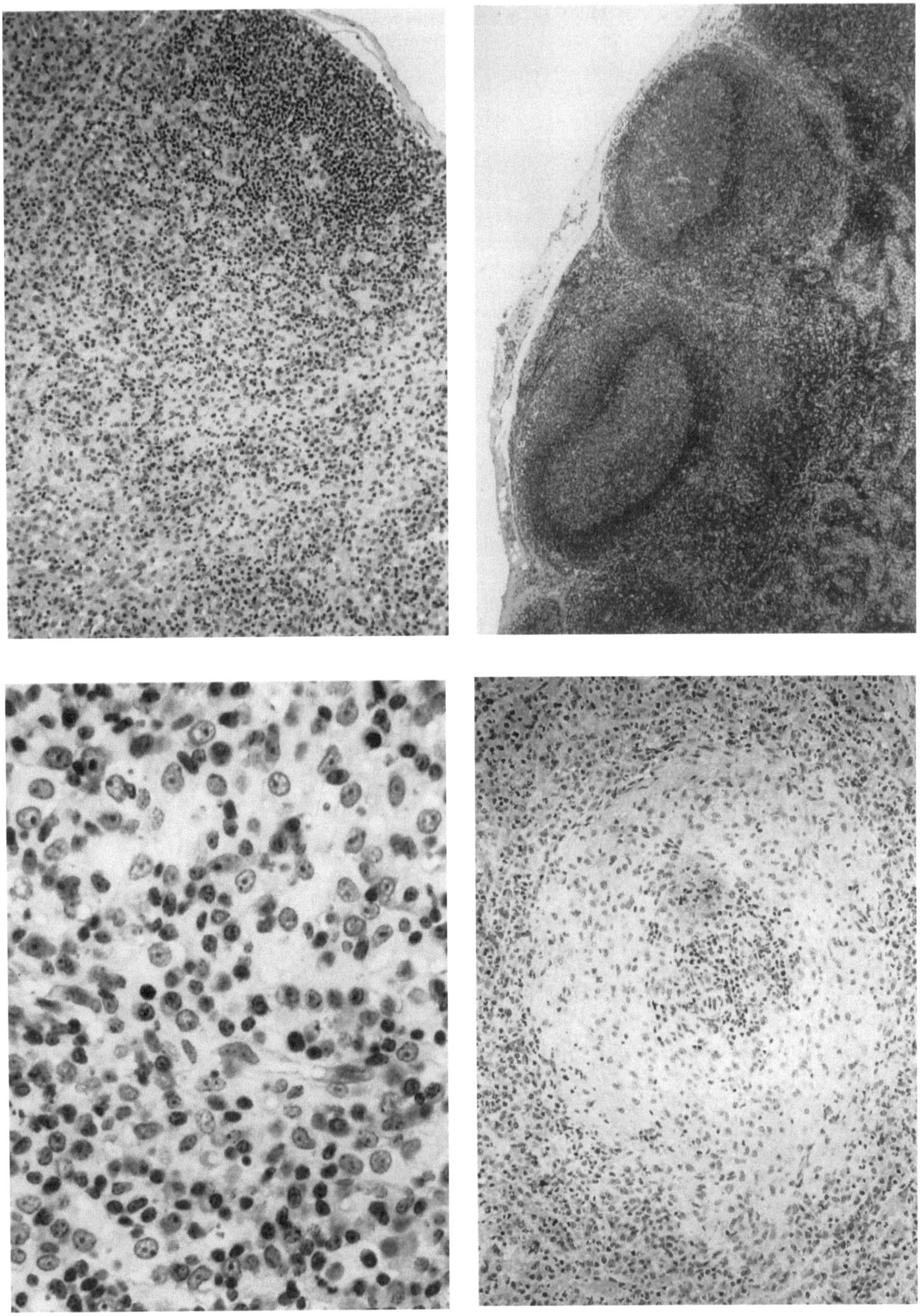

clonal and in none of the cases followed at the California Regional Primate Research Center have they progressed to lymphoma, as may occur in immunosuppressive lentivirus infection in macaques (Baskin et al. 1988; Hunt et al. 1983) (Lowenstine, unpublished data).

The lymph nodes from animals that are stably infected with type D virus or that are progressing toward SAIDS are often morphologically normal with varying degrees of follicular and interfollicular hyperplasia. Immunohistochemistry, however, reveals the presence of abnormal numbers of suppressor/cytotoxic (CD8+) T cells in follicular centers, and there are fewer plasma cells than might be expected (Meyer et al. 1985; Osborn et al. 1985).

In terminally ill animals, there is often profound lymphoid depletion in lymph nodes and spleen. In some cases extensive histiocytosis of cortex, paracortex, cords, and sinusoids may develop. Sinusoidal erythrophagocytosis is commonly observed. In other animals, germinal centers are atrophic and hyalinized, the paracortex is virtually devoid of lymphocytes, and the cords are narrow and hypocellular. In lymph nodes with distensible sinusoids, such as those in the mesenteries, sinusoidal ectasia and lymphoid depletion combine to produce a "cob-web" appearance (Meyer et al. 1985; Chalifoux et al.

◀ **Fig. 20** (*upper left*). Lymph node from a virus and antibody-positive rhesus macaque 1 month after intravenous inoculation with simian retrovirus (SRV)-1. There is expansion and homogenization of the paracortex (interfollicular areas) and a follicle can be seen pushed to the periphery. H and E, ×140

Fig. 21 (*lower left*). Paracortical area from the node of Fig. 20 is expanded by a population of large pale staining cells that were identified by immunohistochemistry to be immature lymphocytes of both CD4 and CD8 subsets. H and E, ×400

Fig. 22 (*upper right*). Lymph node from an antibody-positive/virus-negative rhesus macaque 4 weeks after intravenous inoculation with simian retrovirus (SRV)-1. There is marked follicular hyperplasia with one, large, dumbbell-shaped follicle. H and E, ×48

Fig. 23 (*lower right*). Spleen from a rhesus macaque that was euthanized 109 days after intravenous inoculation with simian retrovirus (SRV)-1. Lymphadenopathy and splenomegaly had been present previously. A periarteriolar lymphoid sheath is nearly devoid of lymphoid cells and is populated by pale staining histiocytic cells. H and E, ×48

1984). In animals that become depleted of lymphoid cells after a period of marked hyperplasia, depletion is also apparent in extranodal lymphoid nodules. In the spleen, white pulp is often lost and replaced by amyloid or fibrin, leaving eosinophilic ghosts of periarteriolar lymphoid sheaths and follicles (Fig. 23).

Histologic changes in other organs and tissues are usually the result of secondary infections or manifestations of persistent antigenic stimulation by the type D retrovirus. In addition to lymphoproliferative disease, amyloidosis involving intestinal lamina propria, spleen, and occasionally liver is seen. Secondary pathogens commonly identified terminally in both natural and experimental type D virus-induced immunodeficiency at the California Primate Center are *Campylobacter jejuni*, cytomegalovirus, gastric trichomonads, *Candida*, and intestinal and biliary *Cryptosporidium* (Osborn et al. 1984; Maul et al. 1986). At other institutions, balantidiasis and amoebiasis are also common (Giddens et al. 1985). The histologic lesions of acute necrotizing ulcerative gingivitis, noma, the fibroproliferative disorders, amyloidosis, and many of the opportunistic infections found in type D virus infected macaques are described in detail elsewhere in this volume.

Lesions that may be the result of direct type D retrovirus infection include nonsuppurative enteritis in which other pathogens are not identified but in which type D antigens can be found in enterocytes by immunohistochemistry (Lackner et al. 1989). Sialoadenitis, involving major and minor salivary glands, is present in many animals and type D retrovirus is abundant in salivary glands (Lackner et al. 1988). Polymyositis has been described in which there is severe perivascular, interstitial, and endomysial mononuclear inflammatory cell infiltration, and proliferation of endomysial fibroblasts. Type D virus antigens, but no other bacterial or viral pathogens including cytomegalovirus, can be detected in these lesions. Primary muscle cell tissue cultures can be infected with the virus (Dalakas et al. 1986). In other studies, however, skeletal muscle consistently served as a negative control for type D virus nucleic acid (Bryant et al. 1986b).

No lesion in the brain specific for type D retrovirus has been identified (Lackner et al. 1989). Encephalitis and meningitis, when present, can usually be attributed to secondary pathogens.

Ultrastructure

The ultrastructural morphology of type D retroviruses is unique (Munn et al. 1985). Immature virions are preformed as intracytoplasmic circular A particles and can be seen in aggregates with vesicles of smooth endoplasmic reticulum (Fig. 24).

The virions bud through the cell membrane as circular or, occasionally, crescentic buds (Fig. 25). The extracellular mature virions are round to hexagonal with a rod-shaped, cylindrical, or occasionally conical core; a diffuse centrally placed nucleoid (best appreciated when the core is seen "end on"); and an external diameter of about 120 nm (Fig. 26). They are most easily found in salivary gland acini, but are present in a variety of lymphoid and nonlymphoid tissues. In lymph nodes they are usually restricted to the germinal centers in the vicinity of dendritic reticular cells. Their presence is not associated with any particular cytopathic effect, but tubulocrystalline arrays and multivesicular bodies have been seen in endothelial and lymphoid cells (Tsai et al. 1985b; R.J. Munn, unpublished).

Differential Diagnosis

In macaques with opportunistic infections consistent with immune deficiency, the differential diagnosis includes infection with the simian immunosuppressive virus (SIV) of the lentivirus subfamily, other immunosuppressive viral infections (especially measles), other noninfectious causes of immunosuppression (e.g., drugs or toxins), and congenital immune deficits (see p. 1). In animals with marked lymphoid hyperplasia or systemic amyloidosis, the differential has to include any persistent antigenic or inflammatory stimulus such as SIV or other chronic infections and catheters or other implanted foreign material.

Infection of macaques with lentivirus is usually experimental (see p. 5), although "naturally occurring" cases have been reported (King 1986; Lowenstine et al. 1992). The types of opportunistic infections may help differentiate between the two diseases (Table 2). Although both the simian type D and lentiviruses cause syncytial giant cell formation in tissue culture, such syncytia are seen in vivo only in SIV infection (Daniel et al. 1984; Baskin et al. 1988; McClure et al. 1989).

Biologic Features

Natural History

Natural infection with exogenous type D retroviruses is found almost exclusively in Asian Old World monkeys of the genus *Macaca*. Serological evidence of infection in African nonhuman primates is restricted to two reports. In one, talapoins (*Miopithecus talapoin*) appeared to have antibodies reactive with type D virus-specific bands recognized by immunoblotting (Lowenstine et al. 1986). In another, antibodies reactive to the Mason-Pfizer monkey virus, the prototype type D retrovirus, were demonstrated in African green monkeys (*Cercopithecus aethiops*) (Fine and Arthur 1981). Monkeys of the guenon group to which these two species belong, however, often have broadly reactive sera of uncertain significance (Pedersen et al. 1986). Squirrel monkeys and African green monkeys are not susceptible to intravenous inoculation (P.A. Marx, unpublished).

Infections in groups of macaques may be either enzootic or epizootic (Lerche et al. 1984, 1987; Tsai et al. 1985a). Epidemics may occur when naive individuals of the same or different species are introduced into a group in which the disease is enzootic or when the virus is accidentally introduced into naive populations by blood transfusions or tissue transplants. The virus is shed in the saliva and transmission requires direct contact. Social interactions such as grooming, licking, and biting are thought to facilitate the spread. Infection in utero may occur (Tsai et al. 1985a; N.W. Lerche, unpublished). In an outdoor-housed population of rhesus in which type D infection was endemic, the mortality rate among introduced juveniles was 85% in the first 9 months. In juveniles born in the cage the earliest mortality from type D virus-related disease was at 3–4 months, and about 45% of each birth cohort died before reaching sexual maturity (Lerche et al. 1987). Reproductive success is often lower in groups of macaques infected with type D virus (Shiigi et al. 1986; N.W. Lerche, unpublished). Some animals, which probably carry the virus from birth, become persistently infected with the virus and may be antibody negative; these animals serve as healthy carriers (Ahmed et al. 1974; Lerche et al. 1986). Some of these animals will eventually succumb from the effects of their infection, with manifestations such as fibrosarcoma, or

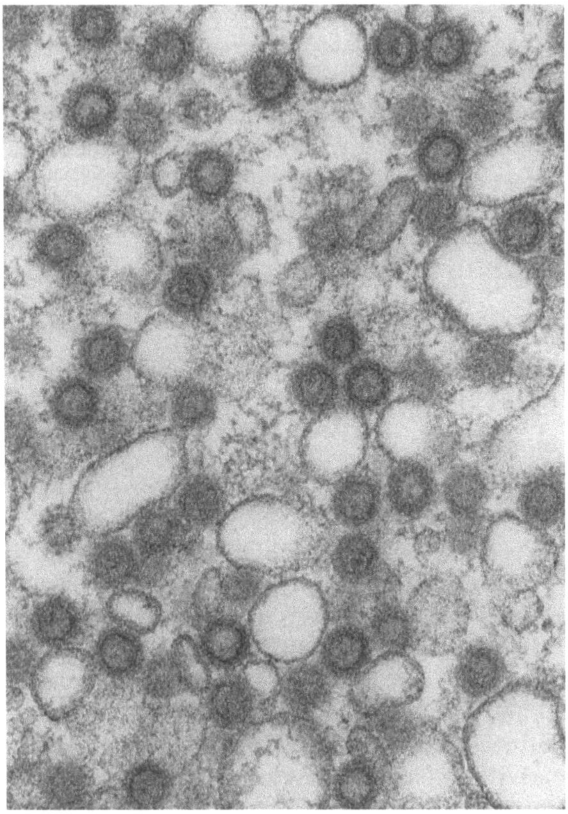

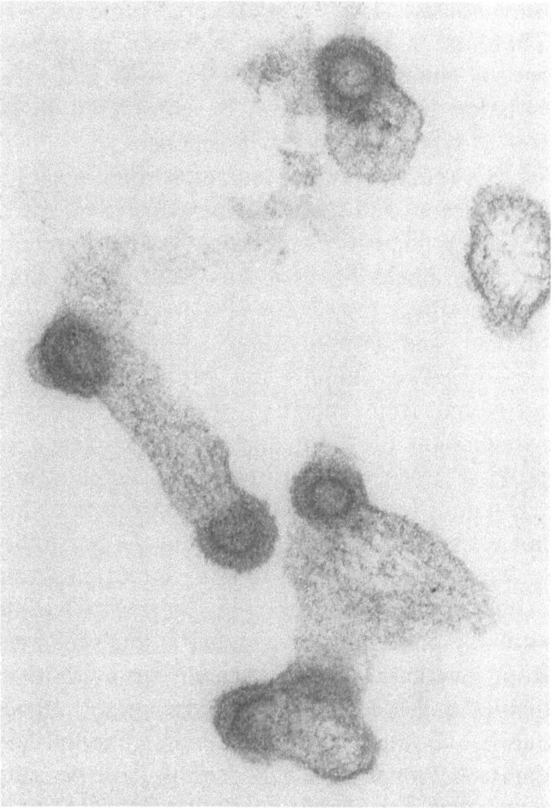

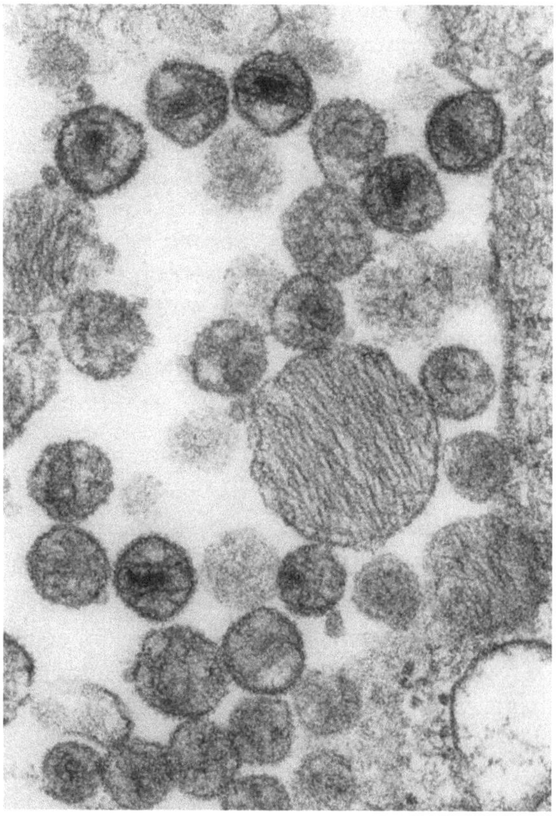

Fig. 24 (*upper left*). Cluster of preformed intracytoplasmic A particles in close association with vesicles of smooth endoplasmic reticulum in a Raji cell culture infected by cocultivation with peripheral blood mononuclear cells from a simian retrovirus (SRV)-1 positive rhesus. TEM, ×80 000. (Courtesy of R.J. Munn)

Fig. 25 (*upper right*). Circular and crescentic buds of simian retrovirus (SRV)-1 in cytoplasmic processes of a Raji cell. TEM, ×80 000. (Courtesy of R.J. Munn)

Fig. 26 (*lower right*). Extracellular mature virions of simian retrovirus (SRV)-1, grown in Raji cells, with characteristic rod-shaped nucleus and centrally placed diffuse nucleoid. ×80 000. (Courtesy of R.J. Munn) amyloidosis (L.J. Lowenstine, unpublished). Therefore it is important to screen individuals several times by both serologic tests and virus isolation before they can be considered to be free of type D retrovirus infections.

ELISA (enzyme-linked immunosorbent assay) is a convenient and relatively inexpensive means of screening populations of monkeys, but immunoblotting should be used for confirmation since false positives may occur (Pedersen et al. 1988; Spinner and Sujipto 1988). Antibody-positive virus-negative animals are presumed to have recovered from infection; however, extensive examination of these animals for evidence of proviral DNA by sensitive gene amplification techniques (e.g., polymerase chain reaction) has not yet been conducted to rule out the possibility of latent infection. In a recently imported group of cynomolgus monkeys (*M. fascicularis*) some antibody-positive, virus-negative animals released from quarantine later became virus positive, highlighting the need to consider these individuals to be latent carriers (Lerche, unpublished data). A formalinized killed whole virus vaccine that induces neutralizing antibodies has been shown experimentally to protect against infection (Marx et al. 1986).

Lesions suggestive of acquired immune suppression have been recognized in captive macaques since the 1960s, when the national primate centers were founded. The importance of retroviruses as the etiologic agents of these immunosuppressive disorders was not recognized until the 1980s, although experimental evidence was published identifying immunosuppression in young macaques inoculated with the Mason-Pfizer monkey virus (Cockrell et al. 1974; Fine et al. 1975; Lowenstine et al. 1992).

Clinical Findings

Infection of rhesus macaques by the type D retroviruses may have a variety of clinical outcomes: (1) development of viremia with antibody response and clearance of the infection (recovery); (2) development of a virus-positive, inapparent, carrier state (with or without detectable antibodies and with inconsistent virus isolation); (3) development of severe immunodeficiency (referred to as simian AIDS or SAIDS) with either an acute course of 1–5 months (often without detectable antibodies) or a more protracted course of 7 or more months; (4) development of persistent lymphadenopathy (referred to as SAIDS-related complex) which usually progresses over a period 2 or more years to overt immunodeficiency; and (5) development of fibroproliferative disorders, retroperitoneal fibromatosis, or subcutaneous fibrosarcomas (often coincident with SAIDS or its related complex) (Kwang et al. 1987; Maul et al. 1986; Osborn et al. 1984; Giddens et al. 1985). There is some evidence that other species of macaques, e.g., cynomolgus (*M. fascicularis*) or bonnet monkeys (*M. radiata*), may have different manifestations of infection (N.W. Lerche et al. unpublished).

A case definition has been developed for simian AIDS caused by type D retrovirus in rhesus. It is: generalized lymphadenopathy and/or splenomegaly with four or more of the following: weight loss (>10% of body weight); anemia (packed cell volume <30%) or neutropenia (<1700/mm^3) in the face of bone marrow hyperplasia; persistent lymphopenia (<1600/mm^3); persistent or intermittent diarrhea poorly responsive to therapy; bacterial infections of skin, oral cavity, or other sites unresponsive to therapy; opportunistic infection such as cytomegalovirus infection, cryptosporidiosis, or oral and esophageal candidiasis. Animals with SAIDS-related complex are those with persistent generalized lymphadenopathy and fewer than four of the criteria used to define SAIDS (Henrickson et al. 1984).

Pathogenesis

The pathogenesis of type D infection has not yet been completely elucidated. Direct contact is needed and inoculation of the virus into the skin by bite wounds or salivary contamination of skin wounds are the most likely natural means of infection (Lerche et al. 1986). Intravenous inoculation is the experimental route most widely studied. Infection can be established using infected blood or tissues, virus grown in nonhuman, primate-derived, cell culture systems, or molecularly cloned virus (Letvin et al. 1984; Marx et al. 1984; Heidecker et al. 1987; Wilson et al. 1986). After intravenous inoculation, viremia is detected in 1–3 weeks by cocultivation of peripheral blood mononuclear cells of the infected animals with susceptible cell lines such as Raji cells in which the virus induces characteristic syncytial cells (Daniel et al. 1984; Maul et al. 1986). The receptor by which the virus attaches to host cells has not yet been identified, but must be present on many cell types. By panning and immunohistochemical techniques to detect virus-encoded antigens and Southern blotting and in situ hybridization to identify viral nucleic acids, it is obvious that type D retroviruses have a broad tropism in vivo which includes B cells, CD4 and CD8 T cells, macrophages, and epithelial cells in salivary gland, oral mucosal Langerhans cells, intestines, and choroid plexus (Maul et al. 1988; Lackner et al. 1988, 1989; Bryant et al. 1986b). Type D retrovirus DNA and RNA can be demonstrated by Southern blotting and in situ hybridization in the brain parenchyma of infected macaques, but viral antigen and virus particles have been demonstrated only in choroid plexus (Lackner et al. 1989). The virus burden and extent of organ involvement appear to increase with time.

Immune suppression in infected rhesus monkeys affects both innate and acquired immunity. Neutropenia often occurs during acute infection and leads to infection with pyogenic bacteria. The pathogenesis of the neutropenia and the anemia and thrombocytopenia which often occur concurrently is not understood. There is no evidence to support autoimmnity or direct lysis of stem cells by the virus (MacKenzie et al. 1986). The marrow is hypercellular and contains many blast cells with little evidence of maturation and no necrosis. Human GM-CSF (immunoglobulin allotype granulocyte-macrophage colony-stimulating factor) has been shown to correct the blood dyscrasia in infected monkeys (Donahue et al. 1986), suggesting that alteration of mediators may be responsible for the hematologic changes.

Macrophages and other antigen-presenting cells, such as Langerhans' cells and dendritic reticular cells, are infected (Lackner et al. 1988, 1989; Maul et al. 1988). Although deficiencies in macrophage killing are suspected from the type of secondary infections, macrophage function in vitro, as measured by phagocytosis and intracellular killing, is not diminished (LeGrand et al. 1985).

Lymphopenia occurs in many type D virus-infected animals especially in the terminal phases of disease and is due to reduction in the number of circulating CD4+ and CE8+ T cells and B cells. Serum immunoglobulins, including IgM, IgG, and IgA, are all decreased in clinically ill animals. There is often evidence of profound lymphoid depletion, suggesting lymphoid necrosis, in the lymph nodes and spleens of these animals, but no histologic evidence of lysis of lymphocytes is seen. Type D retroviruses do cause cytopathic effects in cultured lymphocytes, and there may be a direct cytotoxic effect in vivo as well. More subtle mechanisms are probably involved, however.

Rhesus lymphocytes infected with type D retrovirus do not survive as well in tissue culture as uninfected rhesus lymphocytes. In vitro mitogen blastogenesis responses of macaque peripheral blood monuclear cells are diminished as early as 4 weeks postinfection (Maul et al. 1985; Budzko et al. 1985). Responses to concanavalin A, pokeweed mitogen, phytohemagglutinin, and *Staphylococcus aureus* are all decreased and can be partially or wholly restored with addition of interleukin-2 (IL-2) to the culture system (Maul et al. 1985). This suggests the possibility of alternation of expression of mediators of cellular interactions and lymphocyte proliferation by the integrated type D proviral genome or the production of regulatory factors directly encoded by the virus.

There are two lines of evidence to support these possibilities. One is that a type D retrovirus protein has been shown to have immunosuppressive effects similar to those described for other retroviral proteins. This protein is encoded by the envelope gene (Denner et al. 1985). The other is that type D virus increases the expression of class II antigens in inducible cells such as fibroblasts (C.P. Zaiss, personal communication). Modification of class II major histocompability complex (MHC) expression could seriously impair immunorecognition necessary for antigen presentation and other functions.

Similar alterations in regulatory pathways may occur in other tissues as well. A recent report indicates that macaques with SAIDS, but not healthy carriers, recovered animals, or animals with SAIDS-related complex, have increased levels of quinolinic acid in cerebrospinal fluid and serum. This is a neuroactive metabolite of L-tryptophane that is synthesized in systemic tissues (Heyes et al. 1990). Whether this is a direct effect of type D retroviruses or mediated through other pathways such as interferon induction is unknown. It suggests, however, that mechanisms other than direct cytotoxicity are involved in the pathogenesis of type D retrovirus infections in macaques.

Etiology

Type D retroviruses are placed in the subfamily Oncovirinae. They have a typical retrovirus genome consisting of three major genes: (1) *gag* (which encodes for the core proteins); (2) *pol* (which codes for polymerase or reverse transcriptase and endonuclease enzymes); and (3) *env* (which codes for the transmembrane and external envelope proteins). Gel diffusion of purified whole type D retrovirus reveals several virus-specific bands. Protein products of the envelope (*env*) gene are gp70 and gp20. Structural core protein products of the *gag* gene are p10, p14, p16–18, and p27 (Bradac and Hunter 1984, 1986).

All type D retroviruses isolated thus far have been from nonhuman primates. Both endogenous and exogenous viruses have been described (see reviews by Fine and Schochetman 1978; Lowenstine and Lerche 1988; Gardner et al. 1988). The endogenous viruses are the squirrel monkey (*Saimiri sciureus*) retrovirus and the PO-1-Lu virus isolated from a spectacled langur (*Presbytis obscurus*) (Heberling et al. 1977; Todara et al. 1978). This latter virus is thought to be the progenitor of the exogenous type D retroviruses of macaques. Evidence of related, integrated, endogenous, proviral sequences has been found in many species of colobine monkeys from both Asia and Africa (Benveniste and Todaro 1977). No pathogenic role is known for the endogenous viruses.

The exogenous viruses are found in macaques (genus *Macaca*). The first to be described was the Mason-Pfizer monkey virus which was isolated from a rhesus mammary tumor (Chopra and Mason 1970). Inoculation into infant rhesus did not result in neoplasia; rather, some of the animals died with profound thymic atrophy and evidence of immunosuppression (Cockrell et al. 1974; Fine et al. 1975). The original isolate was resurrected from frozen tumor, inoculated into rhesus, and the earlier observations of immunosuppression were confirmed (Bryant et al. 1986a). The exogenous type D viruses of macaques are all related to the Mason-Pfizer monkey virus and to each other, but not to the human AIDS virus (HIV) or SIV (Desrosiers et al. 1985; Bryant et al. 1985; Thayer et al. 1987; Power et al. 1986). The causative role of type D retroviruses in SAIDS has been irrefutably established by induction of the full spectrum of disease by molecularly cloned type D retrovirus (Heidecker et al. 1987).

Several serotypes of type D retrovirus have been described (Marx et al. 1985 and personal communication; Axthelm, personal communication). The official nomenclature is still in a state of flux. Some authors have used the designation "simian retrovirus" followed by a serotype number, e.g., SRV-1, based on cross neutralization studies (Marx et al. 1985). Others have named their isolates "type D" followed by the name of the laboratory where they were isolated, the species from which they were isolated, and sometimes a laboratory number of the isolate (Hefti et al. 1983). The preferred current nomenclature is D/serotype/laboratory/species, e.g., D/1/California or D/2/Oregon/Celebes. The Mason-Pfizer monkey virus is a distinct serotype, D/3. It is not yet clear whether there are significant differences in pathogenicity of these isolates except in the case of serotype 2. This serotype, which includes type D/Washington and SRV-2 from Oregon, is strongly associated with retroperitoneal fibromatosis and subcutaneous fibroproliferative lesions that some investigators liken to Kaposi's sarcoma (Giddens et al. 1985; Tsai et al. 1985b; Stromberg et al. 1984; Shiigi et al. 1986). Subcutaneous fibrosarcomas and Kaposi's sarcoma-like skin lesions have been seen occasionally in rhesus monkeys infected with SRV-1 (see elsewhere in this volume; London et al. 1983; Osborn et al. 1984; Letvin et al. 1983).

Exogenous type D retrovirus can be isolated by cocultivation of infected macaque peripheral blood mononuclear cells with susceptible cell lines. Raji cells, which are human B lymphocytes transformed by Epstein-Barr virus, form characteristic syncytial cells when infected with type D retroviruses and are the cell line most frequently used for viral isolations (Daniel et al. 1984) CO_2. These cell cultures are best incubated in CO_2, as syncytia do not form well in Hepes buffered media (P.A. Marx, personal communication). Other cell lines used successfully for isolation and cultivation of type D retrovirus include: dog thymus (Stromberg et al. 1984), rhesus foreskin fibroblasts, primary rhesus monkey kidney cells (Marx et al. 1984), and other cells of both epithelial and mesenchymal origin from human, canine, rhesus, bat, and mink (Stromberg et al. 1984). The viruses are enveloped, and not highly cell-associated. They are readily inactivated with 10% household bleach solution, 70% EtOH, and formalin.

Frequency

Type D retroviruses have been isolated from macaques at several of the US regional primate research centers (Hefti et al. 1983; Daniel et al. 1984; Marx et al. 1984, 1985). This infection is the major cause of simian AIDS in colonies of captive macaques. Subclinical type D virus infections occur and may compromise experiments, especially those involving immune function. Infection of macaques with type D retroviruses appears to be fairly common in both captivity and in the wild, but the prevalence of infection varies considerably with the species and population tested and with the method of captive husbandry (e.g., it is less frequent in individually caged than in group-housed animals). Seroepidemiologic surveys of wild caught macaques or those in captivity in their native countries have demonstrated infected cynomolgus (*M. fascicularis*) and pig-tailed macaques (*M. nemestrina*) in Indonesia, bonnet monkeys (*M. radiata*) in India, Tonkeana macaques (*M. tonkeana*) in Sulawesi, and possibly rhesus from Kunming China (Thouless, Marx, and Lerche, personal communication). Interestingly, *M. fascicularis* native to the Philippines and from the Seychelles Islands off the coast of Africa (to which they were transported from Asia by trading vessels several hundred years ago) are seronegative (Houghton 1988). This has been taken as evidence by some investigators that the type D viruses entered *M. fascicularis* some time in the last 200–400 years; however, the "founder effect" is an equally plausible explanation.

Seroepidemiologic surveys of captive populations of macaques for evidence of type D virus infection have yielded disparate results. The earliest survey, using the Mason-Pfizer monkey virus as the antigen, found that 6%–26% of rhesus and 43% of bonnet monkeys in research vivaria were positive (Fine and Arthur 1981). In a more recent survey, the prevalence of antibodies was 30% in several species including rhesus, cynomolgus, and Formosan rock macaques (*M. cyclopis*) (Daniel et al. 1988). In another colony, 70% of Celebes black macaques (*M. nigra*) were seropositive (Shiigi et al. 1986). In a survey of zoo monkeys the overall prevalence was 4%, but among macaques it was 10% and in one group of Tonkeana macaques (*M. tonkeana*) it was 29% (Lowenstine et al. 1986).

Comparison with Other Species

Type D retroviruses are thought to be exclusive to nonhuman primates. Although isolates have been made from human origin tissue cultures, they are considered by most researchers to be laboratory contaminants (Uckert et al. 1987). Most serosurveys of humans have failed to detect antibodies to the type D retroviruses (Charman et al. 1977; N.W. Lerche, unpublished). However, in one study in Belgium of patients with chronic renal disease, neutralizing antibodies to the Mason-Pfizer monkey virus measured by syncytium inhibition were found in 50% of patients with chronic glomerulonephritis (Thirty et al. 1978). In a recent report, a retrovirus highly similar to the Mason-Pfizer monkey virus and SRV-1 was detected in an unusual syncytial variant of B cell lymphoma from a human AIDS patient. Viral sequences were detected in this patient's bone marrow by polymerase chain reaction (Donehower et al. 1990; Bohannon et al., unpublished). Sheep with ovine transmissible pulmonary carcinoma have a 26 kDa protein in the thoracic effusion which is immunologically reactive with the major core protein of Mason-Pfizer monkey virus, but the ovine virus itself has not yet been characterized sufficiently to definitively place it among the type D retroviruses (Rosadio et al. 1988).

SAIDS in macaques induced by type D retrovirus has been used as a model for human AIDS. Although the causative agent of human AIDS, HIV, is a lentivirus not genetically related to the type D retroviruses, many of the clinical manifestations of immune suppression are quite similar. Type D retrovirus infection may also be useful in examining and comparing the consequences of persistent virus infection and pathogenesis of virus-induced blood dyscrasias.

References

Ahmed M, Schidlovsky G, Korol W, Vidrine G, Cicmanec JL (1974) Occurrence of Mason-Pfizer monkey virus in healthy rhesus monkeys. Cancer Res 34:3504–3508

Baskin GB, Murphey-Corb M, Watson EA, Martin LN (1988) Necropsy findings in rhesus monkeys experimentally infected with cultured simian immunodeficiency virus (SIV)/delta. Vet Pathol 25:456–467

Benveniste RE, Todaro GJ (1977) Evolution of primate oncornaviruses: an endogenous virus from langurs (Presbytis spp) with related virogene sequences in other Old World monkeys. Proc Natl Acad Sci USA 74:4557–4561

Bradac J, Hunter E (1984) Polypeptides of Mason-Pfizer Monkey virus: I. Synthesis and processing of gag-gene products. Virology 138:260–275

Bradac J, Hunter E (1986) Polypeptides of Mason-Pfizer monkey virus: II. Synthesis and processing of the env gene products. Virology 150:419–502

Bryant ML, Yamamoto J, Luciw P, Munn R, Marx P, Higgins J, Pedersen N, Levine A, Gardner MB (1985) Molecular comparison of retroviruses associated with human and simian AIDS. Hematol Oncol 3:187–197

Bryant ML, Gardner MB, Marx PA, Maul DH, Lerche NW, Osborn KG, Lowenstine LJ, Bogden A, Aruthur LO, Hunter E (1986a) Immunodeficiency in rhesus monkeys associated with the original Mason-Pfizer monkey virus. J Natl Cancer Inst 77:957–965

Bryant ML, Marx PA, Shiigi SM, Wilson BJ, McNulty WP, Gardner MB (1986b) Distribution of type D retrovirus sequences in tissues of macaques with simian acquired immune deficiency and retroperitoneal fibromatosis. Virology 150:149–160

Budzko DB, Madden DL, London WT, Sever JL (1985) Immunologic alterations in monkeys with simian acquired immunodeficiency syndrome (SAIDS). Proc Soc Exp Biol Med 179:227–231

Califoux LD, King NW, Letvin NL (1984) Morphologic changes in lymph nodes of macaques with an immunodeficiency syndrome. Lab Invest 51:22–26

Charman HP, Rahman R, White MH, Kim N, Gilden RV (1977) Radioimmunoassay for the major structural protein of Mason-Pfizer monkey virus: attempts to detect the presence of antigen or antibody in humans. Int J Cancer 19:498–504

Chopra HC, Mason MM (1970) A new virus in a spontaneous mammary tumor of a rhesus monkey. Cancer Res 30:2081–2086

Cockrell BY, Valerio MC, Garner FM (1974) Cryptosporidiosis in the intestines of rhesus monkeys (Macaca mulatta). Lab Anim Sci 24:881–887

Dalakas MC, London WT, Gravell M, Sever JL (1986) Polymyositis in an immunodeficiency disease in monkeys induced by a Type D retrovirus. Neurology 36:569–572

Daniel MD, King NW, Letvin NL, Hunt RD, Sehgal PK, Desrosiers RC (1984) A new type of D retrovirus isolated from macaques with an immunodeficiency syndrome. Science 223:602–605

Daniel MD, Letvin NL, Sehgal PK, Schmidt DK, Silva DP, Solomon KR, Hodi SF, Ringler DJ, Hunt RD, King NW, Desrosiers RC (1988) Prevalence of antibodies to 3 retroviruses in a captive colony of macaque monkeys. Int J Cancer 31:601–608

Denner J, Wunderlich V, Sydow G (1985) Suppression of human lymphocyte mitogen response by retroviruses of type D: I. Action of highly purified intact disrupted virus. Arch Virol 86:177–186

Desrosiers RC, Daniel MD, Butler CV, Schmidt DK, Letvin NL, Hunt RD, King NW, Barker CS, Hunter E (1985) Retrovirus D/New England and its relationship to Mason-Pfizer Monkey Virus. J Virol 54:552–560

Donahue RE, Wang EA, Stone DK, Kamen R, Wong GG, Sehgal PK, Nathan DG, Clark SC (1986) Stimulation of haematopoiesis in primates by continuous infusion of recombinant human GM-CSF. Nature 321: 872–875

Donehower LA, Bohannon RC, Ford RJ, and Gibbs RA (1990) The use of primers from highly conserved pol regions to identify uncharacterized retroviruses by the polymerase chain reaction. J Virol Methods 28:33–46

Fine DL, Arthur LO (1981) Expression of natural antibodies against endogenous and horizontally transmitted macaque retroviruses in captive primates. Virology 112:49–61

Fine DL, Schochetman G (1978) Type D primate retroviruses: a review. Cancer Res 38:3123–3139

Fine DL, Landon JC, Pienta RJ, Kubicek MT, Valerio MG, Loeb WF, Chopra HC (1975) Responses of infant rhesus monkeys to inoculation with Mason-Pfizer monkey virus materials. J Natl Cancer Int 54:651–658

Gardner MB, Luciw P, Lerche N, Marx P (1988) Nonhuman primate retrovirus isolates and AIDS. Adv Vet Sci Comp Med 32:171–226

Giddens WE Jr, Tsai CC, Morton WR, Ochs HD, Knitter GH, Blakely GA (1985) Retroperitoneal fibromatosis and acquired immunodeficiency syndrome in macaques. Pathologic observations and transmission studies. Am J Pathol 119:253–263

Heberling RL, Barker ST, Kalter SS, Smith GC, Helmke RJ (1977) Oncornavirus: isolation from a squirrel monkey (Saimiri sciureus) lung culture. Science 195: 289–292

Hefti E, Ip J, Giddens WE Jr, Panem S (1983) Isolation of a unique retrovirus, MNV-1, from Macaca nemestrina. Virology 127:309–319

Heidecker G, Lerche NW, Lowenstine LJ, Lackner AA, Osborn KG, Gardner MB, Marx PA (1987) Induction of simian acquired immune deficiency syndrome (SAIDS) with a molecular clone of type D SAIDS retrovirus. J Virol 61:3066–3071

Henrickson RV, Maul DH, Lerche NW, Osborn KG, Lowenstine LJ, Prahalada S, Sever JL, Madden DL, Gardner MB (1984) Clinical features of simian acquired immunodeficiency syndrome (SAIDS) in rhesus monkeys. Lab Anim Sci 34:140–145

Heyes MP, Mefford IN, Quearry BJ, Dedhia M, Lackner AA (1990) Increased ratio of quinolinic acid to kynurenic acid in cerebrospinal fluid of D retrovirus-infected rhesus macaques: relationship to clinical and viral status. Ann Neurol 27:666–675

Houghton P (1988) Virus-free African cynomolgus macaques. Lab Anim 17:36–40

Hunt RD, Blake BJ, Chalifoux LV, Sehgal PK, King NW, Letvin NL (1983) Transmission of naturally occurring lymphoma in macaque monkeys. Proc Natl Acad Sci USA 80:5085–5089

King NW, Hunt RD, Letvin NL (1983) Histopathologic changes in macaques with an acquired immunodeficiency syndrome (AIDS). Am J Pathol 113: 382–388

King NW (1986) Simian models of acquired immunodeficiency syndrome (AIDS): a review. Vet Pathol 23: 345–353

Kwang HS, Pedersen NC, Lerche NW, Osborn KG, Marx PA, Gardner MB (1987) Viremia, antigenemia, and serum antibodies in rhesus macaques infected with simian retrovirus type 1 and their relationship to disease course. Lab Invest 56:591–597

Lackner AA, Rodriguez MH, Bush CE, Munn RJ, Kwang H-S, Moore PF, Osborn KG, Marx PA, Gardner MB, Lowenstine LJ (1988) Distribution of a macaque immunosuppressive type D retrovirus in neural, lymphoid and salivary tissues. J Virol 62: 2134–2142

Lackner AA, Marx PA, Lerche NW, Gardner MB, Kluge JD, Spinner A, Kwang H-S, Lowenstine LJ (1989) Asymptomatic infection of the central nervous system by the macaque immunosuppressive type D retrovirus, SRV-1. J Gen Virol 70:1641–1651

LeGrand EK, Donovan RM, Marx PA, Moulton JE, Cheung ATW, Lewis AE, Gardner MB (1985) Monocyte function in rhesus monkeys with simian acquired immunodeficiency syndrome. Vet Immunol Immunopathol 10:131–146

Lerche NW, Henrickson RV, Maul DH, Gardner MB (1984) Epidemiologic aspects of an outbreak of acquired immunodeficiency in rhesus monkeys (Macaca multatta). Lab Anim Sci 34:146–150

Lerche NW, Osborn KG, Marx PA, Prahalada S, Maul DH, Lowenstine LJ, Munn RJ, Bryant ML, Henrickson RV, Arthur LO, Gilden RV, Barker CS, Hunter E, Gardner MB (1986) Inapparent carriers of simian acquired immune deficiency syndrome type D retrovirus and disease transmission with saliva. J Natl Cancer Inst 77:489–496

Lerche NW, Marx PA, Osborn KG, Maul DH, Lowenstine LJ, Bleviss ML, Moody P, Henrickson RV, Gardner MB (1987) Natural history of endemic type D retrovirus infection and acquired immune deficiency syndrome in group-housed rhesus monkeys. J Natl Cancer Inst 79:847–854

Letvin NL, King NW (1990) Immunologic and pathologic manifestations of the infection of rhesus monkeys with simian immunodeficiency virus of macaques. J Acquir Immune Defic Syndr 3:1023–1040

Letvin NL, Eation KA, Aldrich WR, Sehgal PK, Blake BJ, Schlossman SF, King NW, Hunt RD (1983) Acquired immunodeficiency syndrome in a colony of macaque monkeys. Proc Natl Acad Sci USA 80: 2718–2722

Letvin NL, Daniel MD, Sehgal PK, Chalifoux LV, King NW, Hunt RD, Aldrich WR, Holley K, Schmidt DK, Desrosiers RC (1984) Experimental infection of rhesus monkeys with type D retrovirus. J Virol 52:683–686

London WT, Sever JL, Madden DL, Henrickson RV, Gravell M, Maul DH, Dalakas MC, Osborn KG, Houff SA, Gardner MB (1983) Experimental transmission of simian acquired immunodeficiency syndrome (SAIDS) and Kaposi-like skin lesions. Lancet 2:869–873

Lowenstine LJ, Lerche NW (1988) Retrovirus infections of nonhuman primates: a review. J Zoo Anim Med 19:168–187

Lowenstine LJ, Pedersen NC, Higgins J, Pallis KC, Uyeda A, Marx P, Lerche NW, Gardner MB (1986) Seroepidemiologic survey of captive Old-World primates for antibodies to human and simian retroviruses and isolation of a lentivirus from sooty mangabeys (Cercocebus atys). Int J Cancer 38:563–574

Lowenstine LJ, Lerche NW, Yee JL, Uyeda A, Jennings MB, Munn RJ, McClure HM, Anderson DC, Fultz PN, Gardner M (1992) Evidence for a Leutiviral etiology in an epizootic of immune deficiency and lymphama in stump-tailed macaques (Macaca arctoides). J Med Primatol 21:1–14

MacKenzie M, Lowenstine L, Lalchandani R, Lerche N, Osborn K, Spinner A, Bleviss M, Henrickson R, Gardner M (1986) Hematological abnormalities in simian acquired immune deficiency syndrome. Lab Anim Sci 36:14–19

Marx PA, Maul DH, Osborn KG, Lerche NW, Moody P, Lowenstine LJ, Henrickson RV, Arthur LO, Gravell M, London WT, Sever JL, Levy JA, Munn RJ, Gardner MB (1984) Simian AIDS: isolation of a type D virus and transmission of the disease. Science 223: 1083–1086

Marx PA, Bryant MB, Osborn KG, Maul DH, Lerche NW, Lowenstine LJ, Kluge JD, Zaiss CP, Henrickson RV, Shiigi SM, Wilson BJ, Malley A, Olson L, McNullty WP, Arthur LO, Gilden RV, Barker CS, Hunter E, Munn RJ, Heidecker G, Gardner MB (1985) Isolation of a new serotype of simian acquired immune deficiency syndrome type D retrovirus from Celebes black macaques (Macaca nigra) with immune deficiency and retroperitoneal fibromatosis. J Virol 56:571–578

Marx PA, Pedersen NC, Lerche NW, Osborn KG, Lowenstine LJ, Lackner AA, Maul DH, Kwang H-S, Kluge JD, Zaiss CP, Sharpe V, Spinner A, Allison AC, Gardner MB (1986) Prevention of simian acquired immune deficiency syndrome with a formalin-inactivated type-D retrovirus vaccine. J Virol 60:431–435

Maul DH, Miller CH, Marx PA, Bleviss ML, Madden DL, Henrickson RV, Gardner MB (1985) Immune defects in simian acquired immunodeficiency syndrome. Vet Immunol Immunopathol 8:201–214

Maul DH, Lerche NW, Osborn KG, Marx PA, Zaiss C, Spinner A, Kluge JD, MacKenzie MR, Lowenstine LJ, Bryant ML, Blakeslee JR, Henrickson RV, Gardner MB (1986) Pathogenesis of simian AIDS in rhesus macaques inoculated with the SRV-1 strain of type D retrovirus. Am J Vet Res 47: 863–868

Maul DH, Zaiss CP, MacKenzie MR, Shiigi SM, Marx PA, Gardner MB (1988) Simian retrovirus D serogroup 1 has a broad cellular tropism for lymphoid and non-lymphoid cells. J Virol 62:1768–1773

McClure HM, Anderson DC, Fultz PN, Ansari AA, Lockwood E, Brodie A (1989) Spectrum of disease in macaque monkeys chronically infected with SIV/SMM. Vet Immunol Immunopathol 21:13–24

Meyer PR, Ormerod LD, Osborn KG, Lowenstine LJ, Henrickson RV, Moldin RL, Smith RE, Gardner MB, Taylor CR (1985) An immunopathologic evaluation of lymph nodes from monkey and man with acquired immune deficiency syndrome and related conditions. Hematol Oncol 3:199–210

Munn RJ, Marx PA, Yamamoto JK, Gardner MB (1985) Ultrastructural comparison of the retroviruses associated with human and simian acquired immunodeficiency syndromes. Lab Invest 53:194–199

Osborn KG, Prahalada S, Lowenstine LJ, Gardner MB, Maul DH, Henrickson RV (1984) The pathology of an epizootic of acquired immunodeficiency in rhesus macaques. Am J Pathol 114:94–103

Osborn KG, Lowenstine LJ, Meyer PR, Marx PA, Ormerod LD, Gardner MB (1985) Lymph node-based pathogenesis of experimental simian acquired immune deficiency syndrome. Lab Invest 52:49A

Pedersen NC, Lowenstine LJ, Marx P, Higgins J, Baulu J, McGuire M, Gardner MB (1986) The causes of false positives encountered during the screening of Old World primates for antibodies to human and simian retroviruses by ELISA. J Virol Methods 14:213–228

Power MD, Marx PA, Bryant ML, Gardner MB, Barr PJ, Luciw P (1986) Nucleotide sequence of SRV-1, a type D simian acquired immune deficiency syndrome retrovirus. Science 231:1567–1572

Rosadio RH, Sharp JM, Lairmore MD, Dahlberg JE, De Martini JC (1988) Lesions and retroviruses associated with naturally occurring ovine pulmonary carcinoma (sheep pulmonary adenomatosis). Vet Pathol 25:58–66

Schiodt M, Lackner A, Armitage G, Lerche N, Greenspan JS, Lowenstine Lowenstine L (1988) Oral lesions in rhesus monkeys associated with infection by simian

AIDS retrovirus, serotype-1 (SRV-1). Oral Surg Oral Med Oral Pathol 65:50–55
Shiigi SM, Wilson BJ, Malley A, Chander RA, Howard CF Jr, Olson LC, Palotay JL, McNulty WP, Marx PA (1986) Association of SAIDS/RF-related signs with current or past SAIDS type 2 retrovirus infection in a colony of Celebes black macaques. Lab Anim Sci 36: 20–23
Spinner A, Sutjipto S (1988) Simian AIDS ELISA: sensitivity, specificity and predictive values based on a comparison with Western blot technique. Lab Anim Sci 38:568–572
Stromberg K, Benveniste RE, Arthur LO, Rabin H, Giddens WE Jr, Ochs HD, Morton WR, Tsai C-C (1984) Characterization of exogenous type D retrovirus from a fibroma of a macaque with simian AIDS and fibromatosis. Science 224:289–292
Thayer RM, Power MD, Bryant ML, Gardner MB, Barr PJ, Luciw PA (1987) Sequence relationships of type D retroviruses which cause simian acquired immune deficiency syndrome. Virology 157:317–329
Thiry L, Sprecher-Goldberger S, Bossens M, Cogniaux-Le Clerc J, Vereerstraeten P (1978) Neutralization of Mason-Pfizer virus by sera from patients treated for renal disease. J Gen Virol 41:587–597
Todaro GJ, Benveniste RE, Sherr CJ, Schlom J, Schidlovsky G, Stephenson JR (1978) Isolation and characterization of a new type D retrovirus from the Asian primate, Presbytis obscurus, (spectacled langur). Virology 84:189–194
Tsai CC, Giddens WE Jr, Morton WR, Rosenkranz SL, Ochs HD, Benveniste RE (1985a) Retroperitoneal fibromatosis and acquired immunodeficiency syndrome in macaques: epidemiologic studies. Lab Anim Sci 35:460–464
Tsai CC, Warner TF, Uno H, Giddens WE Jr, Ochs HD (1985b) Subcutaneous fibromatosis associated with an acquired immune deficiency syndrome in pig-tailed macaques. Am J Pathol 120:30–37
Uckert W, Wunderlich V, Fieback H, Hertling I, Stein U, Kraft R, Desrosiers R (1987) Biochemical and immunological characterization of structural proteins from retrovirus-D/New England and comparison to Mason-Pfizer monkey virus and permanent human fibroblast virus. Arch Virol 94:267–282
Wilson BJ, Shiigi SM, Zeigler JL, Olson LC, Malley A, Howard CF Jr (1986) Transmission of simian acquired immunodeficiency syndrome with a type D retrovirus: immunological aspects. Clin Immunol Immunopathol 41:453–460

Cytomegalovirus Infection in Nonhuman Primates

Gary B. Baskin

Synonyms. Inclusion body disease.

Gross Appearance

Lesions due to cytomegalovirus can be recognized grossly in the eye, meninges, intestine, skin, heart, testicle, and lung. Ocular lesions can be observed with the ophthalmoscope and consist of intraretinal hemorrhage and necrosis. In the central nervous system, patchy areas of the meninges covering the brain and spinal cord may be affected. These areas are thickened, edematous, opaque, and discolored yellow-tan (Fig. 27). The covering of the spinal nerve roots is frequently involved. Intestinal lesions consist of focal areas of thickening and sometimes ulceration of the mucosa. Cutaneous and myocardial lesions appear grossly as focal hemorrhages. Affected testicles are enlarged and the cut surface is wet and yellow-tan. Pulmonary lesions can be focal or diffuse. Focal lesions consist of multiple, bright red, 1–3 mm in diameter foci scattered throughout each lobe of the lung (Fig. 28). Diffuse lesions form larger areas of dark red, heavy, wet, consolidated lung.

Microscopic Features

Microscopic changes due to cytomegalovirus may occur in the meninges, brain, spinal cord, spinal and peripheral nerves, arteries, lymph nodes, lung, liver, spleen, skin, heart, kidney, salivary gland, intestine, and testicle. The infections are identified microscopically by characteristic intranuclear and intracytoplasmic inclusion bodies and cytomegaly. Enlarged inclusion-bearing cells may be found in otherwise normal tissues, in small focal areas of necrosis, or associated with extensive necrosis, edema, hemorrhage, and neutrophilic inflammatory infiltration. The cells are usually of mesenchymal origin, although they are often too distorted to identify with certainty. Unlike some herpesviruses, cytomegalovirus does not infect most surface epithelium. Intranuclear inclusions are large, ovoid, dense, homogenous, and may stain eosinophilic to basophilic. They are often surrounded by a clear halo and may be multiple. Cytoplasmic inclusions are less common and are eosinophilic and granular (Fig. 29). Lesions in the meninges are often severe and consist of necrosis, edema, and neutrophil infil-

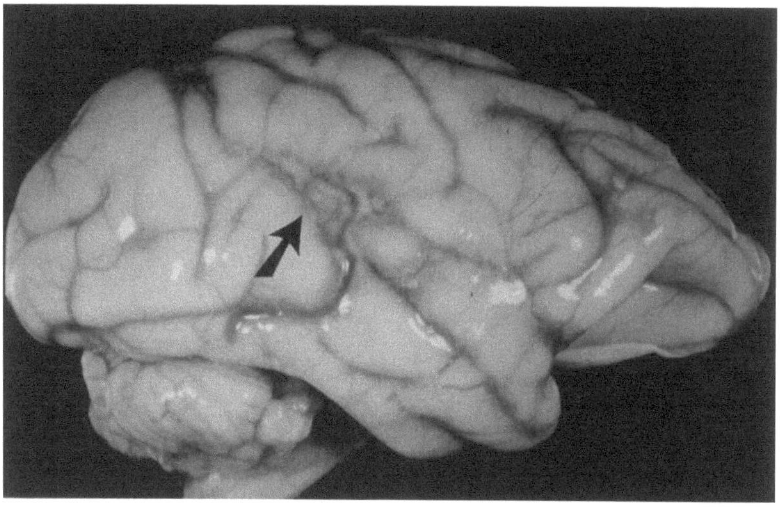

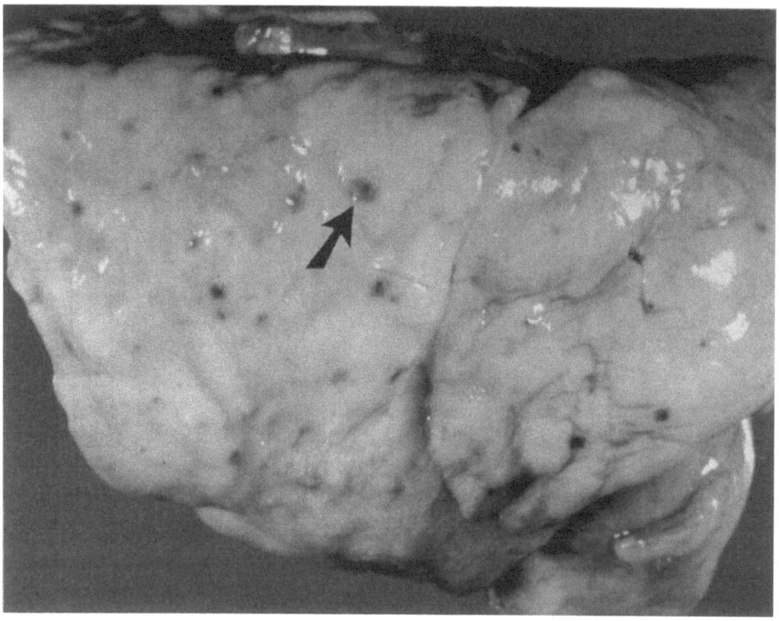

Fig. 27 (*above*). Focal meningitis (*arrow*) caused by cytomegalovirus in a rhesus monkey infected with simian immunodeficiency virus

Fig. 28 (*below*). Multifocal pneumonia (*arrow*) caused by cytomegalovirus in a rhesus monkey infected with simian immunodeficiency virus

tration. The underlying cerebral or spinal cortex may also be infiltrated. The ependyma is sometimes affected, with disruption of the lining cells, necrosis, and inflammation (Fig. 30). Spinal and peripheral nerves may be segementally necrotic, edematous, infiltrated with neutrophils, and contain inclusion-bearing cells. The walls of affected arteries are edematous, necrotic, and infiltrated with neutrophils. Endothelial cells are often enlarged. In lymph nodes, inclusion-bearing cells may be seen within the subcapsular sinus or perinodal tissue of otherwise normal nodes, or they may be associated with extensive areas of necrosis and neutrophilic infiltration within the node. The lung is the most frequently affected organ. The focal and diffuse lesions, similar microscopically except for extent, consist of interstitial pneumonia. Alveolar septae are thickened, hypercellular, and have a hypertrophic lining. Alveolar spaces contain macrophages, neutrophils, proteinaceous fluid, fibrin, and erythrocytes. Inclusion-bearing cells are in alveolar septa, alvelor lining cells, within alveolar spaces, and in the adventitia of arteries and bronchioles (Fig. 31).

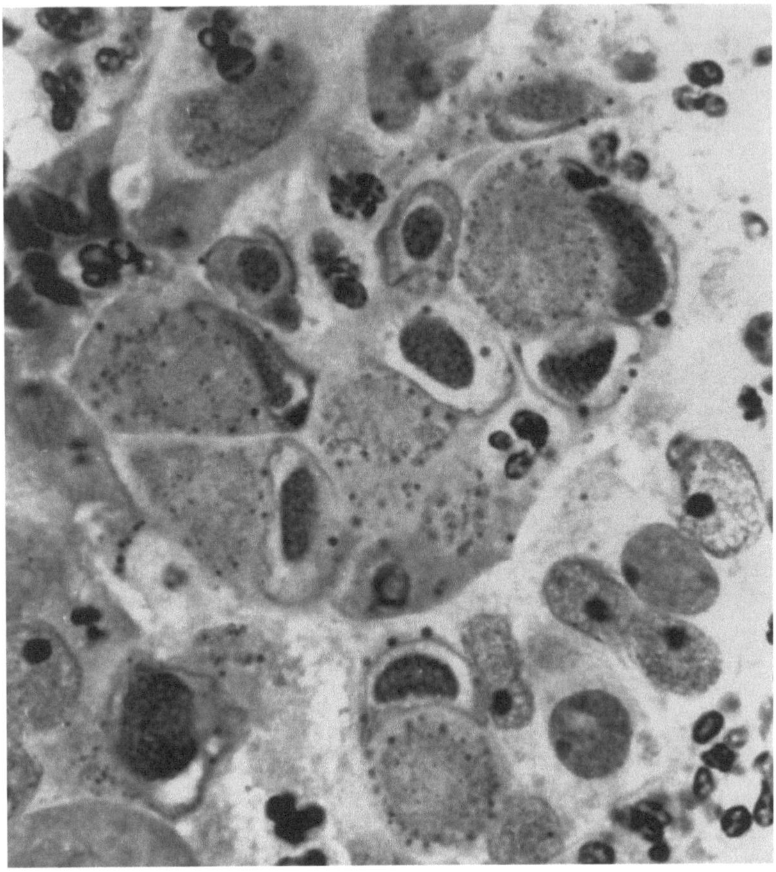

Fig. 29. Intranuclear and intracytoplasmic inclusion bodies and cytomegaly in a seminiferous tubule of a rhesus monkey infected with cytomegalovirus and simian immunodeficiency virus. H and E, ×640

Fig. 30 (*upper left*). Ependymitis in the brain of a rhesus ▶ monkey infected with cytomegalovirus and simian immunodeficiency virus. Note intranuclear inclusion body (*arrow*) and ventricular space (*V*). H and E, ×250

Fig. 31 (*below*). Interstitial pneumonia in a rhesus monkey infected with cytomegalovirus and simian immunodeficiency virus. Note alveolar thickening and hypercellularity and exudate in alveolar spaces. H and E, ×250

Fig. 32 (*upper right*). Intranuclear inclusion body in a hepatocyte of a rhesus monkey infected with cytomegalovirus and simian immunodeficiency virus. H and E, ×400

Hepatic lesions occur within the liver parenchyma and the biliary tract. Parenchymal lesions are usually sparse small foci of necrosis that may contain a few neutrophils. Inclusions occur in hepatocytes and Kupffer's cells (Fig. 32). Biliary lesions are often more severe and consist of enlargement and necrosis of bile duct and gallbladder epithelium, infiltration with neutrophils and mononuclear inflammatory cells, and periductal edema and fibrosis. Inclusions occur in biliary epithelial cells. Cytomegalovirus causes intense neutrophilic infiltration of the lamina propria of the small and large intestine, with inclusion-bearing cells in the lamina propria. These lesions sometimes ulcerate. Inclusion-bearing cells occur in the red and white pulp of the spleen with or without associated foci of necrosis and neutrophilic infiltration. Cutaneous lesions consist of foci of edema, hemorrhage, and mixed inflammatory cell infiltration in the superficial dermis. Inclusions are in inflammatory cells and vascular endothelial cells. Myocardial lesions are likewise composed of patchy areas of interstitial hemorrhage, edema, and mixed inflammatory cell infiltration, with inclusions in inflammatory cells. Rare inclusions may be found within renal glomeruli and in salivary gland epithelium, usually unassociated with other changes. In the testicle there is severe neutrophilic infiltration in the interstitium, although inclusions are found within cells inside the seminiferous tubules (Fig. 33).

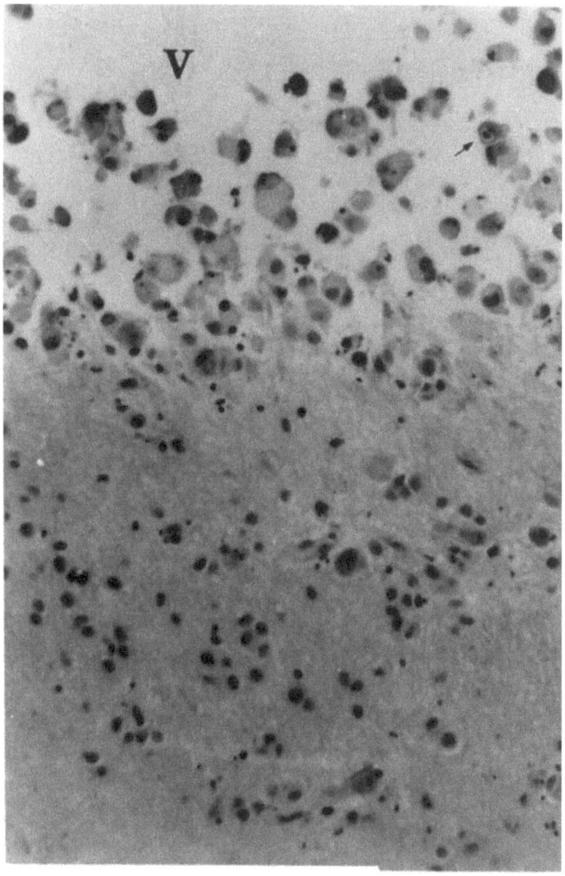
V

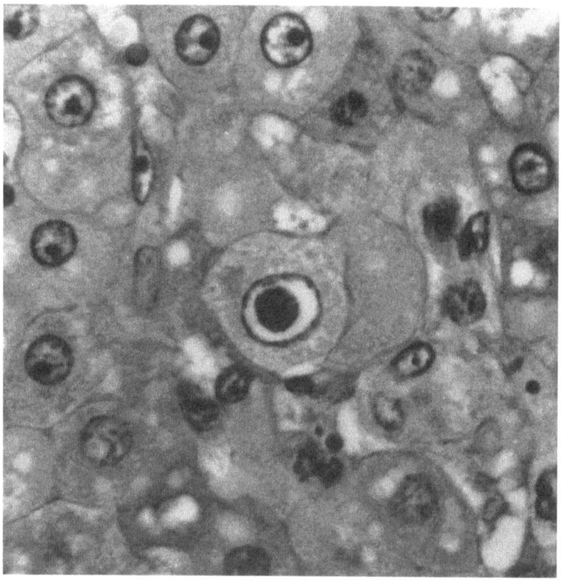

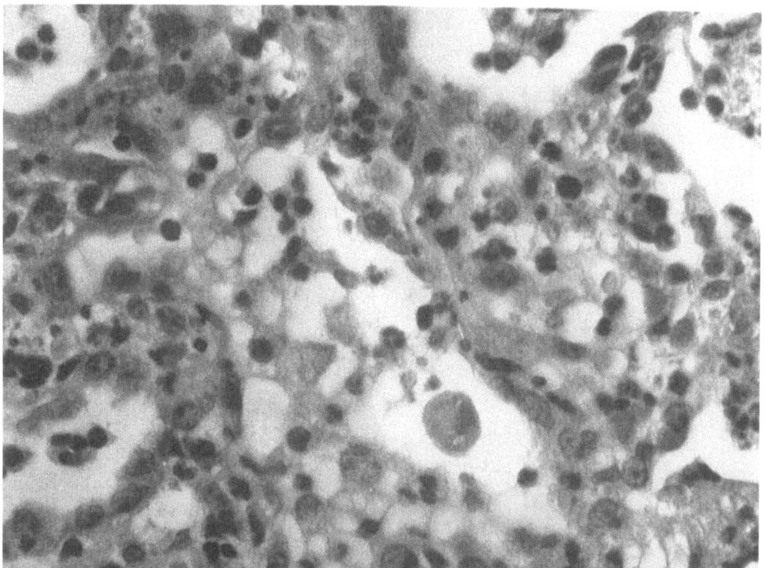

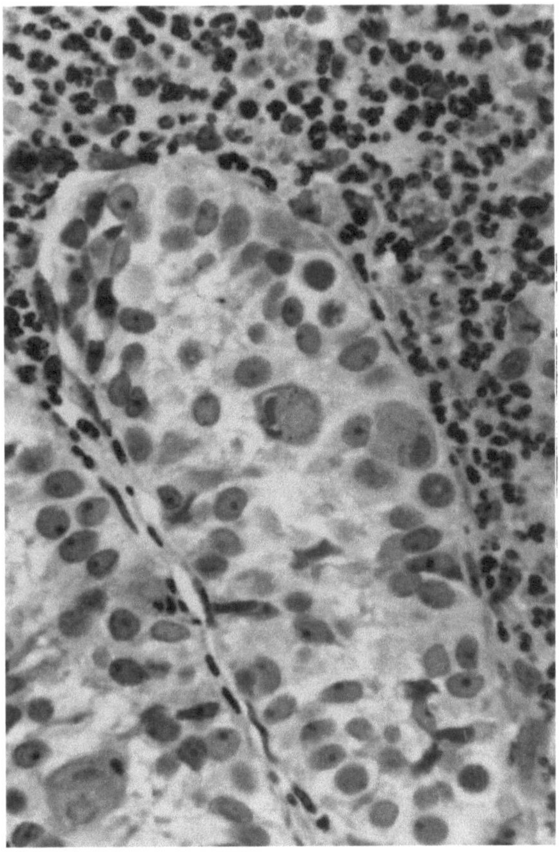

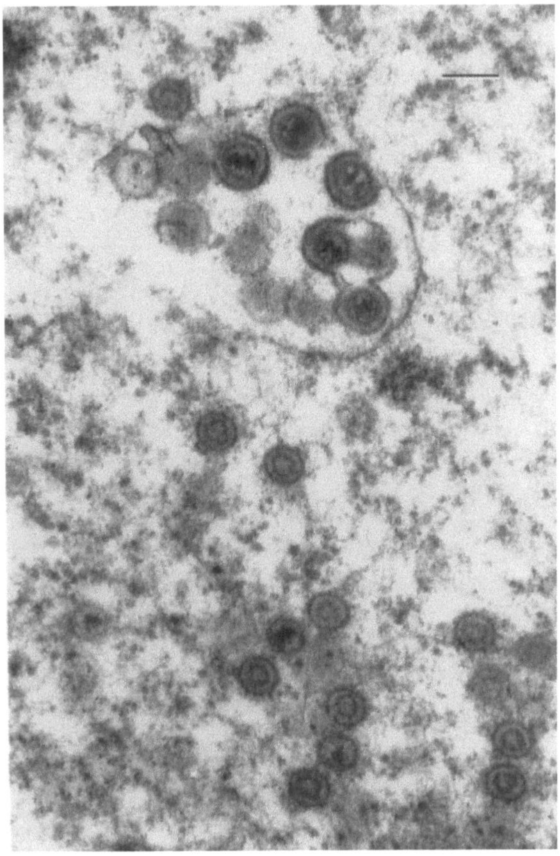

Ultrastructure

Cytomegalovirus has the ultrastructural morphology typical of herpesviruses. The virion measures 120–200 nm in diameter and contains an icosahedral capsid that measures 100–110 nm in diameter. The capsid is surrounded by globular material which is enclosed by a bilayered envelope (Roizman et al. 1981). Both enveloped and unenveloped forms may be observed in infected tissues (Fig. 34). Capsids are assembled in the nucleus and acquire their envelope from the nuclear membrane.

Differential Diagnosis

Lesions of cytomegalovirus must be differentiated from those other viruses that produce intranuclear inclusion bodies. Cytomegalovirus infection is unique in the degree of cytomegaly, the large size of the intranuclear inclusions, and the character of the intracytoplasmic inclusions. Unlike some other herpesviruses, it usually infects mesenchymal rather than epithelial cells. Adenoviruses produce very large intranuclear inclusions, but these are deeply basophilic, "smudgy," and not surrounded by a clear halo. Cytomegalovirus is difficult to isolate in culture from tissues, particularly if they have been frozen.

Biologic Features

Cytomegaloviruses are host-specific herpesviruses belonging to the subfamily Betaherpesvirinae (Roizman et al. 1981) and have been isolated or identified serologically or morphologically in numerous species of apes and monkeys (Rangan and Chaiban 1980; Swack and Hsiung 1982; Vogel and Pinkerton 1955). Cytomegalovirus is usually transmitted horizontally, although in utero infections occasionally occur. Virus is shed

◀ **Fig. 33** (*above*). Orchitis in a rhesus monkey infected with cytomegalovirus and simian immunodeficiency virus. Note necrosis and neutrophilic infiltrate in interstitium; cytomegally and inclusions in seminiferous tubule. H and E, ×250

Fig. 34 (*below*). Virions of cytomegalorivus in the nucleus of a cell in the lung of a rhesus monkey also infected with simian immunodeficiency virus. *Bar*, 200 nm; TEM, ×30 000

in urine and infection is often widespread within a population. Primary infection in all species is usually asymptomatic and the infection becomes chronic or latent after immune responses develop. Clinically significant infections occur only in fetuses and in immunodeficient individuals. A small percentage of human infants who are infected in utero develop damage to the central nervous system. Rhesus and squirrel monkey models of congenital disease have been described (Ordy et al. 1981; London et al. 1986). Immunosuppressed individuals can develop severe disseminated primary or secondary infections. Such infections are common in rhesus monkeys rendered immunodeficient by infection with simian immunodeficiency virus (Baskin 1987) or type D retroviruses, but occur in other situations as well (Ohtaki et al. 1986, 1988) and may occasionally be seen in animals with other debilitating diseases.

Comparison with Other Species

A wide variety of mammalian species, including humans, is susceptible to cytomegaloviruses. The natural history appears to be similar in most species. The infection becomes a serious, often fatal, generalized infection in immunocompromised humans. Inclusion body rhinitis is a cytomegaloviral disease of immunocompetent piglets. Lesions are usually confined to the nasal mucosa, but the infection sometimes becomes generalized.

References

Baskin GB (1987) Disseminated cytomegalovirus infection in immunodeficient rhesus monkeys. Am J Pathol 129:345–352

London WT, Martinez AJ, Houff SA, Wallen WC, Curfman BL, Traub RG, Sever JL (1986) Experimental congenital disease with simian cytomegalovirus in rhesus monkeys. Teratology 33:323–331

Ohtaki S, Kodama H, Hondo R, Kurata T (1986) Activation of cytomegalovirus infection in immunosuppressed cynomolgus monkeys inoculated with varicella-zoster virus. Acta Pathol Jpn 36:1537–1552

Ohtaki S, Hondo R, Kodama H, Kurata T (1988) Experimental activation of latent cytomegalovirus infection of the captive bred (F_1) cynomolgus monkeys by live or killed varicella-zoster virus inoculated under immunosuppression. Acta Pathol Jpn 38:967–978

Ordy JM, Rangan SRS, Wolf RH, Knight C, Dunlap WP (1981) Congenital cytomegalovirus effects on postnatal neurological development of squirrel monkey (Saimiri sciureus) offsping. Exp Neurol 74:728–747

Rangan SRS, Chaiban J (1980) Isolation and characterization of a cytomegalovirus from the salivary gland of a squirrel monkey (Saimiri sciureus). Lab Anim Sci 30:532–540

Roizman B, Carmichael LE, Deinhardt F, de-The G, Nahmias AJ, Plowright W, Rapp F, Sheldrick P, Takahashi M, Wolf K (1981) Herpesviridae. Definition, provisional nomenclature, and taxonomy. The Herpesvirus Study Group, the International Committee of Taxonomy of Viruses. Interviorology 16:201–217

Swack NS, Hsiung GD (1982) Natural and experimental simian cytomegalovirus infections at a primate center. J Med Primatol 11:169–177

Vogel FS, Pinkerton H (1955) Spontaneous salivary gland virus disease in chimpanzees. Arch Pathol Lab Med 60:281–285

Simian Virus 40 Infection

Norval W. King, Jr.

Synonyms. Vacuolating agent infection; simian papovavirus infection; SV 40 infection; simian polyomavirus infection; progressive multifocal leukoencephalopathy; polyomaviral interstitial pneumonia; papovaviral tubulointerstitial nephritis.

Gross Appearance

Macroscopic lesions in macaques with active simian virus 40 infection are subtle and have only been described in the brain or in the lungs or kidneys. (Gribble et al. 1975; Holmberg et al. 1977; Sheffield et al. 1980; King et al. 1983; Horvath et al. 1992). In the brain, multiple, widely scattered, soft, 1–3 mm, gray-pink to translucent foci occur mainly in the subcortical white matter, hypothalamus, medulla oblongata, and adjacent to the cerebral aqueduct (Gribble et al. 1975; Horvath et al. 1992). The gray matter of the cerebrum occasionally may contain gross lesions, but the cerebellum and brain stem are rarely affected. The lungs of animals with interstitial pneumonia induced by this virus contain firm, red, patchy areas that do not collapse when compressed. When cut, a frothy fluid may ooze from the affected areas (Sheffield et al. 1980). Animals with renal lesions have multiple linear petechiae in the inner cortex and medulla.

Microscopic Features

Brain

Microscopically, the brain lesions consist of multifocal and confluent areas of demyelination resulting from injury by the virus to oligodendroglial cells. The areas of demyelination are scattered throughout the cerebral white matter and subependymal regions. These often are associated with multiple foci of microgliosis. Small numbers of glial nodules also may be present in the overlying gray matter. The earliest lesions consist of foci of microgliosis and the presence of intranuclear inclusion bodies in widely scattered oligodendroglial cells and astrocytes. The inclusions fill and sometime cause enlargement of the nuclei and vary from early forms that are lightly basophilic and granular to later forms that are condensed and intensely basophilic (Fig. 35). In advanced lesions, large numbers of foamy macrophages (gitter cells) containing phagocytized myelin debris and inclusion body-bearing astrocytes are present throughout the areas of demyelination. These brain lesions resemble the subacute demyelinating condition of humans referred to as progressive multifocal leukoencephalopathy (Gribble et al. 1975).

Lung

The parenchyma contains patchy areas of interstitial pneumonia in which there is variable thickening of alveolar due to congestion or infiltration by histiocytes and to a lesser extent fibroblasts. Affected septa are lined by hypertrophied and hyperplastic type II pneumocytes, some of which contain prominent basophilic inclusion bodies similar to those found in the brain (Fig. 36). The alveoli in these areas are often atelectatic and contain increased numbers of macrophages. Bronchial and bronchiolar epithelia are generally not affected.

Kidney

The inner cortex and medulla contain scattered collecting tubules lined by hypertrophied epithelial cells containing finely granular to intensely basophilic intranuclear inclusion bodies. These necrotic inclusion body-bearing cells desquamate into the lumens of affected tubules to form prominent cellular casts (Fig. 37).

Ultrastructure

The intranuclear inclusions are composed of numerous 45 nm simian virus 40 virions that fill the nucleoplasm and displace the chromatin to the periphery (Fig. 38).

Differential Diagnosis

Other viral infections of macaque monkeys in which there are intranuclear inclusion bodies

Fig. 35 (*upper left*). Simian virus 40 infection of the ▶ cerebral cortex in a rhesus monkey. Multiple oligodendroglial cells contain early, finely granular (*open arrows*) and late, dense (*closed arrow*) basophilic intranuclear inclusion bodies. H and E, ×450

Fig. 36 (*lower left*). Simian virus 40 infection of the lung in a rhesus monkey. Alveolar septa are markedly thickened; hypertrophied type II pneumocytes contain dense, basophilic, intranuclear inclusion bodies. H and E, ×250

Fig. 37 (*upper right*). Simian virus 40 infection of the kidney in a rhesus monkey. Nuclei of epithelial cells of the collecting tubules contain early granular and late intensely basophilic intranuclear inclusion bodies. Some of the infected cells have desquamated into the lumen of the tubules. H and E, ×450

Fig. 38 (*lower right*). Simian virus 40 infection. One of the intranuclear inclusion bodies in the kidney illustrated in Fig. 37. The nucleoplasm is filled with numerous SV 40 virions that displace much of the chromatin to the periphery. TEM, ×15 000. *Inset*, Higher magnification of the SV 40 virions revealing their cubic symmetry and 45 nm diameter. TEM, ×100 000

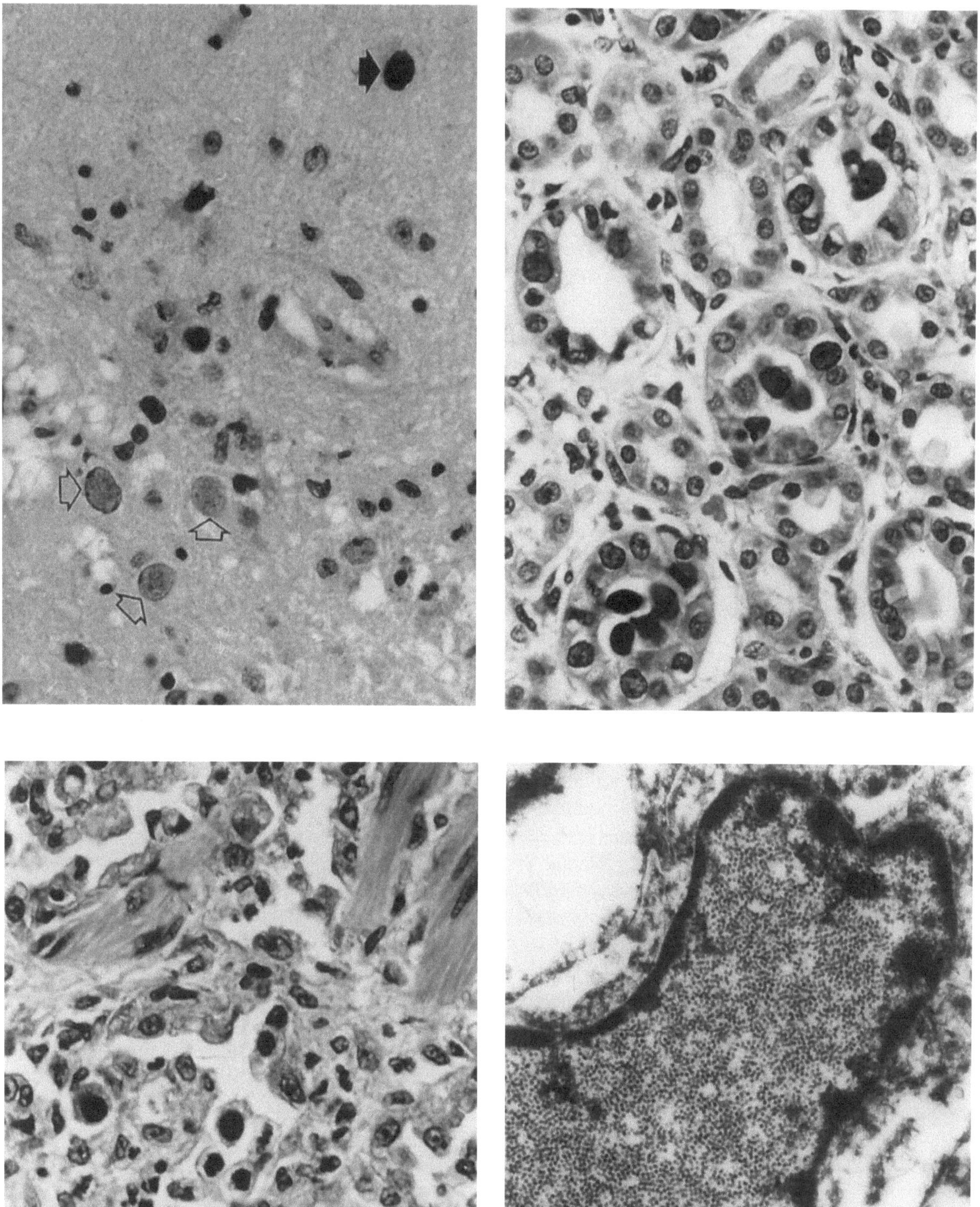

that must be distinguished from those caused by simian virus 40 include adenovirus, cytomegalovirus, *Herpesvirus simiae*, and other herpesvirus infections. In fixed tissues, virions of simian virus 40 can be easily distinguished from these other viral agents by their size, using electron microscopy, or by their antigenic or genetic composition, using virus-specific immunohistochemistry and *in situ* hybridization, respectively. Virus isolation can be used if fresh tissues are available.

Biologic Features

Natural History and Epidemiology

Simian virus 40 occurs commonly as a latent infection of feral and captive Asian monkeys, especially *Macaca mulatta*, *M. fuscata*, *M. cyclopis*, and to a lesser extent *M. fascicularis*, but rarely causes clinical illness in these species (Shah and Southwick 1965). The virus was originally isolated from normal appearing rhesus and cynomolgus kidney cell cultures and from seed stocks of poliovirus and several strains of adenovirus grown in these cells for human vaccine production (Sweet and Hilleman 1960a,b). Rhesus and cynomolgus monkey kidney cell cultures infected with simian virus 40 exhibit no cytopathic effect even though the virus replicates to high titers in these cultures. When the medium from such cultures is used to inoculate cultures of African green monkey (*Cercopithecus aethiops*) kidney cells, the latter develop prominent cytoplasmic vacuolization as the result of simian virus 40 infection. This property led to its original designation as "vacuolating agent" or "vacuolating virus" (Sweet and Hilleman 1960a,b; Hsiung and Gaylord 1961). As mentioned, this virus is not a significant cause of clinical illness in macaque monkeys. In fact, the number of reported cases of active simian virus 40 infection in macaque monkeys is extremely small and has only involved animals with suspected or confirmed immune dysfunction (Gribble et al. 1975; Holmberg et al. 1977; Sheffield et al. 1980; King et al. 1983; Horvath et al. 1992). The cause of the immune compromise, resulting in reactivation of latent simian virus 40 infection in these macaques, has been confirmed as simian immunodeficiency virus (SIV) infection in at least two instances (Horvath et al. 1992; Lowestine et al. 1992; see p. 20, this volume).

Etiology

Simian virus 40 is a 45 nm, unenveloped, icosahedral, DNA-containing virus classified in the Polyomavirinae subfamily of the family Papovaviridae. Its genome consists of superhelical, double-stranded, circular DNA of approximately 5000 base pairs. Its genome contains early and late coding regions and a noncoding regulatory region. The early region encodes for two DNA binding proteins, referred to as large and small T (tumor) antigens, which bind to the viral regulatory sequences to facilitate transcription of the late viral genes. The late genes encode for three capsid proteins (VP1, VP2, VP3) which contain genus-specific antigenic determinants (Eckhart 1990). The virus can be propagated in a variety of different cell types, including those of monkey, human, and murine origin. In vitro infection may be lytic or nonlytic, depending upon the cell cultures employed. In addition, the virus has been shown to transform a variety of rodent cells in vitro and to be oncogenic in several rodent species in vivo (see p. 5, this volume).

Pathogenesis

Natural transmission of simian virus 40 infection in macaque monkeys is thought to occur via the respiratory route. The virus has been experimentally transmitted to rhesus and African green monkeys via the intranasal, intragastric, and subcutaneous routes (Fabiyi et al. 1967; Shah et al. 1969). In these experimental studies, none of the animals became clinically ill during their primary infection, but a number of them did develop mild to moderate azotemia and red blood cells and casts in their urine, indicating that the virus induces a mild, but transient, impairment of renal function (Fabiyi et al. 1967). Virus isolation studies have indicated that simian virus 40 replicates efficiently in the renal medulla, spleen, and lymph nodes and that it is excreted in high concentrations in the urine (Ashkenazi and Melnick 1962). Infected urine may serve as the principal source of infection of susceptible animals. During the first week of infection animals become viremic. Neutralizing antibodies to the virus occur in the blood by 21 days postinfection and in the urine by 9 weeks after infection. Thereafter, the virus enters a latent state in which it persists in the renal epithelium

indefinitely. Latent infection may become reactivated when the host's cell-mediated immune response is impaired, thereby allowing the virus to cause progressive multifocal leukoencephalopathy or polyomaviral interstitial pneumonia and/or papovaviral nephritis. Interestingly, animals with this infection of the brain have not had concomitant simian virus 40-induced pneumonia or nephritis, although the latter two conditions have been observed in the same animal (Sheffield et al. 1980; King et al. 1983).
Intracerebral inoculation of fetal rhesus monkeys with this virus has been associated with postnatal hydrocephalus, seizures, and incoordination (London et al. 1983).

Comparison with Other Species

Early studies with simian virus 40 revealed that it is highly oncogenic when inoculated into newborn Syrian hamsters (Eddy et al. 1961, 1962). Hamsters become progressively more resistent to the oncogenic effects of the virus with advancing age (Eddy et al. 1962). Like other polyomaviruses, simian virus 40 induces a variety of tumors including lymphocytic leukemia, lymphosarcoma, reticulum cell sarcoma, osteogenic sarcoma, and undifferentiated sarcomas when inoculated into hamsters (Diamandopoulos 1972). Intracerebral inoculation of this virus into newborn hamsters and mastomys (*Mastomys natalensis*) induces ependymomas (Gerber and Kirschstein 1962; Rabson et al. 1962). Interestingly, the type of tumor induced varies depending upon the dose and route of inoculation of the virus (Diamandopoulos and McLane 1975). Tumors have also been experimentally induced in mice inoculated with this virus (Abramczuk et al. 1984; Hargis and Malkiel 1979). Due to its ability to readily transform cells in vitro and produce malignant neoplasms in vivo, simian virus 40 is one of the most extensively studied viruses of nonhuman primate origin. Its genome has been cloned and sequenced and the molecular basis for its oncogenicity in laboratory rodents is reasonably well understood. This subject is beyond the scope of this manuscript.
Following the discovery, in 1961, that simian virus 40 is oncogenic in hamsters, there was great concern that this virus might also cause neoplasms in humans. Between 1954 and 1961, hundreds of thousands of people were exposed to simian virus 40 through ingestion of the widely distributed, live, oral poliomyelitis vaccine. This vaccine was subsequently found to be contaminated with simian virus 40 from the rhesus monkey kidney cells in which the poliovirus was grown. Children who ingested this vaccine were found to excrete small amounts of simian virus 40 in their stools for up to 5 weeks without evidence of clinical illness. Large-scale epidemiologic studies of people exposed during this period have failed to reveal convincing evidence of neoplasia induced by this virus or other illness in this large population. There is a recent report describing virus sequences similar to those of simian virus 40 in ependymomas and choroid plexus tumors of children (Bergsagel et al. 1992).
A high percentage of humans harbor one or two papovaviruses related to simian virus 40 that have been associated with clinical disease primarily in immunocompromised patients. These agents, referred to as JC virus and BK virus (the initials of the patients from which they were originally isolated), are acquired during childhood, presumably as a mild or subclinical respiratory infection. In these patients, following this primary subclinical infection, the viruses become latent and persist indefinitely in the kidneys. During states of immune compromise, such as those resulting from cancer chemotherapy, long-term steroid treatment, certain forms of neoplasia, inherited immunologic disorders, and, more recently, the acquired immunodeficiency syndrome (AIDS), latent JC virus may become activated and cause the demyelinating disease progressive multifocal leukoencephalopathy (Padgett et al. 1971; Weiner et al. 1972). Reactivation of latent BK virus infections has been associated with kidney infection in patients given renal transplants and cystitis in others receiving bone marrow transplants. In addition, BK virus genomic sequences have been associated with pancreatic islet cell tumors, a variety of different brain tumors, and with Kaposi's sarcoma, but its role in the induction of these tumors has not been established.

Acknowledgement. This work was supported by Public Health Service Award RR00168 from the National Institutes of Health.

References

Abramczuk J, Pan S, Maul G, Knowles BB (1984) Tumor induction by simian virus 40 in mice is controlled by long-term persistence of the viral genome and the immune response of the host. J Virol 49:540–548

Ashkenazi A, Melnick JL (1962) Induced latent infection of monkeys with vacuolating SV40 papovavirus. Virus in kidneys and urine. Proc Soc Exp Med III:367–372

Bergsagel DJ, Finegold MJ, Butel JS, Kupsky WJ, Garcea RL (1992) DNA sequences similar to those of simian virus 40 in ependymomas and choroid plexus tumors of childhood. N Engl J Med 326:988–993

Diamandopoulos GT (1972) Leukemia, lymphoma, and osteosarcoma induced in the Syrian golden hamster by simian virus 40. Science 176:173–175

Diamandopoulos GT, McLane MF (1975) Effect of host age, virus dose, and route of inoculation on tumor incidence, latency, and morphology in Syrian hamsters inoculated intravenously with oncogenic DNA simian virus 40. J Natl Cancer Inst 55:479–482

Eckhart W (1990) Polyomavirinae and their replication. In: Knipe DM et al. (eds) Virology, 2nd edn. Raven, New York, pp 1593–1623

Eddy BE, Borman GS, Berkeley WH, Young RD (1961) Tumors induced in hamsters by injection of rhesus monkey kidney cell extracts. Proc Soc Exp Biol Med 107:191–197

Eddy BE, Borman GS, Grubbs GE, Young RD (1962) Identification of the oncogenic substance in rhesus monkey kidney cell cultures as simian virus 40. Virology 17:65–75

Fabiyi A, Calcagno PL, Sever JL, Antonovych T, Wolman F (1967) Isolation of SV40 virus and its effect on renal function in African green monkeys. Nature 215:88–89

Gerber R, Kirschstein RL (1962) SV40 induced ependymomas in newborn hamsters: I. Virus tumor relationships. Virology 18:582–588

Gribble DH, Haden CC, Schwartz LW, Henrickson RV (1975) Spontaneous progressive multifocal leukoencephalopathy (PML) in macaques. Nature 254:602–604

Hargis BJ, Malkiel S (1979) Sarcomas induced by injection of simian virus 40 into neonatal CFW mice. J Natl Cancer Inst 63:965–968

Holmberg CA, Gribble DH, Takemoto KK, Howley PM, Espana C, Osburn BI (1977) Isolation of simian virus 40 from rhesus monkeys (Macaca mulatta) with spontaneous progressive multifocal leukoencephalopathy. J Infect Dis 136:593–596

Horvath CJ, Simon MA, Bergsagel DJ, Pauley DR, King NW, Garcea RL, Ringler DJ (1992) Simian virus 40 (SV40)-induced disease in macaques with simian AIDS. Am J Pathol (in press)

Hsiung GD, Gaylord WH (1961) The vacuolating virus of monkeys: I. Isolation, growth characteristics and inclusion body formation. J Exp Med 114:975–986

King NW, Hunt RD, Letvin NL (1983) Histopathologic changes in macaques with an acquired immunodeficiency syndrome (AIDS). Am J Pathol 113:382–388

London WT, Gravell M, Curfman B, Martinez AJ, Sever JL (1983) Hydrocephalus, seizures, and incoordination in rhesus monkeys infected in utero with SV40 virus. Lab Anim Sci 33:482–483

Lowenstine LJ, Lerche NW, Yee JA, Uyeda A, Jennings MB, Munn RJ, McClure HM, Anderson DC, Fultz PN, Gardner MB (1992) Evidence for a lentiviral etiology in an epizootic of immune deficiency and lymphoma in stump-tailed macaques (Macaca arctoides). J Med Primatol (in press)

Padgett BL, Walker DL, ZuRhein GM, Eckroade RJ, Dessel BH (1971) Cultivation of papova-like virus from human brain with progressive multifocal leukoencephalopathy. Lancet 1:1257–1260

Rabson AS, O'Conor GT, Kirschstein RL, Branigan WJ (1962) Papillary ependymomas produced in Rattus (Mastomys) natalensis inoculated with vacuolating virus (SV40). J Natl Cancer Inst 29:765–787

Shah KV, Southwick CH (1965) Prevalence of antibodies to certain viruses in sera of free-living rhesus and captive monkeys. Indian J Med Res 53:488–500

Shah KV, Willard S, Myers RE, Hess DM, DiGiacomo R (1969) Experimental infection of rhesus with simian virus 40 (SV40). Proc Soc Exp Biol Med 130:196–203

Sheffield WD, Strandberg JD, Braun L, Shah K, Kalter SS (1980) Simian virus 40-associated fatal interstitial pneumonia and renal tubular necrosis in a rhesus monkey. J Infect Dis 142:618–622

Sweet BH, Hilleman MR (1960a) Detection of a "non-detectable" simian virus (vacuolating agent) present in rhesus and cynomolgus monkey kidney cell culture material. 2nd international conference on live poliovirus vaccines. Pan American Health Organization and the World Heath Organization, Washington, pp 79–85

Sweet BH, Hilleman MR (1960b) The vacuolating virus SV40. Proc Soc Exp Biol Med 105:420–427

Weiner LP, Herndon RM, Narayan O, Johnson RT, Shah K, Rubinstein LJ, Perziosi TJ, Conley FK (1972) Isolation of virus related to SV40 from patients with progressive multifocal leukoencephalopathy. N Engl J Med 286:385–390

Malignant Lymphoma of Unknown Cause

Ronald D. Hunt

Synonym. Lymphosarcoma.

Gross Appearance

Lymphomas are extremely variable in their gross appearance. They may occur as: (1) enlargement of one or more lymph nodes; (2) one to multiple nodules or masses of variable size either in or adjacent to various viscera such as the heart, liver, spleen, kidneys, gastrointestinal tract, thymus, or in such locations as the subcutaneous tissues, muscles, orbit, bone, or central nervous system; or (3) diffuse infiltration of visceral organs. The tumor tissue is yellow, gray, or off-white and often contains foci of hemorrhage.

Microscopic Features

Most lymphomas in nonhuman primates are characterized by a uniform population of poorly differentiated lymphocytes occurring in sheets. The neoplastic cells have round, oval, or angular nuclei, usually containing a single distinct nucleolus. Chromatin is uniformly dispersed or clumped. By Rappaport's (1966) classification these tumors are diffuse, poorly differentiated, lymphocytic lymphomas. Other patterns and degrees of differentiation also occur in nonhuman primates and include lymphocytic, well differentiated; nodular, differentiated; nodular, poorly differentiated; diffuse, mixed lymphocytic and histiocytic; diffuse, histiocytic; and undifferentiated (Figs. 39–45). Many reported examples of lymphoma describe phagocytic macrophages throughout the neoplastic cells giving rise to the "starry-sky" pattern so typical of Burkitt's lymphoma, associated with the Epstein-Barr virus, in humans. This pattern, however, does not imply any other similarity or a common cause such as herpesvirus. Lymphomas resembling Hodgkin's disease have been reported

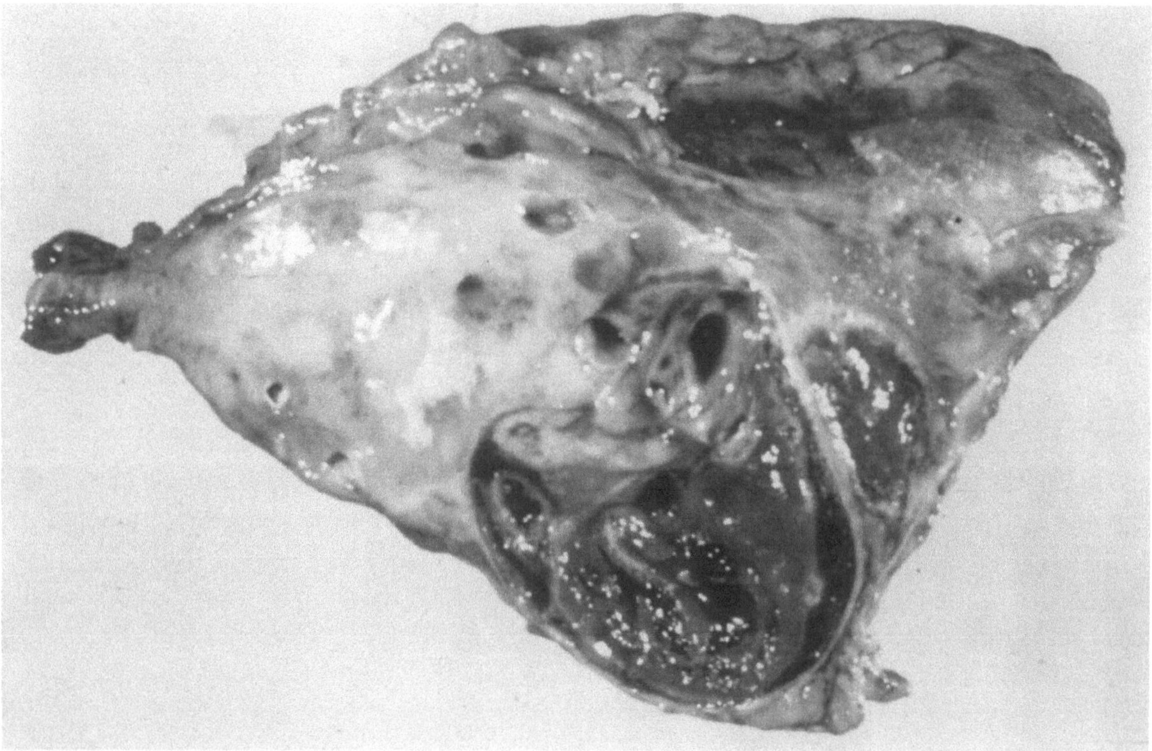

Fig. 39. A spontaneous lymphoma in the mediastinum of a rhesus monkey replacing the thymus and invading the heart

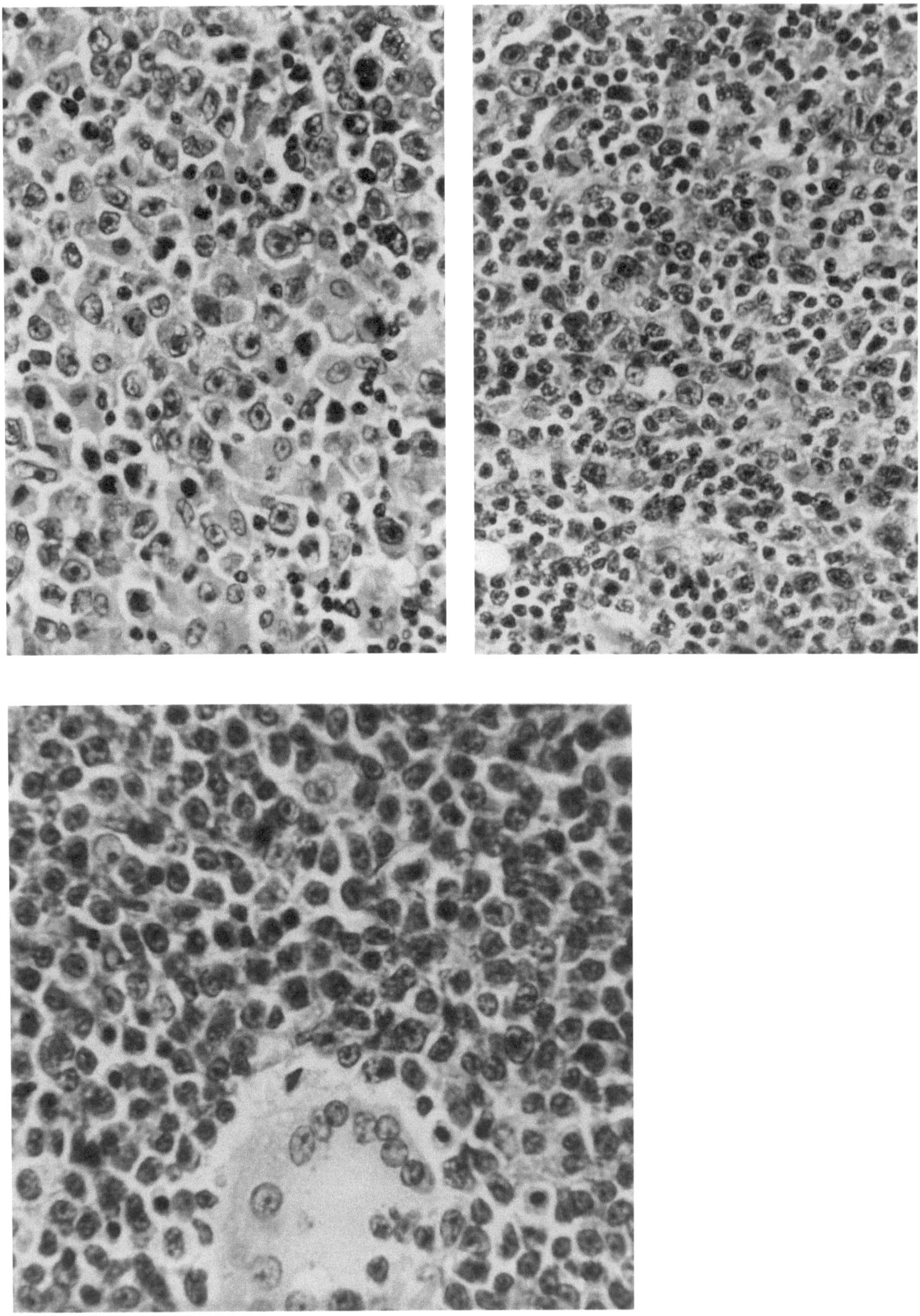

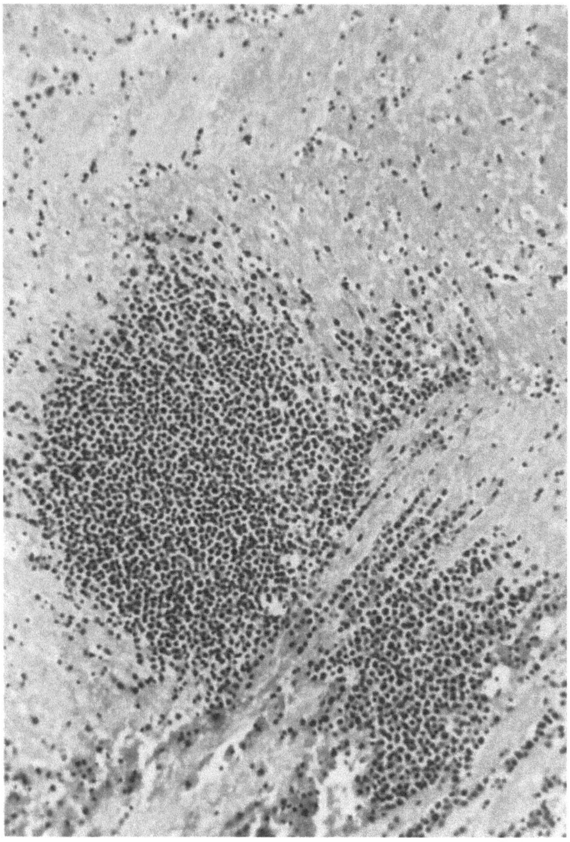

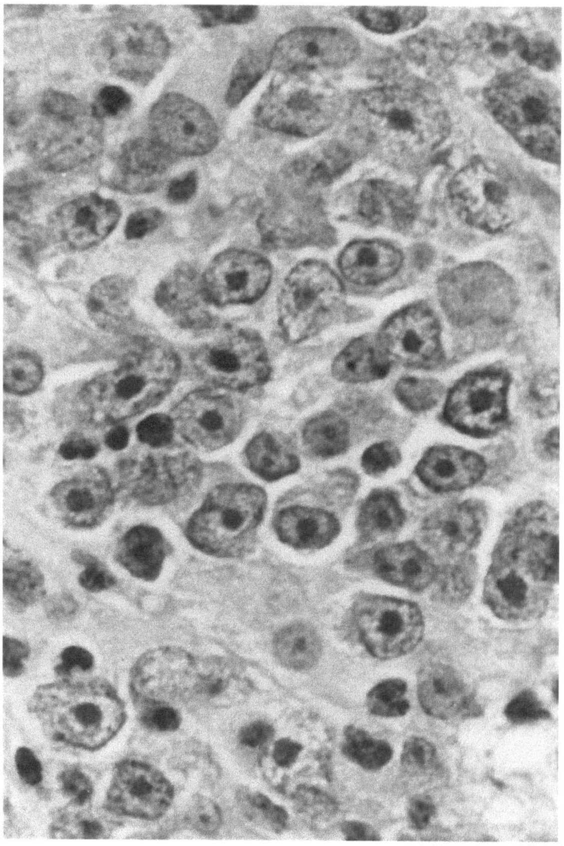

◀ **Fig. 40** (*upper left*). A spontaneous lymphoma arising in the mandible of a rhesus monkey. The neoplasm composed of large cells with prominent nucleoli and is classified as a diffuse histiocytic lymphoma or immunoblastic sarcoma. H and E, ×400

Fig. 41 (*upper right*). Undifferentiated lymphoma of the intestine composed of pleomorphic cells intermixed with more mature lymphocytes. H and E, ×400

Fig. 42 (*below*). Spontaneous poorly differentiated lymphoma containing a multinucleated giant cell in a rhesus monkey naturally infected with simian immunodeficiency virus. H and E, ×400

Fig. 43 (*left*). Spontaneous lymphoma invading the brain of a rhesus monkey. H and E, ×100

Fig. 44 (*right*). A spontaneous histiocytic lymphoma in a rhesus monkey. H and E, ×400

in nonhuman primates (Hofmann et al. 1981; Gleiser et al. 1984; Wadsworth et al. 1985; Lowenstine 1986).

Biologic Features

Natural History

In colonies of nonhuman primates, lymphomas are usually seen as an isolated occurrence. The neoplastic disease is typically described in young adults, but this may be a reflection of the composition of most colonies of nonhuman primates. There are no specific clinical signs. Leukemia is rarely detected. Although very few reported examples of spontaneous lymphoma in nonhuman primates have been characterized with techniques which identify the functional category of the cell type, both B and T cell types have been described (Holmberg et al. 1978). Based upon morphologic features alone, most resemble B cell lymphomas.

Etiology and Frequency

By definition, the malignant lymphomas considered in this report represent lymphomas of unknown or unidentified etiology. More than likely some, if not many, are caused by viruses

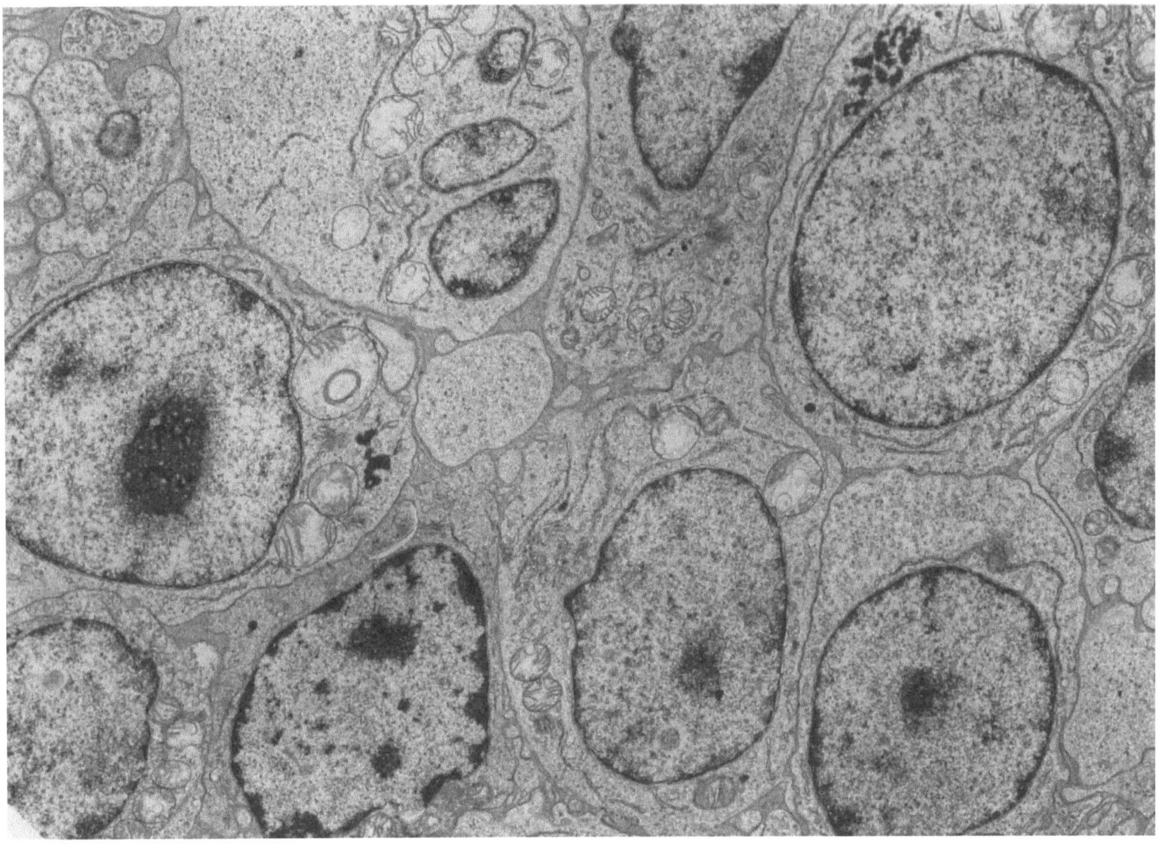

Fig. 45. Spontaneous lymphoma in a rhesus monkey. The fine detail does not allow differentiation of T or B cell origin. TEM, ×5000

such as: Epstein-Barr type viruses, *Herpesvirus saimiri*, *Herpesvirus ateles*, simian T-lymphotropic virus type 1 (STLV-1), and gibbon ape leukemia virus (GALV) (Rangan and Gallagher 1979; Rangan et al. 1986; Soike et al. 1984; Desrosiers and Hunt 1982; Sakakibara et al. 1986). In contrast to the usual sporadic occurrence of lymphoma, "outbreaks" of lymphoma have been reported from the California Regional Primate Research Center (Manning and Griesemer 1974). In one interval between February 1969 and December 1977, lymphoma was seen in 42 rhesus monkeys (*M. mulatta*) and three stumptail macaques (*M. arctoides*) (Terrell et al. 1980). These developed in conjunction with the occurrence of a simian acquired immunodeficiency syndrome caused by either a type D retrovirus or simian immunodeficiency virus. Immunosuppression apparently allowed an unidentified infectious agent to express itself. A similar occurrence on a much smaller scale was seen at both the New England Regional Primate Research Center and the Delta Regional Primate Research Center where the lymphomas were shown to be transmissible (Hunt et al. 1983; Baskin et al. 1986). Lymphomas have also been observed with a higher than expected frequency in cynomolgus monkeys (*M. fascicularis*) immunosuppressed for transplantation studies (Reitz et al. 1981, 1982; Garnett et al. 1983). It was suggested to be related to activation of a virus resembling the Epstein-Barr virus (Heberling et al. 1982). An unusually high incidence of lymphoma, believed to be infectious, has been occurring for several years in hamadryas baboons at the Sukhumi Primate Center. This purportedly followed the introduction of leukemic blood from human beings into several animals which were released into the colony (Lapin 1973, 1987).

Lymphoma has been described in many species of nonhuman primates: prosimians; Old World monkeys; New World monkeys; and great apes (Beniashvili 1989; Lowenstine 1986; Hofmann et al. 1981; Holmberg et al. 1985; Lingeman et al. 1969; McClure 1973; Manning and

Griesemer 1974; Snyder et al. 1973). With the exception of experiences at the California Primate Research Center and the Russian Center of Sukhumi, lymphoma is not seen as the most common neoplastic disease in any nonhuman primate species.

Comparison with Other Species

The morphology and distribution of lymphoma in nonhuman primates are comparable to that seen in other animals and human beings; however, in many species the number of reported examples is far too small to make any meaningful comparisons. The relatively small number of reported lymphomas in so many species in the order Primates has precluded the development of a completely satisfactory classification system for lymphomas in nonhuman primates.

Acknowledgement. This work was supported by Public Health Service Award RR00168 from the National Institutes of Health.

References

Baskin GB, Martin LN, Rangan SRS, Gormus BJ, Murphey-Corb M, Wolf RH, Soike KF (1986) Transmissible lymphoma and simian acquired immunodeficiency syndrome in rhesus monkeys. J Natl Cancer Inst 77:127–139

Beniashvili DS (1989) An overview of the world literature on spontaneous tumors in nonhuman primates. J Med Primatol 18:423–437

Desrosiers RC, Hunt RD (1982) Herpesvirus-induced malignant lymphoma in nonhuman primates. In: Capen CC, Hackel DB, Jones TC, Magaki G (eds) Handbook: animal models of human disease, Fasc 11 Model No 60, Suppl update. AFIP, Washington

Garnett NL, Taylor HR, Hoffman GJ, Heberling RL, Ambinder RF, Huber AC, Valentine BA (1983) Malignant lymphoma in nontransplanted cynomolgus monkeys receiving cyclosporine. Transplant Proc 15 [Suppl 1]:2808–2812

Gleiser CA, Carey KD, Heberling RL (1984) Malignant lymphoma and Hodgkin's disease in baboons (Papio sp.). Lab Anim Sci 34:286–289

Heberling RL, Bieber CP, Kalter SS (1982) Establishment of a lymphoblastoid cell line from a lymphomatous cynomolgus monkey. In: Yohn DS, Blakeslee JR (eds) Advances in comparative leukemia research. Elsevier/North Holland, New York, pp 385–386

Hofmann W, Möller P, Schwaier A, Flügel RM, Zöller L, Darai G (1981) Malignant tumors in Tupaia (tree shrew). J Med Primatol 10:155–163

Holmberg CA, Osburn BI, Terrell TG, Manning JS (1978) Cellular immunologic studies of malignant lymphoma in rhesus macaques. Am J Vet Res 39:469–472

Holmberg CA, Henrickson R, Anderson J, Osburn BI (1985) Malignant lymphoma in a colony of Macaca arctoides. Vet Pathol 22:42–45

Hunt RD, Blake BJ, Chalifoux LV, Sehgal PK, King NW, Letvin NL (1983) Transmission of naturally occurring lymphoma in macaque monkeys. Proc Natl Acad Sci USA 80:5085–5089

Lapin BA (1973) The epidemiologic and genetic aspects of an outbreak of leukemia among hamadryas baboons of the Sukhumi monkey colony. In: Dutcher RM, Chieco-Bianchi L (eds) Unifying concepts of leukemia. Bibl Haematol Basel 39:263–268

Lapin BA (1987) Nonhuman primate models of human hematological malignancies. Haematol Bluttransfus 31: 440–444

Lingeman CH, Reed RE, Garner FM (1969) Spontaneous hematopoietic neoplasms of nonhuman primates. Review, case report, and comparative studies. Natl Cancer Inst Monogr 32:157–170

Lowenstine LJ (1986) Neoplasms and proliferative disorders in nonhuman primates. In: Benirschke K (ed) Primates: the road to self-sustaining populations. Springer, Berlin Heidelberg New York, pp 781–814

Manning JS, Griesemer RA (1976) Spontaneous lymphoma of the nonhuman primate. Lab Anim Sci 24:204–210

McClure HM (1973) Tumors in nonhuman primates: observations during a six-year period in the Yerkes Primate Center colony. Am J Phys Anthropol 38:425–429

Rangan SRS, Gallagher RE (1979) Tumors and viruses in nonhuman primates. Adv Virus Res 24:1–123

Rangan SRS, Martin LN, Bozelka BE, Wang N, Gormus BJ (1986) Epstein-Barr virus-related herpesvirus from a rhesus monkey (Macaca mulatta) with malignant lymphoma. Int J Cancer 38:425–432

Rappaport H (1966) Tumors of the hematopoietic system. Atlas of tumor pathology, Section III, Fasc 8. AFIP, Washington, pp 91–206

Reitz BA, Bieber CP, Raney AA, Pennock JL, Jamieson SW, Oyer PE, Stinson EB (1981) Orthotopic heart and combined heart and lung transplantation with cyclosporin-A immune suppression. Transplant Proc 13:393–396

Reitz BA, Bieber CP, Pennock JC (1982) Cyclosporin A and lymphoma in nonhuman primates. In: White DJ (ed) Cyclosporin A. Elsevier, Amsterdam, pp 317–328

Sakakibara I, Sugimoto Y, Sasagawa A, Honjo S, Tsujimoto H, Nakamura H, Hayami M (1986) Spontaneous malignant lymphoma in an African green monkey naturally infected with simian T-lymphotropic virus (STLV). J Med Primatol 15:311–318

Snyder SP, Dungworth DL, Kawakami TG, Callaway E, Lau DT-L (1973) Lymphosarcomas in two gibbons (Hylobates lar) with associated C-type virus. J Natl Cancer Inst 51:89–94

Soike KF, Rangan SRS, Gerone PJ (1984) Viral disease models in primates. Adv Vet Sci Comp Med 28:151–199

Terrell TG, Gribble DH, Osburn BI (1980) Malignant lymphoma in macaques: a clinicopathologic study of 45 cases. J Natl Cancer Inst 64:561–568

Wadsworth PF, Jones DM, Pugsley SL (1985) A survey of mammalian and avian neoplasms at the Zoological Society of London. J Zoo Anim Med 16:73–80

Fibromatosis in Macaques Infected with Type D Retroviruses

Che-Chung Tsai

Synonyms. Retroperitoneal fibromatosis; subcutaneous fibromatosis; Kaposi's sarcoma-like lesion; mesenchymal neoplasm; mesenchymal tumor-like lesions.

Gross Appearance

The two anatomic locations of these lesions led to the designations retroperitoneal and subcutaneous fibromatosis. The course of the disease also occurs in two forms: localized and progressive. In the localized form, a nodular or multinodular proliferation of fibromatous tissue is found beneath the peritoneum at the mesenteric attachment of the ileocecal junction and/or adjacent lymph nodes (Fig. 46). These lesions vary from 1 to 4 cm in diameter, are very firm and pale, and resist cutting. Most localized lesions remain at one site but some become progressive, particularly in monkeys that manifest simian acquired immunodeficiency syndrome (SAIDS) (Giddens et al. 1985; Tsai et al. 1985a,b, 1986). In a few affected monkeys the localized lesions undergo spontaneous regression.

In the progressive form, which is uniformly fatal, fibromatous tissue encircles the intestines (Fig. 47) and adjacent lymph nodes. Expansion of lesions throughout the abdominal cavity involves either multiple discrete nodules of fibromatous tissue or fusion into a large mass. The lesions usually extend up the mesentery, cover the dorsal and ventral walls of the abdominal cavity, and cause a marked fibromatous thickening of the omentum. The small and large intestines are often encased in a solid mass of fibrous tissue. The fibromatous tissue may extend through the diaphragm into the pleural cavity, covering the parietal and visceral pleura of the lungs and the pericardium. Fibromatous tissue also may extend into the inguinal canal and surround the testicles.

A subcutaneous or oral mucous form of fibromatosis may be observed in a small number of affected monkeys. This is manifest by single or multiple nodules in the skin or oral mucosa (Fig. 48).

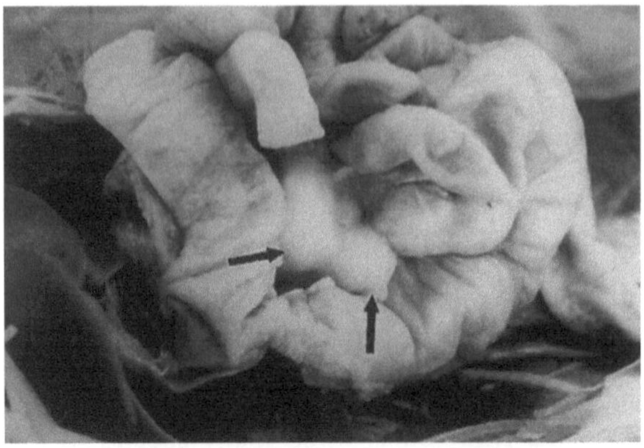

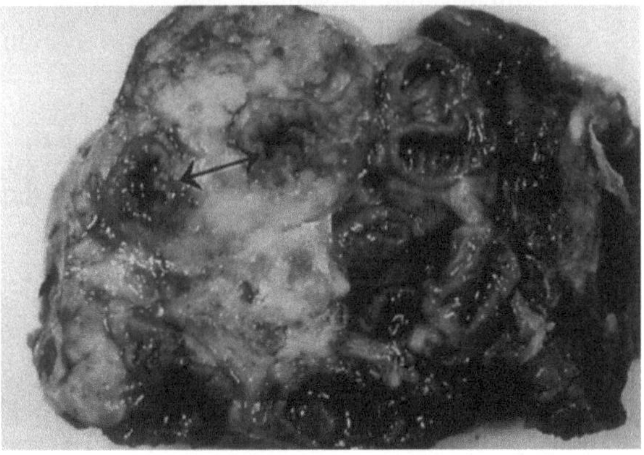

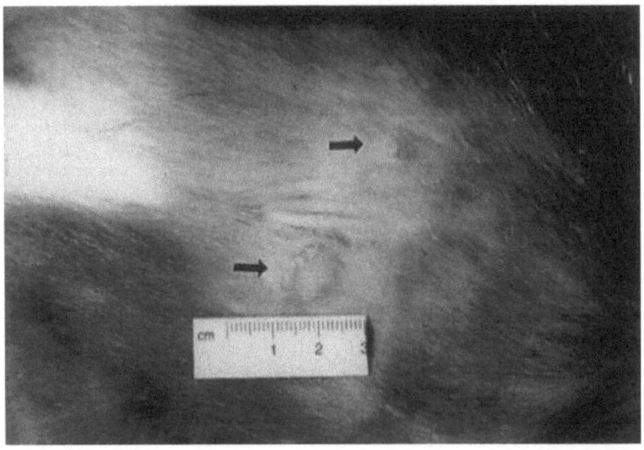

Fig. 46 (*above*). Localized form of retroperitoneal fibromatosis, *Macaca nemestrina*. Multiple enlarged nodules (*arrows*) surrounding mesenteric lymph nodes adjacent to the ileocecal junction

Fig. 47 (*middle*). Progressive form of retroperitoneal fibromatosis. A cross-section of intestinal loop (*arrow*) is circumscribed by fibromatous tissue

Fig. 48 (*below*). Two subcutaneous nodules (*arrows*) of fibromatosis over the abdomen of a *Macaca nemestrina*

Microscopic Features

Regardless of where the lesions occur, two basis morphologic patterns are recognizable: proliferative and sclerotic. Both types are occasionally observed at different sites in the same affected individual. In the proliferative type, the tissue is composed of large ovoid or spindle-shaped cells in a disorganized stromal network of collagen and reticulum fibers (Fig. 49). The nuclei of these cells are vesicular and contain up to two nucleoli and occasional mitotic figures (Fig. 50). Although these lesions do not appear congested, both well-formed blood vessels and immature capillaries are often present. Erythrophagocytosis is not a common feature. Lymphocytes and plasma cells in varying numbers are almost always present in the lesions. The shape and arrangement of the cells and their production of collagen suggest that they are fibroblasts. Despite the proliferative behavior of these fibroblasts, they lack the densely cellular and anaplastic features of fibrosarcoma cells (Fig. 51).

In the sclerotic lesion there are fewer fibroblasts and more collagenous stroma. The nuclei of the fibroblasts are smaller and less vesicular, and mitotic figures are absent. In most lesions lymphocytic infiltration is less prominent. Parts of the nodule or mass may consist of sparsely scattered mature fibroblasts in dense collagen.

In the localized form of the disease, the lesion usually is found in the lymph nodes adjacent to the ileocecal junction. It appears to originate in the capsule of the lymph node (Fig. 52), expanding outward along the lymphatic channels and mesentery toward the adjacent intestine or centrally toward the mesenteric root. The fibromatosis remains retroperitoneal and a distinct serosal lining is usually present over it.

In the progressive form of the disease, fibromatosis occurs as multiple to small or large solitary nodules or as large sheaths or masses that appear to result from expansion and coalescence of multiple focal lesions in the peritoneum and sometimes in the pleura. Fibromatous tissue usually circumscribes the intestinal loops (Fig. 53) without encroaching on the mucosa or lumen, but it occasionally invades the tunica of the gut and forms small submucosal nodules (Fig. 54). Fibromatous tissue rarely invades parenchymal organs such as liver, kidney, spleen, or lung, but it often forms small nodules or plaques on the capsular or serosal surface. There is often local invasion of mesenchymal tissues, particularly the diaphragm (Fig. 55). Progressive cases with involvement of the diaphragm often have lesions in the pleural cavity. As in the abdominal cavity, lesions consist of multiple plaques or nodules on both the parietal and visceral pleura and the pleural covering of the pericardium. In several cases the lesions extend down the inguinal canals to surround the epididymis or testes.

Fibromatous nodules or plaques in the skin and/or oral cavity have features very similar to those lesions with a progressive pattern in the abdominal cavity. In another form the disease occurs as an obliterative vasculopathy or an exuberant intimal fibromuscular dysplasia (or hyperplasia).

Ultrastructure

The predominant cells are spindle-shaped fibroblasts (Fig. 56) surrounded by collagen fibers, organelles of disrupted cells, and intracellular matrix. These cells contain abundant dilated rough endoplasmic reticulum, clusters of ribosomes, small vesicles, and scattered filaments. Capillaries and larger vessels with an investment of pericytes are found; occasionally two or three endothelial cells form structures with a slit-shaped lumen. Fascicles of mature collagen fibers are plentiful in most fields. Occasionally lymphocytes, macrophages, and plasma cells may be identified.

Virions may be found among fibroblasts in some cases. Their average diameter is 102 nm. They possess an electron-dense eccentric nucleoid, and a cylindrical nucleoid may be seen in a few viral particles.

Differential Diagnosis

Regardless of anatomic site and type of fibromatous lesions, fibromatosis should be distinguished from fibrosarcoma. The fibroblastic cells of fibromatosis lack the densely cellular and anaplastic features and high mitotic activity of fibrosarcoma. Fibromatosis is an unusual fibroproliferative disorder in macaques infected with SAIDS and has many similarities to Kaposi's sarcoma in human patients with acquired immunodeficiency syndrome (AIDS) (Gottlieb et al. 1983; Fauci et al. 1985; Ziegler and Levy 1985). The principal difference is that Kaposi's

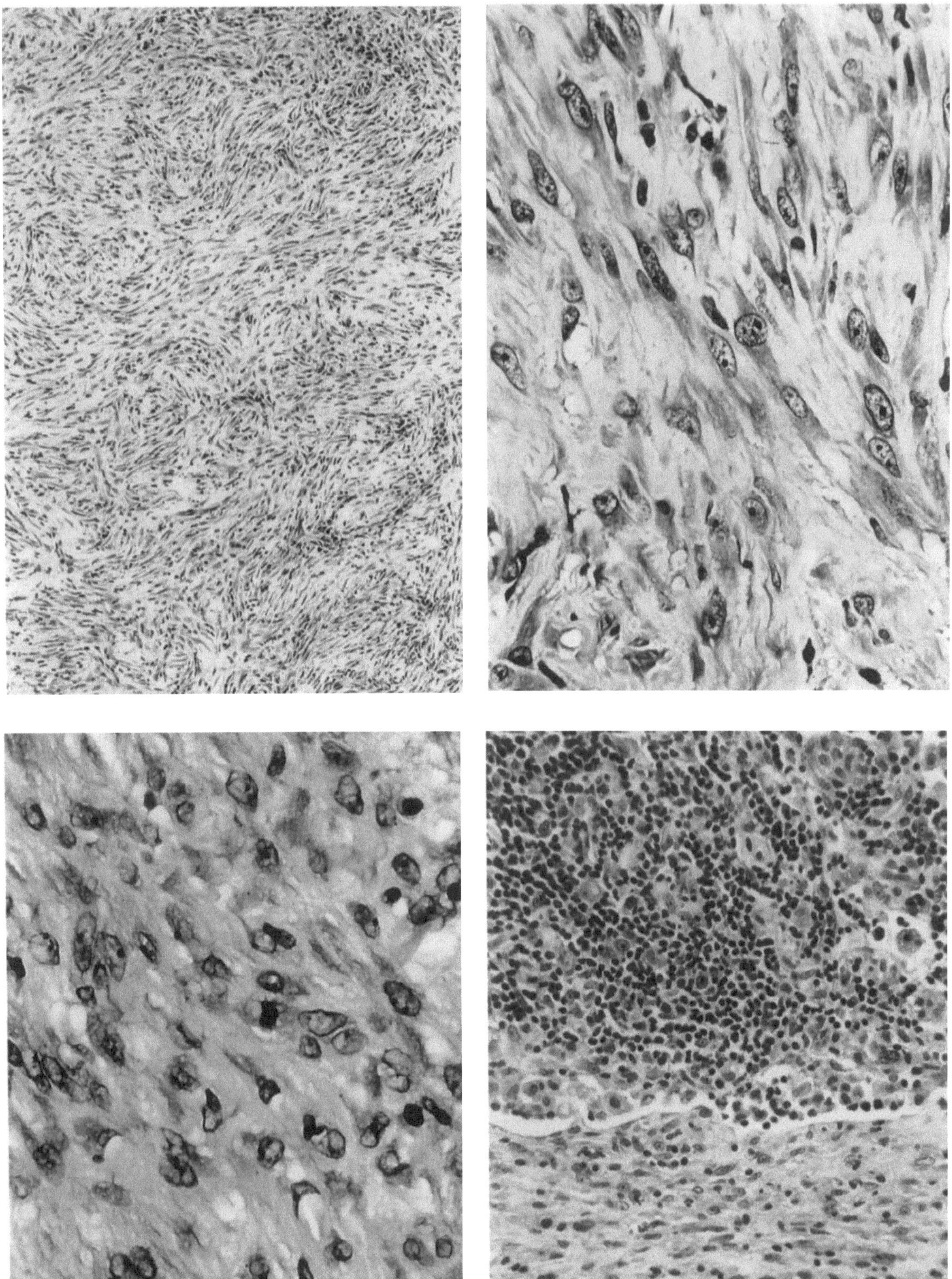

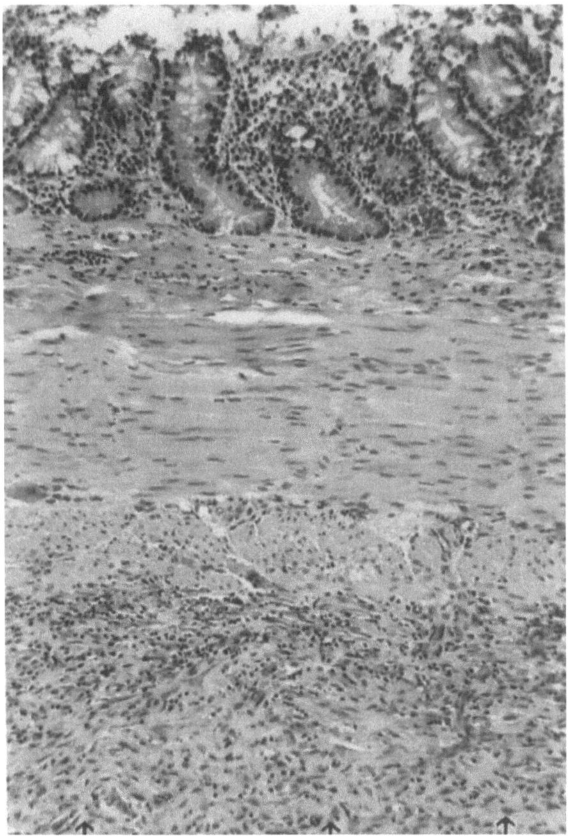

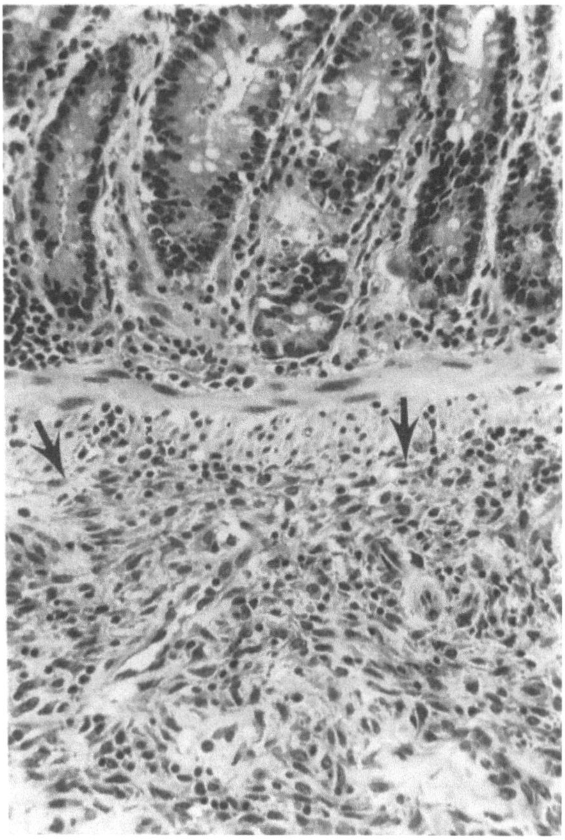

◀ **Fig. 49** (*upper left*). Spindle-shaped cells of retroperitoneal fibromatosis, *Macaca nemestrina*. Note multifocal interwoven or whorled pattern. H and E, ×100

Fig. 50 (*lower left*). Progressive form of retroperitoneal fibromatosis, *Macaca nemestrina*. Many spindle-shaped cells, with vesicular nuclei, prominent nucleoli, and occasionally mitotic figures, surrounded by collagen. H and E, ×400

Fig. 51 (*upper right*). Proliferating spindle-shaped cells of retroperitoneal fibromatosis, *Macaca nemestrina*. Note pleomorphic nuclei and tapering ends of cytoplasm resembling smooth muscle cells. These cells are immuochemically positive for desmin, α-actin, and myosin. H and E, ×400

Fig. 52 (*lower right*). Early lesion of retroperitoneal fibromatosis in a mesenteric node, *Macaca nemestrina*. Proliferative fibromatous tissue is in the capsule (*arrows*). H and E, ×200

Fig. 53 (*upper left*). Fibromatous mass of in the serosa ▶ (*arrows*) of the small intestine, *Macaca nemestrina*. H and E, ×100

Fig. 54 (*upper right*). Fibromatous tissue invading the tunica muscularis (*arrows*) of the small intestine, *Macaca nemestrina*. H and E, ×200

Fig. 55 (*lower right*). Fibromatous tissue invading skeletal muscle of the diaphragm, *Macaca nemestrina*. H and E, ×200

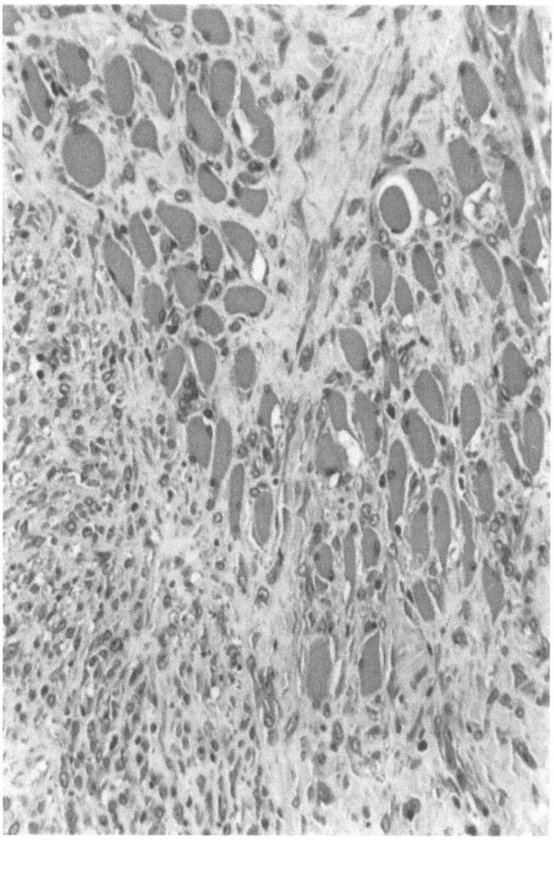

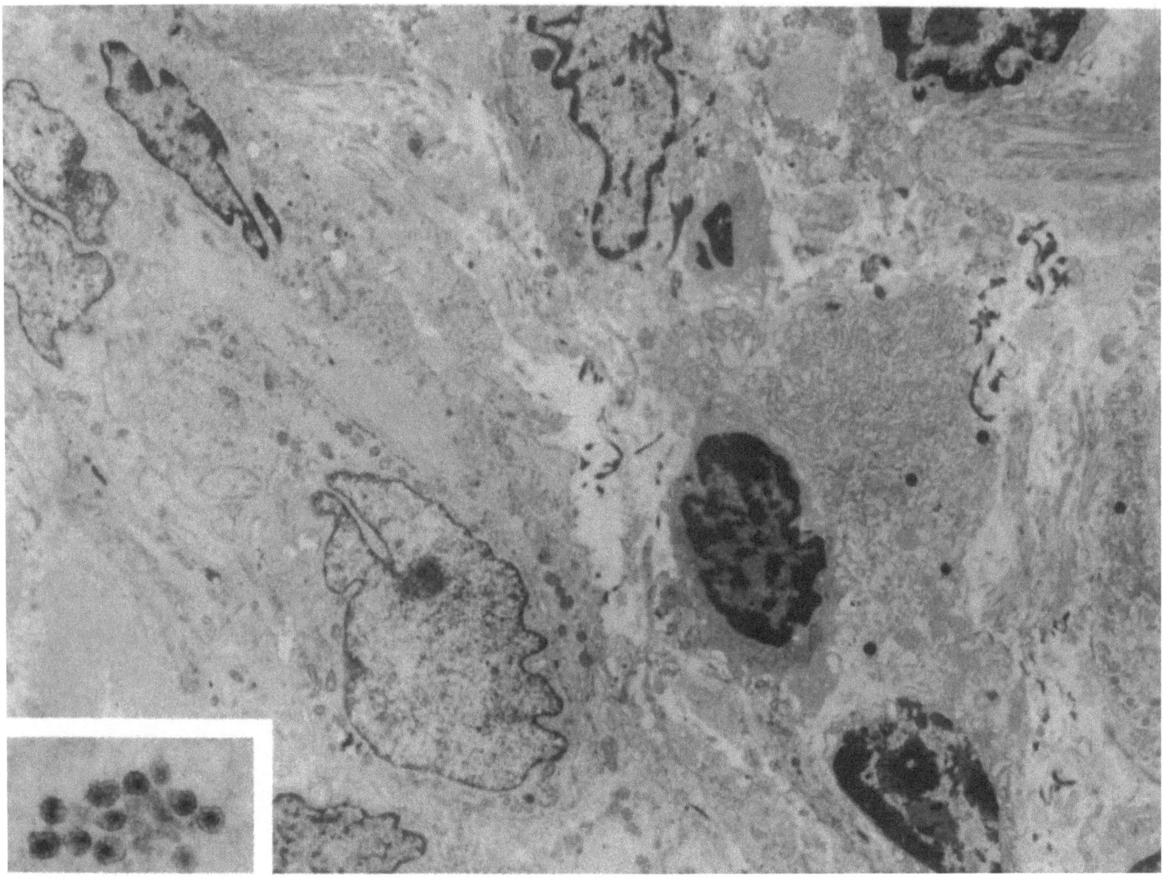

Fig. 56. Fibromatous tissue of spindle cells with rough endoplasmic reticulum and fascicles of collagen, *Macaca nemestrina*. Note the lymphocyte *right of center*. TEM, ×5400. *Insert*, cluster of virions. ×41 250. (From Tsai et al. 1985a)

sarcoma is characteristically a multicentric cutaneous lesion that is more vascular and more hemorrhagic than fibromatosis, which typically arises in the lymph nodes of the ileocecal region and in the peritoneum covering the intestines. The next most frequent site of fibromatous involvement is the pleural cavity and lung. However, more than 50% of AIDS patients with skin involvement by Kaposi's sarcoma have one or more similar lesions in the gastrointestinal tract (Friedman-Klein et al. 1982; Saltz et al. 1984). About 50% of AIDS patients also have pulmonary involvement by Kaposi's sarcoma (Meduri et al. 1986; Purdy et al. 1986). Other evidence leading to a differential diagnosis can be obtained by immunohistochemistry, in situ hybridization, or polymerase chain reaction specific for viral antigens or cell markers.

Biologic Features

Natural History and Epidemiology

A peculiar fibroproliferative syndrome has been recognized since 1976 at the University of Washington Regional Primate Research Center in pig-tailed macaques (*Macaca nemestrina*) and was subsequently diagnosed in cynomolgus (*M. fascicularis*), rhesus (*M. mulatta*), and Japanese macaques (*M. fuscata*) at this center (Giddens et al. 1979, 1983, 1985; Tsai et al. 1984, 1985a). It was also reported in Celebes black macaques (*M. nigra*) at the Oregon Regional Primate Research Center (Marx et al. 1985). The principal mesenchymal lesions in macaques are retroperitoneal and subcutaneous fibromatosis. The retroperitoneal form is an endemic syndrome which occurs predominantly in pig-tailed (*Macaca nemestrina*) and Celebes macaques (*Macaca nigra*). The subcutaneous form has

Table 3. Incidence of retroperitoneal fibromatosis (RF) in *M. nemestrina*

	1980–81	1982–83	1984–85	1986–87
Animals at risk per year	1120	1145	1193	1249
Progressive RF	16 (1.4%)	12 (1.05%)	9 (0.7%)	8 (0.6%)
Localized RF	6 (0.5%)	12 (1.05%)	26 (2.2%)	25 (2.0%)
Total	22 (1.9%)	24 (2.1%)	35 (2.9%)	33 (2.6%)

been observed in some cases in pig-tailed macaques (Tsai et al. 1985b) and two rhesus macaques (Osborn et al. 1984). Subcutaneous fibromatosis has also been called a cutaneous Kaposi's sarcoma-like tumor (Warner et al. 1984). In pig-tailed macaques, a few animals have either subcutaneous or nodular fibromatosis in the oral cavity. In others, fibromatous lesions occur only in the visceral or parietal pleura.

At the Washington Regional Primate Research Center, endemic cases of retroperitoneal and subcutaneous fibromatosis and other mesenchymal lesions have been observed in more than 190 macaques since 1976. The incidence rates in these animals are listed in Table 3. In localized and progressive forms of retroperitoneal fibromatosis immunodeficiency accompanies a proliferative lesion, primarily affects young monkeys (1–3 years), and has a high fatality rate. The average age at death is 25 months. Based on the occurrence of retroperitoneal fibromatosis in colony-born (earliest age when lesions observed was 285 days) and imported macaques (earliest case observed was 272 days following transfer), the minimum incubation period was considered to be about 9 months. The mean incidence of the disease in pig-tailed macaques was about 1% in all age groups and about 10% in colony-born juvenile macaques (0.8–3 years). The frequency of retroperitoneal fibromatosis is not significantly different in either sex. Epidemiologic observations revealed no common genetic, environmental, chemical, or dietary factors that were considered to be etiologically significant.

Based on immunohistochemistry in proliferative lesions using monoclonal antibody specific for factor VIII (Giddens et al. 1985), most of the proliferative fibroblasts were found to contain factor VIII-related antigen. This implies that endothelial cells are the progenitors of retroperitoneal fibromatosis. In the less cellular or sclerotic lesions, only small numbers of spindle cells and almost all vascular endothelial cells are positive for factor VIII. However, it may be difficult to distinguish the abnormal spindle cells from the normal proliferative vessels within them. Furthermore, the most recent evidence has suggested that retroperitoneal fibromatosis may not arise from endothelial cells (Tsai et al. 1990a) but from vascular smooth muscle-like cells (or pericytes) (Tsai et al. 1990b).

Retroperitoneal and subcutaneous fibromatosis do not behave like soft tissue sarcoma, which grows locally and metastasizes to other organs. In the early stage of fibromatosis, small foci can be detected in the lymph node capsule and sinuses, often accompanied by angioproliferation. For these reasons retroperitoneal or subcutaneous fibromatosis is not considered to be a true neoplasm but a proliferative disorder possibly originating in endothelial cells. Over time and with the possible influence of other factors, localized endothelial proliferative lesions may develop into the more characteristic proliferative fibroblasts of fibromatosis.

Affected macaques often manifest SAIDS. Chronic immunosuppression appears to be an important factor. The variations in histologic and clinical forms indicate that the histogenetic cell may be multipotential, resulting in lesions with the characteristics of various mesenchymal tumors.

Etiology and Pathogenesis

Type D retroviruses associated with SAIDS have been found in several species of macaques at various primate centers in the USA. These type D viruses are presently designated simian retroviruses (SRV). Among serotype isolates, SRV-1 and SRV-2 are predominantly found in infected captive macaques (Marx et al. 1985; Lowenstine and Lerche 1988; p. 20, this volume). At the Washington Regional Primate Research Center, four naturally occurring disease patterns of

SAIDS associated with SRV-2 have been recognized in macaque colonies (Tsai et al. 1988). These are: (1) SAIDS without neoplasms; (2) retroperitoneal fibromatosis, a fibroproliferative lesion associated with SAIDS; (3) subcutaneous fibromatosis, a multifocal lesion associated with SAIDS; and (4) spontaneous lymphomas. These have been observed in one cynomolgus and 11 pig-tailed macaques in association with SAIDS. In addition, 21 pig-tailed macaques naturally infected with SRV-2 have developed a lymphoproliferative disorder associated with SAIDS. The first pattern, SAIDS without neoplasms, is the most frequent and is characterized by lymphadenopathy beginning as interfollicular hyperplasia of T helper (CD4) and T suppressor (CD8) cells and follicular hyperplasia of predominantly B cells. Terminally the lymph nodes are markedly depleted. Immune dysfunction is indicated by lymphocytopenia (less than 2000 lymphocytes/µl), reduced responses of peripheral blood lymphocytes to mitogens, and impaired ability to produce specific antibody to T cell-dependent antigens (Tsai et al. 1986). Enterocolitis with recurrent diarrhea, weight loss, anemia, and opportunistic infections commonly leads to death or euthanasia.

The data obtained from epidemiologic observations on spontaneous retroperitoneal fibromatosis in pig-tailed and Celebes black macaques (*Macaca nemestrina* and *M. nigra*) indicate a strong association between this disease and SRV-2 (Benveniste et al. 1986). SRV-2 has been isolated, and proviral DNA from SRV-2 has been detected in tissues taken from retroperitoneal fibromatosis lesions (Bryant et al. 1986). Retroviral particles typical of type D viruses have been seen in subcutaneous fibromatosis lesions by transmission electron microscopy (Tsai et al. 1985a). SRV-1 and its proviral DNA have also been demonstrated in tissues from subcutaneous fibromatosis lesions (Gardner and Marx 1985). In transmission studies of SRV-1-induced SAIDS, two of 25 juvenile rhesus macaques developed the retroperitoneal lesion (London et al. 1983). Three (19%) of 16 juvenile macaques inoculated with suspension of tissue from retroperitoneal fibromatosis developed progressive disease (Giddens et al. 1985). In transmission studies of SRV-2-induced SAIDS, in four separate experiments, 30 juvenile pig-tailed macaques were inoculated intravenously with approximately $1 \times 10^4 - 1 \times 10^6$ cell culture infectious doses of SRV-2. Five (17%) of these 30 inoculated macaques developed retroperitoneal fibromatosis, a frequency not significantly different from the incidence (10%) in the general population of juvenile macaques (1–3 years old) at risk (Tsai et al. 1985b). Although SRV-2 as an etiologic factor alone does not increase the incidence of retroperitoneal lesion, the data obtained from naturally occurring and experimentally induced SAIDS indicate a role for SRV. Although studies on the genomes of SRV-1 and SRV-2 indicated a lack of oncogenes (Power et al. 1986; Thayer et al. 1987), two possible mechanisms of pathogenesis must be considered: (1) certain unidentified regions of the SRV-2 genome may be directly involved in induction of the lesion, (2) the immunosuppressive property of SRV may allow other agents to induce tumors (Marx and Lowenstine 1987).

Pathogenesis by Other Experimental Methods

Histogenesis and Tissue Culture

Three continuous culture cell lines were established from pig-tailed macaques with both forms of fibromatosis, retroperitoneal and subcutaneous (RF/SF) (Tsai et al. 1989, 1990b). The growth characteristics of these cells at different passages (over 3 years) were similar. SRV-2 has been isolated from the original tumor tissues and from various passages of the cultured cells by Raji cell coculture technique and identified by electron microscopy and immunocytochemistry using monoclonal antibodies to SRV-2 (Tsai et al. 1989; 1990). Another subcutaneous tumor (SF) cell line, with biologic features similar to the cultured retroperitoneal (RF) cell lines, has also been established from a pig-tailed macaque with SAIDS. Both types of cells contain diploid DNA based on flow cytometric analysis (DNA index = 1.03) and have high proliferative activity (S = 20%–30%, GT_2/M = 25%–35%). The supernatants from the culture media in which the fibromatosis cells were grown were tested for mitogenic effect on several mesenchymal cells, including normal monkey pulmonary fibroblasts, NIH 3T3, endothelial cells, and melanocytic cells. The data suggest that the conditioned media contains a fibroblast growth factor-like agent which stimulates the proliferation of mesenchymal cells.

Cells from the continuous cell lines were examined during different passages for cell morphology, growth pattern, and immunocytochemical studies. Cells harvested from different passages were also fixed in glutaraldehyde for ultrastructural identification of viral particles and intracellular substructures. A large panel of monoclonal and polyclonal antibodies to various human cell differentiation antigens and markers were used to characterize the cells. Heterogeneous populations of endothelial, pericytic, and fibroblastic phenotypes were identified (Tsai et al. 1989, 1990a). The continuous cell lines were positive for SRV-2, but no cross-reactivity was detected to human immunodeficiency virus type 1, cytomegalovirus, herpes simplex type I and II, or Epstein-Barr virus.

Experimental Transmission of Cultured RF Cells in Nude Mice

A xenograft of RF cells in 12 athymic nude mice produced a transient, self-limited, nodular lesion at the site of subcutaneous injection (Tsai et al. 1990). The early (after 2–3 days) lesion at the site of infection consists of an intense host inflammatory response with polymorphonuclear neutrophil, eosinophil, and histocyte admixtures accompanying some degenerating grafted RF cells. The later (6–9 days) lesions underwent fibrovascular proliferation involving dermis, subcutis, and regional lymph nodes. Vascular formations by injected RF cells and erythrophagocytosis were evident. The grafted lesions were histologically indistinguishable from those of original retroperitoneal fibromatosis tissues. The control injection sites of normal monkey lung fibroblasts or human perinodal fibroblasts produced no visible or microscopic lesions.

Experimental Transmission of Retroperitoneal Fibromatosis

Four juvenile macaques (2–2.5 years old) were persistently viremic with SRV-2 and negative for retroperitoneal or subcutaneous fibromatosis prior to experimental transmission. A subcutaneous allograft of RF cells in these four viremic macaques produced a transient self-limited nodule in the subcutaneous injection site over 2 weeks duration. With intraperitoneal allograft, all macaques developed a large retroperitoneal fibromatosis type mass in the peritoneum and adjacent abdominal wall about 2–3 months after injection. Histologically, the allografted tumors were similar to the original lesions. Using exploratory laparotomy for observation of gross morphology of the tumor and for biopsy of tumor tissues, four offspring cell lines have been successfully established for the grafted tumors of four experimental macaques. Biologic features including the immunophenotypes of these grafted cells lines are identical to the original cells. These cell lines continue to harbor SRV-2.

Based on the same clinicopathologic morphology and biologic features of the original retroperitoneal fibromatosis tumors and their cultured cells and the grafted RF tumors and their cells, these findings indicate successful transmission using cultured RF cells as the inoculum. Moreover, the persistent association with SRV-2 in both the original and grafted tumors and their respectively cultured cells would prove the etiologic significance of SRV-2. Further studies on the immunosuppressive nature of SRV-related tumor induction must be considered.

Comparison with Other Species

Retroperitoneal or subcutaneous fibromatosis in macaques with SAIDS has a vascular fibroproliferative nature with many similarities to epidemic Kaposi's sarcoma in AIDS patients infected with human T-lymphotrophic retrovirus (now called HIV) (Barre'-Sinoussi et al. 1983; Gallo et al. 1984). Each of these lesions consists of a proliferation of spindle-shaped mesenchymal cells, chiefly fibroblasts and endothelial cells. In early stage lesions, more spindle-shaped cells contain factor VIII-related antigen. In late stage lesions, only a few cells contain antigen. The lesions in macaques and human patients show a strong association between retrovirus infection and immunosuppression (Safai et al. 1985). Retroperitoneal and subcutaneous fibromatosis is associated with SRV-2, and Kaposi's sarcoma is associated with HIV. An increase in frequency of a benign form of Kaposi's sarcoma in homosexual men without HIV infection indicates that it may be caused by another sexually transmitted infectious agent more readily expressed in HIV infection (Beral et al. 1990).

Recent developments in tissue culture methods indicate that growth of AIDS-Kaposi's sarcoma (KS) cells is stimulated in culture by a growth factor released by CD_4+ T cells infected with human retrovirus (Nakamura et al. 1988). These

AIDS-KS cells can produce autocrine and paracrine growth factors for in vitro proliferation and can induce transient lesions resembling Kaposi's sarcoma in mice (Salahuddin et al. 1988; Ensoli et al. 1989). However, the AIDS-KS cells cannot sustain long-term growth without the specific growth factor supplement. These cells do not contain HIV-1 or cytomegalovirus (CMV) DNA (Nakamura et al. 1988). The histogenesis of the AIDS-KS cells remains uncertain and may arise from spindle cells of lymphatic or vascular origin (Dorfman et al. 1988; Salahuddin et al. 1988).

The establishment of continuous RF/SF cells lines in tissue cultures has provided evidence of cellular differentiation of mesenchymal cells into a heterogenous population. The RF/SF cultured cells are persistently associated with SRV-2 and produce autocrine and paracrine growth factors (Tsai et al. 1990a; Roodman et al. 1991). In preliminary xenograft and allograft studies, RF/SF cultured cells produce *SF-like* lesions in nude mice and *RF-like* lesions in immunosuppressed monkeys.

References

Barre'-Sinoussi F, Chermann J-C, Rey F, Nugeyre MT, Chamaret S, Greust J, Dauguet C, Axler-Blin C, Vezimet-Brum F, Rouzioux C, Rozenbaum W, Montagnier L (1983) Isolation of a T-lymphotropic retrovirus from a patient at risk for acquired immune deficiency syndrome (AIDS). Science 220:868–871

Benveniste RE, Stromberg K, Morton WR, Tsai C-C, Giddens WE Jr (1986) Association of retroperitoneal fibromatosis with type D retroviruses. In: Salzman LA (ed) Animal models of retrovirus infection and their relation to AIDS. Academic, San Diego, pp 335–353

Beral V, Peterman TA, Berkelman RL, Jaffe HW (1990) Kaposi's sarcoma among persons with AIDS: a sexually transmitted infection? Lancet 335I:L123–128

Bryant ML, Marx PA, Shiigi SM, Wilson BJ, McNulty WP, Gardner MB (1986) Distribution of type D retroviral sequences in tissues of macaques with simian acquired immune deficiency and retroperitoneal fibromatosis. Virology 150:149–160

Dorfman RF, Wood GS, Beckstead JH (1988) Histogenesis. In: Ziegler JL, Dorfman RF (eds) Kaposi's sarcoma. Dekker, New York, pp 71–112

Ensoli B, Nakamura S, Salahuddin SZ, Biberfield P, Larsson L, Beaver B, Wong-Staal F, Gallo RC (1989) AIDS-Kaposi's sarcoma-derived cells express cytokines with autocrine and paracrine growth effects. Science 243:223–226

Fauci AS, Masur H, Gelmann EP, Markham PD, Hahn BH, Lane HC (1985) The acquired immunodeficiency syndrome: an update. Ann Intern Med 102:800–813

Friedman-Kien AE, Laubenstein LJ, Rubinstein P, Buimovici-Klein E, Marmor M, Stahl R, Spigland I, Kim KS, Zolla-Pazner S (1982) Disseminated Kaposi's sarcoma in homosexual men. Ann Intern Med 96:693–700

Gallo RC, Salahuddin SZ, Popovic M, Shearer GM, Kaplan M, Haynes BF, Palker TJ, Redfield R, Oleske J, Safai B, White G, Foster P, Markham PD (1984) Frequent detection and isolation of cytopathic retroviruses (HTLV-III) from patients with AIDS and at risk for AIDS. Science 224:500–503

Gardner MB, Marx PA (1985) Simian acquired immunodeficiency syndrome. In: Klein G (ed) Viruses as the causative agents of naturally occurring tumors. Advances in viral oncology, vol 5. Raven, New York, pp 57–81

Giddens WE Jr, Bielitzki JT, Morton WR, Ochs HD, Meyers MS, Blakely GA, Boyce JT (1979) Idiopathic retroperitoneal fibromatosis: an enzootic disease in the pigtailed monkey Macaca nemestrina. Lab Invest 40:294 (abstr)

Giddens WE Jr, Morton WR, Hefti E, Panem S, Ochs H (1983) Enzootic retroperitoneal fibromatosis in Macaca spp. In: Kalter SS (ed) Viral and immunological diseases in nonhuman primates. Liss, New York, pp 249–253

Giddens WE Jr, Tsai C-C, Morton WR, Ochs HD, Knitter GH, Blakely GA (1985) Retroperitoneal fibromatosis and acquired immunodeficiency syndrome in macaques. Pathologic observations and transmission studies. Am J Pathol 119:253–263

Gottlieb MS, Groopman JE, Weinstein WM, Fahey JL, Detels R (1983) The acquired immunodeficiency syndrome. Ann Intern Med 99:208–220

London WI, Madden DL, Gravell M, Dalakas MC, Houff SA, Sever JL, Henrickson RV, Maul DH, Osborn KG, Gardner MB (1983) Experimental transmission of simian acquired immunodeficiency syndrome (SAIDS) and Kaposi-like skin lesions. Lancet 2:869–873

Lowenstine LJ, Lerche NW (1988) Retrovirus infections of nonhuman primates: a review. J Zoo Anim Med 19:168–187

Lowenstine L, Lerche N, Jennings M, Marx P, Gardner M, Pedersen N (1988) An epizootic of simian AIDS caused by SIV in captive macaques in the 1970's. In: Girard M, Valette L (eds) Retroviruses of human A.I.D.S. and related animal diseases. 2nd Colloquium of the Cent Gardes, Foundation Marcel Merieux, Lyon, pp 174–176

Marx PA, Lowenstine LJ (1987) Menechyman neoplasms associated with type D retroviruses in macaques. Cancer Surv 6:101–115

Marx PA, Bryant ML, Osborn KG, Maul DH, Lerche NW, Lowenstine LJ, Kluge JD, Zaiss CP, Henrickson RV, Shiigi SM, Wilson BJ, Malley A, Olson LC, McNulty WP, Arthur LO, Gilden RV, Barker CS, Hunter E, Munn R, Herdecker G, Gardner MB (1985) Isolation of a new serotype of simian acquired immune deficient syndrome type D retrovirus from celebes black macaques (Macaca nigra) with immune deficiency and retroperitoneal fibromatosis. J Virol 56: 571–578

Meduri GU, Stover DE, Lee M, Myskowski PL, Caravelli JF, Zaman MB (1986) Pulmonary Kaposi's sarcoma in the acquired immune deficiency syndrome. Clinical, radiographic and pathologic manifestations. Am J Med 81:11–18

Nakamura S, Salahuddin SZ, Biberfield P, Ensoli B, Markham PD, Wong-Staal F, Gallo RC (1988) Kaposi's sarcoma cells: long-term culture with growth factor from retrovirus-infected CD4+ T cells. Science 242: 426–430

Osborn KG, Prahalada S, Lowenstine LJ, Gardner MB, Maul DH, Henrickson RV (1984) The pathology of a enzootic acquired immunodeficiency in rhesus macaques. Am J Pathol 114:94–103

Power MD, Marx PA, Bryant ML, Gardner MB, Barr PJ, Luciw PA (1986) The necleotide sequence of a type D retrovirus (SRV-1) etiologically linked with the simian acquired imunodeficiency syndrome. Science 231:1567–1572

Purdy LJ, Colby TV, Yousem SA, Battifora H (1986) Pulmonary Kaposi's sarcoma: premortem histologic diagnosis. Am J Surg Pathol 10:301–311

Roodman ST, Woon MD, Hoffman JW, Theodorakis P, Tsai CC, Wu H-N, Tsai C-C (1991) Interleukin-6 and retroperitoneal fibromatosis from SRV-2-infected macaques with simian AIDS. J Med Primatol 20:201–205

Safai B, Johnson KG, Myskorvski PL, Koziner B, Yang SY, Cunningham-Rundles S, Godbold JH, Dupont BO (1985) The natural history of Kaposi's sarcoma in the acquired immunodeficiency syndrome. Ann Intern Med 103:744–750

Salahuddin SZ, Nakamura S, Biberfeld P, Kaplan MH, Markham PD, Larsson L, Gallo RC (1988) Angiogenic properties of Kaposi's sarcoma-derived cells after long-term culture in vitro. Science 242:430–433

Saltz RK, Kurtz RC, Lightdale CJ, Myskowski KP, Cunningham-Rundles S, Urmacher C, Safai B (1984) Kaposi's sarcoma. Gastrointestinal involvement, correlation with skin findings and immunologic function. Dig Dis Sci 29:817–823

Thayer RM, Power MD, Bryant ML, Gardner MB, Barr PJ, Luciw PA (1987) Sequence relationships of type D retroviruses which cause simian acquired immunodeficiency syndrome. Virology 157:317–327

Tsai C-C, Blakely GA, Uno H, Frohna B, Warner TFCS (1984) Retroperitoneal fibromatosis (Kaposi's sarcoma-like tumor) with lymph node lesions in pigtaled macaques (Macaca nemestrina). Lab Invest 50:61A

Tsai C-C, Warner TFCS, Uno H, Giddens WE Jr, Ochs HD (1985a) Subcutaneous fibromatosis associated with an acquired immunodeficiency syndrome in pigtailed (Macaca nemestrina) macaques. Am J Pathol 120:30–37

Tsai C-C, Giddens WE Jr, Morton WR, Rosenkranz SL, Ochs HD, Benveniste RE (1985b) Retroperitoneal fibromatosis and acquired immunodeficiency syndrome in macaques: epidemiologic studies. Lab Anim Sci 35:460–464

Tsai C-C, Giddens WE Jr, Ochs HD, Morton WR, Knitter GH, Blakely GA, Benveniste RE (1986) Retroperitoneal fibromatosis and acquired immunodeficiency syndrome in macaques: clinical and immunologic studies. Lab Anim Sci 35:119–125

Tsai C-C, Thouless ME, Morton WR, Benveniste RE (1988) Changing tumorigenesis of simian acquired immunodeficiency syndrome-associated type D retrovirus. Lab Invest 58:95A

Tsai CC, Tsai C-C, Sligh J, Roodman S (1989) Kaposi's sarcoma-like cell line from simian AIDS with retroperitoneal fibromatosis (RF) associated with type D retrovirus (SRV-2) infection. Lab Invest 60:98A

Tsai CC, Tsai C-C, Roodman ST, Woon M-D (1990a) Mesenchymal tumor-like lesions (MTL) in simian AIDS associated with SRV-2 infection. J Med Primatol 19:189–202

Tsai C-C, Wu H-N, Tsai CC (1990b) Phenotypic characterization of Kaposi's sarcoma-like tumor associated with SAIDS-D retrovirus in pigtailed macaques. 8th annual symposium on nonhuman primate models for AIDS, New Orleans, November 1990, p 121

Warner TFCS, Uno H, Tsai C-C, Giddens WE Jr (1984) Cutaneous Kaposi's sarcoma-like tumor in pigtail macaques (Macaca nemestrina). Lab Invest 50:65–66A (abstr)

Ziegler JL, Levy JA (1985) Acquired immunodeficiency syndrome and cancer. In: Klein G (ed) Viruses as the causative agents of naturally occurring tumors. Advances in viral oncology, vol 5. Raven, New York, pp 239–255

Mycobacterium avium-intracellulare Infection

Norval W. King, Jr.

Synonyms. Avian tuberculosis, avian mycobacteriosis, nontuberculous mycobacteriosis, atypical mycobacteriosis, Battey bacillus infection, Runyon type III mycobacterial infection.

Gross Appearance

M. avium-intracellulare infection in nonhuman primates begins as a primary infection of the small and large intestines and draining lymph nodes. Early in the course of the infection, lesions may not be discernible grossly. In advanced cases, however, lesions may become widely disseminated and involve the spleen, liver, kidney, heart, lungs, and other organs. The affected segments of small and large bowel have roughened serosal surfaces and raised, yellow-white, 0.5–2.0 mm, granular nodules within the serosal and mesenteric lymphatics. The intestinal wall is firm, thickened, and the mucosa often contains coarse yellow folds. The

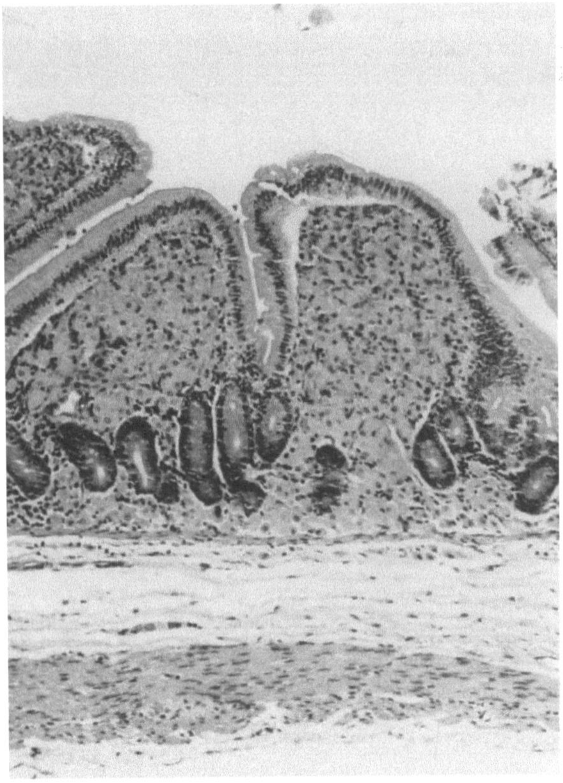

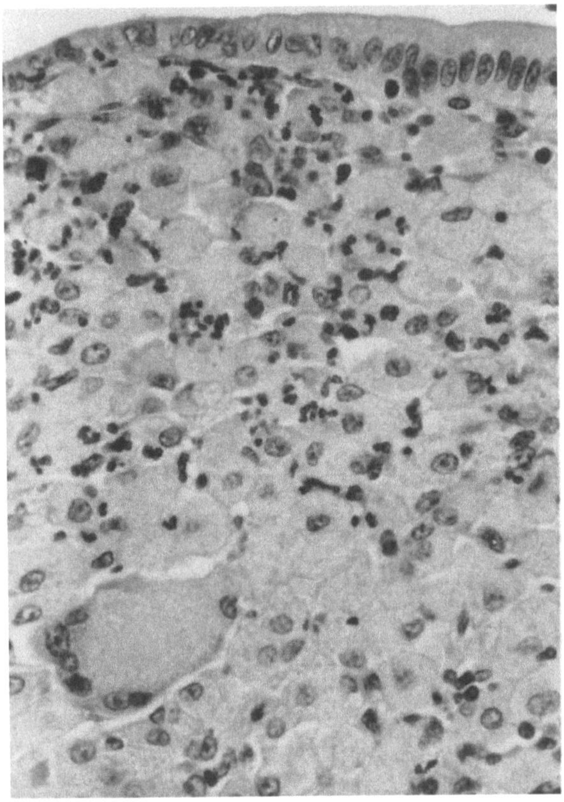

mesenteric and ileocolic lymph nodes are diffusely enlarged, firm, yellow-white, and contain no gross evidence of necrosis. The liver may be somewhat enlarged and have areas of yellow-tan discoloration. In advanced cases the spleen is enlarged, firm, and contains less than the normal amount of blood. Although *Mycobacterium avium-intracellulare* infections in nonhuman primates are typically characterized by lack of tubercle formation, in several reports tubercle formation has been described (Holmberg et al. 1982; Goodwin et al. 1988; Bellinger and Bullock 1988). In one of these cases, the disease was also unusual in that the lesion involved the skin only (Bellinger and Bullock 1988).

Microscopic Features

Microscopically, the lamina propria of the small intestine, cecum, and colon and infiltrated with diffuse sheets of large epithelioid histiocytes with finely vacuolated cytoplasm (Fig. 57). A feature that can often be used to distinguish *M. avium-intracellulare* infection from tuberculous forms of mycobacteriosis is the lack of caseous necrosis in the granulomatous infiltrates. There are, however, a few reports in which typical tubercles have been found in monkeys from which *M. avium-intracellulare* has been cultured (Holmberg et al. 1982; Goodwin et al. 1988; Bellinger and Bullock 1988). Lymphocytes are usually not a prominent feature of this form of granulomatous enterocolitis either. Multinucleate giant cells are rarely found in the typical epithelioid cell infiltrates, but neutrophils may be present in areas where the surface epithelium has been breached or in areas where epithelioid cells

◀**Fig. 57** (*above*). Ileum, *Macaca mulatta*, *Micobacterium avium-intracellulare* infection. The lamina propria has been extensively infiltrated by diffuse sheets of large epithelioid histiocytes that have caused blunting and fusion of intestinal villi. Note the lack of caseous necrosis and the intact surface epithelium. H and E, ×100

Fig. 58 (*below*). Higher magnification of Fig. 57. Note the large, foamy, epithelioid histiocytes that expand the lamina propria of the intestinal mucosa, the single multinucleate giant cell, and the scattered neutrophils within the infiltrate. H and E, ×250

have ruptured (Fig. 58). The intestinal villi are thickened, fused, and blunted due to the expanding masses of histiocytes in their lamina propria, but the surface epithelium remains mostly intact. The epithelioid cells may extend through the submucosa and tunica muscularis and involve the serosa and lymphatics where they form small noncaseous granulomas. Mesenteric and ileocolic lymph nodes are massively infiltrated by similar large, foamy epithelioid cells that obliterate their normal architecture (Figs. 59–61). The liver and spleen, in advanced cases, may contain numerous focal and confluent noncaseous microgranulomas that replace the functional parenchyma (Fig. 62). Other organs, including the lung, heart, and kidney, may contain similar but smaller numbers of microgranulomas. Acid-fast stains reveal numerous acid-fast bacilli filling the cytoplasm of virtually every epithelioid cell in these infiltrates (Figs. 59, 61). In contrast, in those exceptional cases in which *M. avium-intracellulare* infection has been associated with tubercle formation, the numbers of organisms demonstrable in the lesions have been more variable.

Differential Diagnosis

Animals with *M. avium-intracellulare* infection may manifest an equivocal or weakly positive tuberculin reaction to the mammalian old tuberculin (OT) used routinely in tuberculosis testing programs in nonhuman primate colonies. If retested with tuberculin prepared from *M. avium*, these animals will often manifest a more intense positive reaction. However, false negative reactions may result from anergy in animals with overwhelming infection or reflect a severe immunodeficiency such as that caused by simian immunodeficiency virus (SIV) infection that predisposes the monkey to this opportunistic infection. Pathologically, *M. avium-intracellulare* infection must be distinguished from other granulomatous diseases in which acid-fast bacilli are demonstrable. Features useful in distinguishing *M. avium-intracellulare* infection from tuberculous forms of mycobacterioses are: its extensive intestinal involvement, the massive numbers of acid-fast bacilli within the infiltrating histiocytes, and the lack of tubercle formation and caseous necrosis. The reader should be aware, however, that there are several reported cases of simian *M. avium-intracellulare* infection in which tubercles with caseous necrosis were found. Another mycobacterial disease of nonhuman primates that is virtually indistinguishable grossly and microscopically from *M. avium-intracellulare* infection is paratuberculosis, caused by *Mycobacterium paratuberculosis*. Paratuberculosis has only recently been described and confirmed by culture in stumptail macaques at the Yerkes Primate Center (McClure et al. 1987). It, like *M. avium-intracellulare* infection, is also primarily an enteric disease that resembles Johne's disease of cattle. Differentiation between these two diseases can only be achieved through culture and identification of the causative organism.

Biologic Features

Natural History and Epidemiology

M. avium is a natural pathogen of wild birds, domestic poultry, and, rarely, swine, cattle, sheep, goats, and horses. In other mammalian species it occurs only as an opportunistic pathogen. Likewise, *M. intracellulare*, which exists in nature as a common environmental saprophyte of soil, water, and plants, can, under the appropriate circumstances, become an opportunistic pathogen. The first description of *Mycobacterium avium-intracellulare*-associated disease in nonhuman primates was in 1973 by investigators at the California and New England Regional Primate Research Centers (Smith et al. 1973), although others had earlier reported the isolation of *M. avium* from tuberculin-positive pig-tailed macaques (*Macaca nemestrina*) with no evidence of disease (Sedgwick et al. 1970). The outbreak of *M. avium-intracellulare* infection at the two centers occurred over a 2 year period and involved a total of 63 macaques of four different species (*Macaca mulatta*, *M. cynomologi* (*fascicularis*), *M. speciosa* (*arctoides*), and *M. nemestrina*) (Smith et al. 1973). Since then, other outbreaks, as well as sporadic cases, of *M. avium-intracellulare* infection have been described in these same species (Sesline et al. 1975; Latt 1975; Fleischmann et al. 1982; Holmberg et al. 1982) and in an island siamang (*Hylobates syndactylus*) (Munson et al. 1991).

Nonhuman primates presumably acquire *M. avium-intracellulare* infection through ingestion of contaminated foodstuffs. Susceptibility of humans and nonhuman primates to infection by

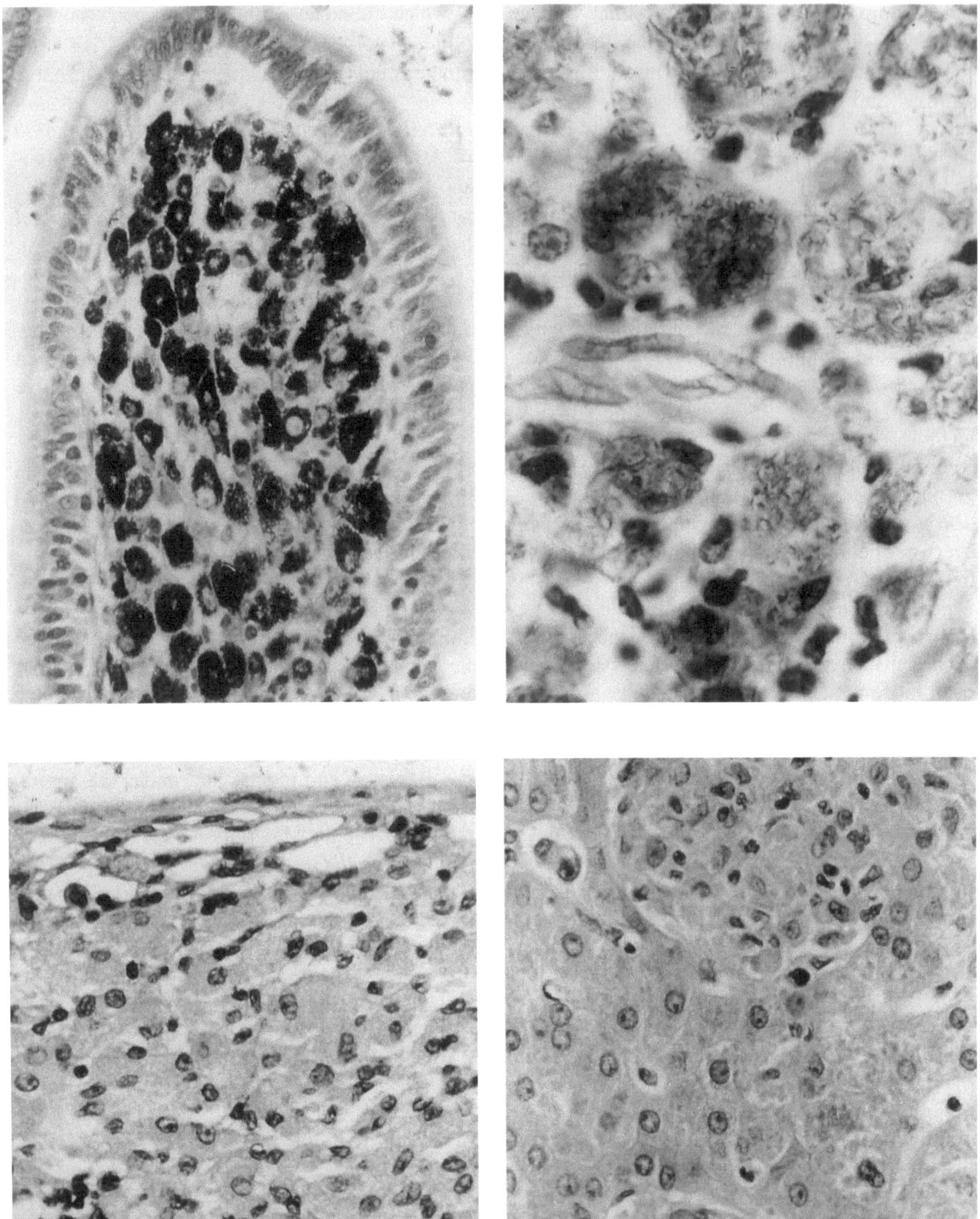

M. avium-intracellulare requires that the host be immunocompromised; hence infection with these organisms is not regarded as a contagious disease in these species. In humans, infections with *M. avium-intracellulare* have been associated with steroid therapy, cytotoxic drugs, and, more recently, with the acquired immunodeficiency syndrome (AIDS) caused by the human immunodeficiency virus (HIV). Healthy nonhuman primates are highly resistant to *M. avium-intracellulare* infection (Karlson 1978). In fact, in their original report, Smith et al. (1973) indicated that attempts to transmit *M. avium-intracellulare* to healthy, uninfected, rhesus monkeys via the oral route were unsuccessful. There is now evidence, based upon retrospective serologic testing and limited virus isolation studies using frozen serum and tissues from monkeys included in the earlier reports of spontaneously occurring outbreaks of *M. avium-intracellulare* infections at the California and New England Primate Centers, that those animals were also infected with SIV and probably immunosuppressed. (Lowenstine et al. 1987, 1992; King et al. 1990). Moreover, *M. avium-intracellulare* infection occurs as one of the more common opportunistic infections of rhesus monkeys experimentally infected with SIV_{MAC} and SIV_{SMM} (Baskin et al. 1988; King et al. 1990). Interestingly, exposure to outdoor environments where there is access to birds and bird droppings is not essential for infection by these agents. This author has seen *M. avium-intracellulare* infection in experimentally SIV-infected, colony-born rhesus monkeys that were never housed outdoors.

◀ **Fig. 59** (*upper left*). Ileum, *Macaca mulatta*, *M. avium-intracellulare* infection. Lamina propria of a villus is filled with histiocytes containing massive numbers of acid-fast organisms. Ziehl-Neelsen stain, ×150

Fig. 60 (*lower left*). Mesenteric lymph node, *M. avium-intracellulare* infection, *Macaca mulatta*. The parenchyma has been infiltrated and replaced by solid sheets of epithelioid histiocytes. A small aggregate of residual lymphocytes is at the *lower left*; capsule at the *top*. H and E, ×250

Fig. 61 (*upper right*). Higher magnification of lymph node in Fig. 60. The cytoplasm of the infiltrating epithelioid cells is filled with acid-fast bacilli. Ziehl-Nelsen stain, ×1500

Fig. 62 (*lower right*). Liver, *Macaca mulatta*, *M. avium-intracellulare* infection. Note two noncaseous microgranulomas composed of collections of pleomorphic epithelioid histiocytes. H and E, ×250

Etiology

M. avium (the avian tubercle bacillus) is a slow-growing, aerobic, nonpigment-producing (Runyon's group III), acid-fast bacillus. Like all mycobacteria, it has a high lipid content that includes esterified mycolic acid residues that account for its acid-fastness and poor stainability with Gram's stain. Currently there are at least 20 different serotypes of *M. avium*, including some organisms that were previously referred to as *M. intracellulare* or Battey bacilli (Thoen 1978). Since *M. avium* and *M. intracellulare* are so similar and exceedingly difficult to distinguish from one another, they have been recently referred to as the *M. avium-intracellulare* complex. Serotypes 1, 2, and 3 are commonly isolated from birds and are pathogenic for chickens, swine, and cattle (Thoen et al. 1981). Serotype 8 has been associated with severe outbreaks in swine in the United States and other countries (Thoen et al. 1975). Serotypes 4–20, by contrast, are common environmental saprophytes that are generally not considered pathogenic for chickens or immunocompetent mammalian hosts (Thoen 1978). In the outbreaks originally reported in immunocompromised macaques at the California and New England Primate Centers, *M. intracellulare*, serotype Davis, and *M. avium*, serotypes 1, 2, 4, 9, 18, 20, and double serotype 1 and 8, were isolated (Sesline 1978). These so-called atypical mycobacteria are generally resistant to the common forms of mycobacterial therapy.

Pathogenesis

Nonhuman primates are naturally resistant to acquired and experimental infection by *M. avium-intracellulare* but may be rendered susceptible by factors that impair immune function, especially T cell activity. These factors include certain immunosuppressive drugs and viral infections, such as SIV, that cause severe clinical immunosuppression. The earliest lesions of *M. avium-intracellulare* infection in monkeys occur in the intestinal tract, and thus the organism is thought to be acquired through ingestion. During passage through the gut, these acid-fast bacilli come in contact with the lamina propria, presumably through small defects in the mucosal epithelium, and are phagocytized by mucosal macrophages that become transformed into epithelioid histiocytes. The waxy adjuvant component of the mycobacterial cell walls of *M. avium-intracellulare* attracts additional antigen-presenting macrophages to the site of infection where they accumulate in solid sheets. Due to the host's compromised immune system, these histiocytes are incapable, either alone or in concert with the host's T lymphocytes, of killing and eliminating the organism from the gut. Instead, *M. avium-intracellulare* undergoes extensive multiplication within phagocytic vacuoles of the epithelioid cells eventually leading to their death, rupture, and release of numerous bacilli that attract additional macrophages. Neutrophils may be present among the epithelioid cells presumably in response to either chemical mediators, or mycobacteria released from dying epithelioid cells, or enteric bacteria that enter the lamina propria through small breaks in the overlying epithelium. Lymphocytes are not a prominent feature of the granulomatous inflammation caused by *M. avium-intracellulare* in nonhuman primates. This possibly reflects the deficiency of CD4+ lymphocytes and the related T cell dysfunction that originally predisposes these animals to this opportunistic infection. Moreover, this immune dysfunction may also explain the absence of caseous necrosis in *M. avium-intracellulare*-induced granulomas in nonhuman primates, since cytotoxic T cells play a critical role in the pathogenesis of caseous necrosis in tubercle formation in tuberculous forms of mycobacteriosis. The infection becomes disseminated from the primary site in the gut to the regional lymph nodes and other organs either by migration of infected macrophages to these other sites, or by the mycobacteria entering venules or lymphatics directly, or both.

Comparison with Other Species

Serotypes 1, 2, and 3 of *M. avium* are naturally occurring pathogens of wild birds and domestic poultry and on rare occasions may infect swine, cattle, sheep, goats, horses, dogs, cats, and cold blooded animals (Thoen et al. 1981). It is not known whether these mammalian species, like human and nonhuman primates, require some form of immunosuppression to be susceptible to *M. avium* infection. In birds, *M. avium* causes typical tuberculous lesions in which early epithelioid granulomas progress to typical tubercles with central cores of caseous necrosis surrounded by histiocytes, prominent multinucleate giant cells, lymphocytes, and varying amounts of fibrosis. In most domestic farm animals, *M. avium-intracellulare* causes diffuse granulomatous lesions similar to those described in nonhuman primates, the exception being that there may be small foci of caseous necrosis present in some of the epithelioid granulomas. Unlike the lesions caused by *M. tuberculosis* and *M. bovis* in these species, calcification is not a feature of *M. avium-intracellulare*-induced granulomas.

Acknowledgement. This work was supported by Public Health Service Award RR00168 from the National Center for Research Resources, National Institutes of Health, to the New England Regional Primate Research Center.

References

Baskin GB, Murphey-Corb M, Watson EA, Martin LN (1988) Necropsy findings in rhesus monkeys experimentally infected with cultured simian immunodeficiency virus (SIV)/delta. Vet Pathol 25:456–467

Bellinger DA, Bullock BC (1988) Cutaneous Mycobacterium avium infection in a cynomolgus monkey. Lab Anim Sci 38:85–86

Fleischman RW, du Moulin GC, Esber HJ, Ilievski V, Bogden AE (1982) Nontuberculous mycobacterial infection attributable to Mycobacterium intracellulare serotype 10 in two rhesus monkeys. J Am Vet Med Assoc 181:1358–1362

Goodwin BT, Jerome CP, Bullock BC (1988) Unusual lesion morphology and skin test reaction for Mycobacterium avium complex in macaques. Lab Anim Sci 38:20–24

Holmberg CA, Hendrickson RV, Malaga C, Schneider R, Gribble D (1982) Nontuberculous mycobacterial disease in rhesus monkeys. Vet Pathol 19 [Suppl 7]: 9–16

Karlson AG (1978) Avian tuberculosis. In: Montali RJ (ed) Mycobacterial infections of zoo animals. Smithsonian Institution, Washington, pp 21–24

King NW, Chalifoux LV, Ringler DJ, Wyand MS, Sehgal PK, Daniel MD, Letvin NL, Desrosiers RC, Blake BJ, Hunt RD (1990) Comparative biology of natural and experimental SIV_{MAC} infection in macaque monkeys: a review. J Med Primatol 19:109–118

Latt RH (1975) Runyon group III atypical mycobacteria as a cause of tuberculosis in a rhesus monkey. Lab Anim Sci 25:206–209

Lowenstine LJ, Lerche N, Jennings M, Marx P, Gardner M, Pedersen N (1987) An epizootic of simian AIDS caused by SIV in captive macaques in the 1970s. In: Giarard M, Vallette L (eds) Retroviruses of human AIDS and related animal viruses. Pasteur Vaccines, Paris, pp 174–176

Lowenstine LJ, Lerche NW, Yee JA, Uyeda A, Jennings MB, Munn RJ, McClure HM, Anderson DC, Fultz PN, Gardner MB (1992) Evidence for a lentiviral etiology in an epizootic of immune deficiency and lymphoma in stump-tailed macaques (Macaca arctoides). J Med Primatol (in press)

McClure HM, Chiodini RJ, Anderson DC, Swenson RB, Thayer WR, Coutu JA (1987) Mycobacterium paratuberculosis infection in a colony of stumptail macaques (Macaca arctoides). J Infect Dis 155:1011–1019
Munson L, Luibel FJ, Van Kruiningen HJ (1991) Siderophilic bodies associated with hemosiderosis and atypical mycobacterial infection in an island siamang (Hylobates syndactylus). J Med Primatol 20:265–270
Sedgwick C, Parcher J, Durham R (1970) Atypical mycobacterial infection in the pig-tailed macaque (Macaca nemestrina). J Am Vet Med Assoc 157:724–725
Sesline DH (1978) Mycobacterium avium enteritis in nonhuman primates. In: Montali RJ (ed) Mycobacterial infections in zoo animals. Smithsonian Institution, Washington, pp 157–159
Sesline DH, Schwartz LW, Osburn BI, Thoen CO, Terrell T, Holmberg C, Anderson JH, Hendrickson (1975) Mycobacterium avium infection in three rhesus monkeys. J Am Vet Med Assoc 167:639–645
Smith EK, Hunt RD, Garcia FG, Fraser CEO, Merkal RS, Karlson AG (1973) Avian tuberculosis in monkeys: a unique mycobacterial infection. Am Rev Respir Dis 107:469–471
Thoen CO (1978) Isolation and identification of mycobacteria from tissues of exotic animals. In: Montali RJ (ed) Mycobacterial infections of zoo animals. Smithsonian Institution, Washington, pp 99–101
Thoen CO, Jarnagin JL, Richards WD (1975) Isolation and identification of mycobacteria from porcine tissues: a three-year summary. Am J Vet Res 36:1383–1386
Thoen CO, Karlson AG, Himes EM (1981) Mycobacterial infections in animals. Rev Infect Dis 3:960–972

Toxoplasmosis

Daniel C. Anderson and Harold M. McClure

Synonyms. Infection with *Toxoplasma gondii.*

Gross Appearance

Animals with toxoplasmosis are frequently found dead without previous signs of illness and are often in very good physical condition at death, reflecting the very acute nature of infection. Clinical signs, when observed, are nonspecific and usually consist of lethargy or severe depression. Animals generally die shortly after the observed onset, frequently within minutes or hours, although illness of 1–5 days and weight loss over several months in one zoo have been described (McKissick et al. 1968).

The lungs are consistently involved and are usually firm, uniformly dark red, and congested and edematous on section. The trachea frequently contains red-tinged froth. Lymphadenopathy is common with mesenteric lymph nodes often enlarged with hemorrhage and necrosis. Pinpoint white foci may be seen in the liver and the spleen; however, this is uncommonly reported. Gastrointestinal lesions are much less frequently reported; they have been described in natural and experimental infections in owl monkeys (*Aotus trivirgatus*) and marmosets (*Sanguinus oedipus*) (Seibold and Wolf 1971; Werner et al. 1969).

Microscopic Features

Solitary tissue cysts of *T. gondii* have been described in monkeys in the absence of other lesions and were likely incidental findings (Dubey and Beattie 1988). In active infections, reported lesions and tissue distribution are remarkably similar in most nonhuman primate species. The characteristic lesion is necrosis, with or without an inflammatory response, and is usually seen in multiple organs. Lesions are most frequently reported in the liver, spleen, mesenteric lymph nodes, lungs, heart, and adrenal gland. The liver is invariably involved with numerous randomly scattered necrotic foci throughout the parenchyma (Fig. 63). Tissue cysts (bradyzoites) or free forms (tachyzoites) are often found at the periphery of the lesion and in Kupffer's and hepatic cells. Necrosis may be less apparent in the spleen; however, an acute inflammatory response is usually seen with numerous organisms in macrophages or free forms throughout the red pulp (Fig. 64). The mesenteric lymph nodes are often severely affected with extensive hemorrhage and necrosis resulting in severe effacement or the virtual destruction of nodal architecture (Fig. 65); similar lesions are less frequently reported in thoracic lymph nodes. Organisms may be found in affected lymph nodes or in nodes with little or no pathologic change (Fig. 66). The lungs are usually severely

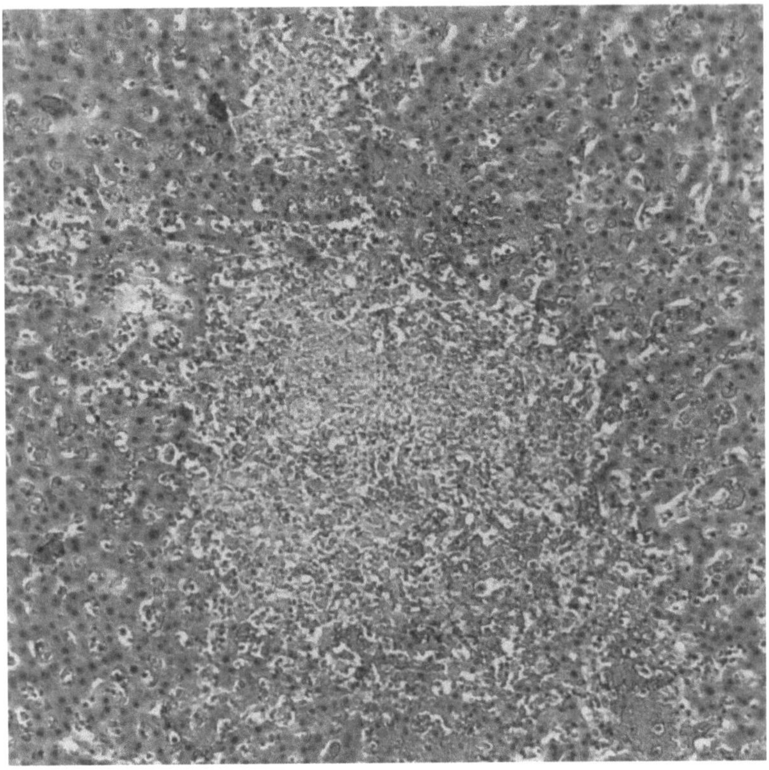

Fig. 63. Liver of squirrel monkey, toxoplasmosis with multifocal necrotizing hepatitis. H and E, ×117

congested and edematous with multifocal hemorrhage and serofibrinous exudate within the alveolar spaces. An acute inflammatory response may be minimal and an early interstitial pneumonia may be observed. When found, random multifocal areas of necrosis may be severe. Organisms may be numerous or be found with some difficulty. Small randomly scattered foci of necrosis are also frequently noted in the heart (Fig. 67) and in the adrenal gland. These tissues are usually less severely affected and cysts or free forms may be seen in either tissue with little or no inflammatory response (Fig. 68).

Organisms are frequently seen in the kidney; however, they may or may not be associated with inflammation or necrosis. Lesions are often absent in the brain, but may be found widely scattered throughout and consist either of small foci of hemorrhage and necrosis or gliosis (Fig. 69). Organisms may be found both in affected and normal areas and have been described in vessels in association with small infarcts (McKissick et al. 1968). Gastrointestinal lesions range from superficial erosion to extensive inflammation and necrosis that extend deeply into the submucosa (Werner et al. 1969). Tissue examination may reveal *T. gondi* protozoa in virtually any organ. Encysted forms may be found in the absence of inflammation. With disseminated infection, organisms have been infrequently reported in other tissues, including bone marrow, skeletal muscle, thyroid, pancreas, and the ovary and have usually been observed without associated inflammation.

Ultrastructure

Toxoplasma is classified in the protozoan subphylum Apicomplexia which includes all coccidia of veterinary and medical importance (Dubey and Beattie 1988). Ultrastructural characteristics of this subphylum are useful in distinguishing them from other protozoa with which they may be confused at the light microscopic level (Figs. 70, 71). The electron microscopic features of the organism have been described in detail (Dubey and Beattie 1988). In general, features of significance include an outer layer (pellicle) which consists of two membranes, a polar ring which encircles a conid, microtubules (subpellicular fibrils) which extend from the polar ring for most of the length of the

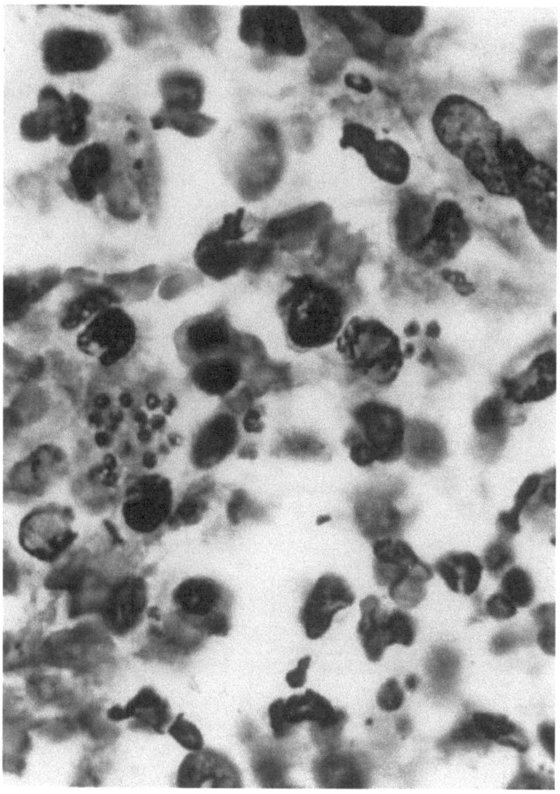

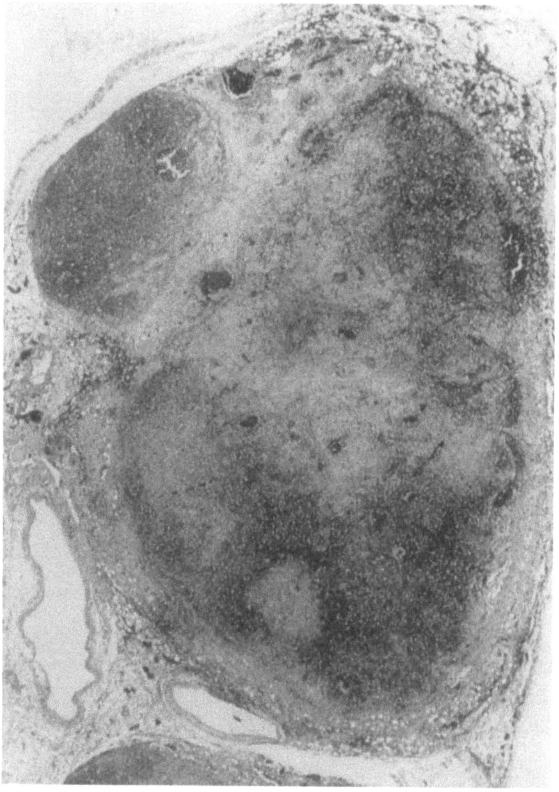

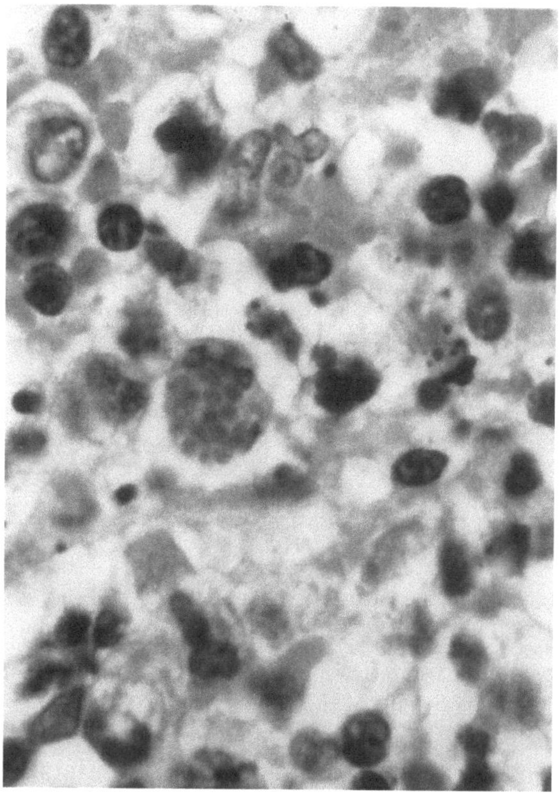

Fig. 64 (*upper left*). Spleen from squirrel monkey with numerous *Toxoplasma* tachyzoites scattered throughout the parenchyma. Wilder's stain, ×1320

Fig. 65 (*upper right*). Mesenteric lymph node of squirrel monkey with diffuse necrotizing lymphadenitis. Note virtually total effacement of the lymph node by hemorrhage and necrosis. H and E, ×16

Fig. 66 (*lower right*). High magnification of lymph node in Fig. 65. Note *Toxoplasma* organisms in necrotic focus. H and E, ×528

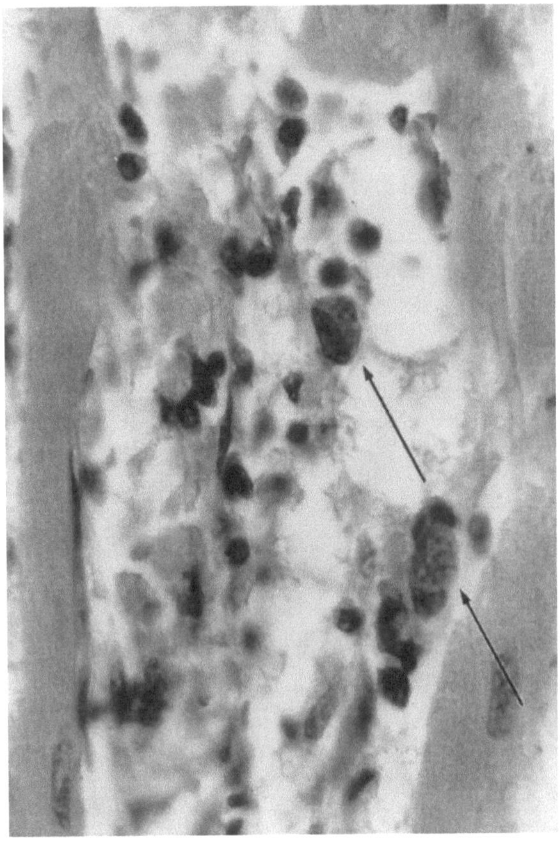

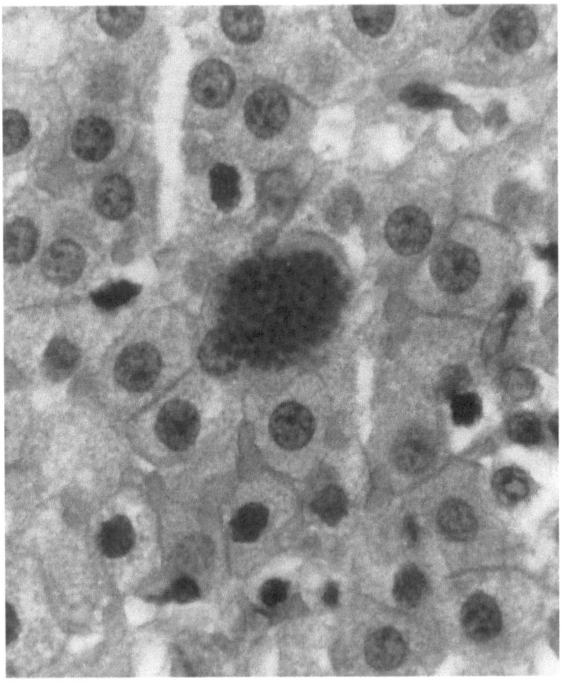

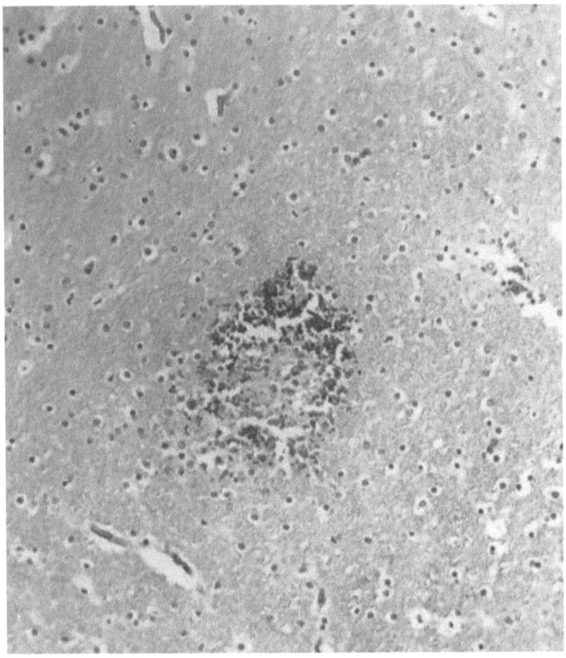

Fig. 67 (*upper left*). Heart of a lemur with multifocal necrotizing myocarditis. Note necrotic focus containing *Toxoplasma* organisms (*arrows*). H and E, ×720

Fig. 68 (*upper right*). Adrenal cortex from squirrel monkey. Toxoplasmal cyst containing numerous bradyzoites. Note lack of inflammatory reaction. H and E, ×760

Fig. 69 (*lower right*). Brain, squirrel monkey. Note focal hemorrhage and necrosis. H and E, ×150

cell, and rhopteries and micronemes which are in the anterior end of the cell. Cytoplasmic organelles include mitochondria, endoplasmic reticulum, Golgi apparatus, and ribosomes. Structurally, bradyzoites differ only slightly from tachyzoites. Bradyzoites are always found in tissue cysts and are surrounded by a single unit membrane. The nucleus is well defined and is usually located centrally (tachyzoites) or near the posterior end (bradyzoites). Bradyzoites are also smaller and contain a few PAS-positive glycogen granules which are indistinct or absent in tachyzoites.

Differential Diagnosis

Toxoplasma may resemble several protozoal or fungal agents (Frenkel 1976). Tachyzoites are about the same size as *Histoplasma* and the leishmanial forms of *Leishmania* sp. in tissue macrophages or *Trypanosoma* sp. in myocytes. The latter can be distinguished by the presence of an intracytoplasmic kinetoplast and nucleus. *Histoplasma* may be readily differentiated by capsular staining using the PAS or GMS method. *Toxoplasma* bradyzoites in tissue cysts must be differentiated from pseudocysts of *Encephalitozoon cuniculi* in multiple tissues and from *Sarcocystis* in muscle. *Encephalitozoon cuniculi* can be differentiated from *Toxoplasma* by spore and cyst size, morphology, and special stains (Zeman and Baskin 1985). *E. cuniculi* is gram-positive and has larger pseudocysts (60–120 μm in diameter) with small, oval, rod-shaped spores (1.5 × 2.5 μm). *Toxoplasma* cysts are less than 60 μm in diameter with individual organisms 2 × 6 μm in width and crescent-shaped on impression smears; they appear round in tissue sections (Frenkel 1976). *Sarcocystis* is in the same subphylum with *Toxoplasma* and shares many characteristics. In general, *Sarcocystis* bradyzoites and cysts are usually larger and PAS-positive granules are seen within the cytoplasm. In addition, bradyzoites contain numerous micronemes (Dubey 1976) with cysts divided into compartments evident by electron microscopy. *Toxoplasma* bradyzoites are not septated and micronemes are fewer.

Biologic Features

Natural History

Toxoplasma gondii was first described in a North African rodent species (*Ctenodactylus gundi*) in 1908 (Nicolle and Manceaux 1909). It was also described in a laboratory rabbit in Brazil that same year and has been reported in numerous mammals and birds since that time. It was first reported in a howler monkey (*Alouatta seniculus*) in 1916 and has been reported in at least 15 New World and seven Old World nonhuman primate species (Anderson and McClure 1982). New World species are considered to be more susceptible to *Toxoplasma* infection than Old World primate species, based on both the frequency of reported cases and on experimental studies. Although most reports of spontaneous toxoplasmosis in nonhuman primates have documented fatal cases in one or two animals, there have been reports of endemic or epizootic infec-

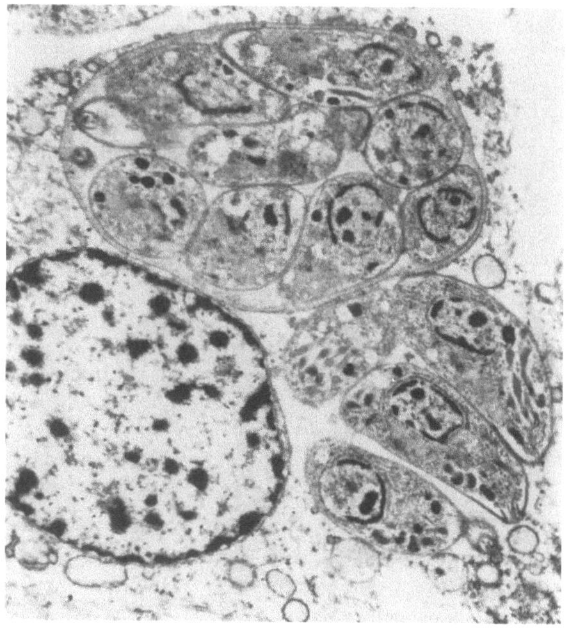

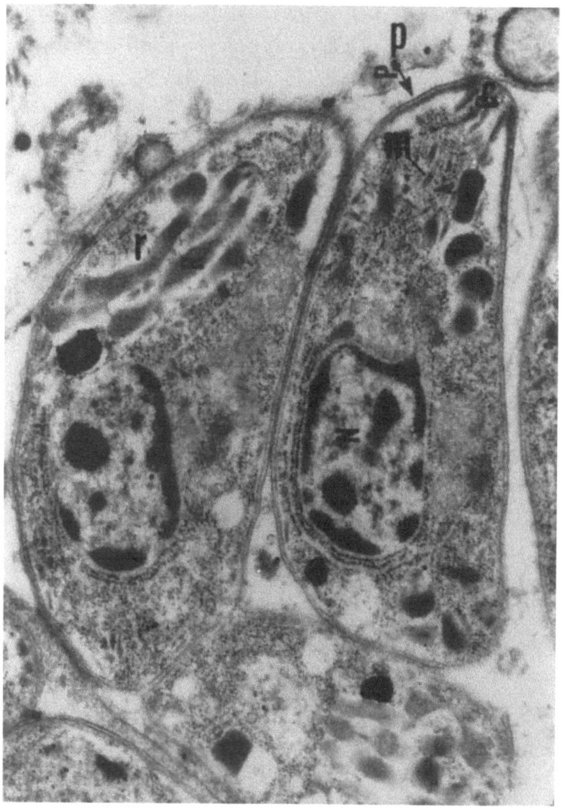

Fig. 70 (*above*). Kidney, squirrel monkey. Note tachyzoites within cells in interstitium of the renal cortex. (Reprinted by courtesy of the JAVMA, Anderson and McClure 1982) TEM, ×3150

Fig. 71 (*below*). Higher magnification of kidney in Fig. 70. Note a double-layered pellicle (*p*), conoid (*c*), rhoptries (*r*), micronemes (*m*), and a nucleus (N). (Reprinted by courtesy of the JAVMA, Anderson and McClure 1982) TEM, ×8400 tions which have resulted in multiple deaths in New World species or in prosimians (McKissick et al. 1968; Dickson et al. 1983; Borst and van Knapen 1984; Dubey et al. 1985). In these outbreaks, the source of infection was likely the ingestion of raw horse meat or sheep hearts which were added to the food rations (McKissick et al. 1968; Dickson et al. 1983; Borst and van Knapen 1984) or by the ingestion of oocysts excreted in the feces of infected cats that were housed in an adjacent cage (Dubey et al. 1985).

Pathogenesis

The life cycle of *T. gondii* is well characterized (Dubey and Beattie 1988; Frenkel 1990a). Cats are the only known definitive host with other mammalian species serving as intermediate hosts. A coccidian cycle with sexual development of *Toxoplasma* in the cat was described in 1970 (Frenkel et al. 1970). Only *Felidae* have an enteroepithelial phase of development; an extraintestinal cycle also occurs and is similar to the cycle in nonfeline hosts. Cats probably become infected in nature by eating infected animals or tissues; however, they may become infected by ingesting any of the three infectious parasitic stages (tachyzoites in pseudocysts; bradyzoites in tissue cysts; sporozoites in oocysts) (Dubey and Beattie 1988). Fecal shedding of oocysts varies with ingestion of the different stages; the prepatent period is 3–10 days after ingesting cysts and 19–20 days or longer after ingesting tachyzoites or oocysts (Dubey 1976). Infection of humans and other animals occurs primarily by the ingestion of food which is contaminated with oocysts from cat feces or by the ingestion of tissue cysts in raw or undercooked meat. Congenital transmission, although serious, is less frequent and primarily occurs by transplacental infection of the fetus following primary infection of the mother during pregnancy (Frenkel 1990b). The causative organism is an intracellular parasite that can infect virtually any cell in the body with the exception of nucleated red blood cells (Feldman 1968). Following acute intestinal infection, tachyzoites disseminate via the blood stream throughout the body and rapidly multiply within intracellular vacuoles of host cells with destruction of parasitized cells, resulting in lesions in multiple tissues. Immunity usually develops by the third week of infection and coincides with the disappearance of tachyzoites (Frenkel 1990a). Bradyzoites continue to multiply slowly and are tightly packed in cysts which may remain for the life of the host. These are usually of little significance unless immunosuppression results in reactivation of the infection.

Etiology

Toxoplasma gondii has a worldwide distribution and a wide range of intermediate hosts (Ratcliffe and Worth 1951; Dubey and Beattie 1988; Dreesen 1990). The organism is a cyst-forming coccidian which is in the protozoan phylum Apicomplexia. *Toxoplasma* develops asexually by endodyogeny in tissue stages as tachyzoites or bradyzoites. Tachyzoites divide rapidly within host cells until the cell ruptures. Individual tachyzoites measure about $2 \times 6\,\mu m$ and are usually crescent-shaped in fresh preparations and are oval or round in sections (Frenkel 1976). Bradyzoites multiply slowly and persist in a chronic state in host cells for the life of the animal. Cysts range from 5 to $50\,\mu m$ in diameter with larger mature cysts containing hundreds of organisms (Dubey and Beattie 1988).

Frequency

The presence of antibodies in the serum indicates that *T. gondii* infection is widespread in humans and animals. Studies in humans have revealed infection rates ranging from less than 1% to greater than 80%–90% in some areas of the world, with estimates in the United States ranging from 30% to 60% (Krogstad et al. 1991; Dubey and Beattie 1988). Similar studies have revealed high rates of antibodies in serum of domestic animals and in wildlife, including marine and zoo animal populations (Dreesen 1990). Serologic studies in nonhuman primates suggest that significant levels of *T. gondii* antibodies are rarely found in New World or in Old World monkeys (Dubey and Beattie 1988). Naturally occurring fatal toxoplasmosis in these animals appears to be an unusual sporadic event.

Comparison with Other Species

Toxoplasmosis is a common infection in most animal species; however, clinical disease is infrequent or mild. The clinical diagnosis is often difficult due to the large numbers of seropositive individuals in both human and animal populations. Significant antibody levels are often absent. Low anti-*Toxoplasma* IgG antibody only indicates past exposure and may reflect infection years earlier. Very high IgG levels are needed to be considered evidence of active infection. The question of active or chronic infection can often be determined by the presence of IgM antibodies; however, they also may persist for up to 12–18 months (Wilson et al. 1990; Krogstad et al. 1991). IgM antibody determinations are needed to diagnose congenital infection; IgG antibodies will cross the placental barrier while IgM antibodies are considered to be of fetal origin. The diagnosis is much more difficult if the patient is immunocompromised, when antibody levels are often low or absent. In addition, the organism may be isolated from chronically infected individuals and both tachyzoites and bradyzoites may be found in the absence of clinical disease (Remington and Cavanaugh 1965; Feldman 1968; Krogstad et al. 1991). While the isolation of the parasite or the demonstration of tissue cysts are not considered absolute evidence of disease, the demonstration of tachyzoites or antigen in tissue is considered to be diagnostic of active infection (Krogstad et al. 1991; Dubey and Beattie 1988). Unfortunately, tachyzoites may be difficult to recognize by routine staining methods and result in underdiagnosis (Conley et al. 1981). Recently, peroxidase-antiperoxidase staining of tissues has been useful in documenting active infection by demonstrating free tachyzoites and infected cells (Conley et al. 1981; Hauser et al. 1982; Borst and van Knapen 1984).

Acknowledgements. This work was supported in part by NIH grant RR-00165 from the National Center for Research Resources to the Yerkes Regional Primate Research Center. The Yerkes Center is fully accredited by the American Association for Accreditation of Laboratory Animal Care.

References

Anderson DC, McClure HM (1982) Acute disseminated fatal toxoplasmosis in a squirrel monkey. J Am Vet Med Assoc 181:1363–1366

Borst GHA, van Knapen F (1984) Acute acquired toxoplasmosis in primates in a zoo. J Zoo Anim Med 15:60–62

Conley FK, Jenkins KA, Remington JS (1981) Toxoplasma gondii infection of the central nervous system. Use of the peroxidase-antiperoxidase method to demonstrate Toxoplasma in formalin fixed, paraffin embedded tissue sections. Hum Pathol 12:690–698

Dickson J, Fry J, Fairfax R, Spence T (1983) Epidemic toxoplasmosis in captive squirrel monkeys (Saimiri sciureus). Vet Rec 112:302

Dreesen DW (1990) Toxoplasma gondii infections in wildlife. J Am Vet Med Assoc 196:274–276

Dubey JP (1976) A review of Sarcocystis of domestic animals and of other coccidia of cats and dogs. J Am Vet Med Assoc 169:1061–1078

Dubey JP, Beattie CP (1988) Toxoplasmosis of animals and man. CRC Press, Boca Raton

Dubey JP, Kramer LW, Weisbrode SE (1985) Acute death associated with Toxoplasma gondii in ring-tailed lemurs. J Am Vet Med Assoc 187:1272–1273

Feldman HA (1968) Toxoplasmosis. N Engl J Med 279:1370–1375, 1431–1437

Frenkel JK (1976) Toxoplasmosis. In: Binford CH, Connor DH (eds) Pathology of tropical and extraordinary diseases. Armed Forces Institute of Pathology, Washington, pp 284–300

Frenkel JK (1985) Toxoplasmosis, lung, mouse and hamster. In: Jones TC, Mohr U, Hunt RD (eds) Respiratory system. Springer, Berlin Heidelberg New York, pp 227–230 (Monographs on pathology of laboratory animals)

Frenkel JK (1990a) Transmission of toxoplasmosis and the role of immunity in limiting transmission and illness. J Am Vet Med Assoc 196:233–240

Frenkel JK (1990b) Toxoplasmosis in human beings. J Am Vet Med Assoc 196:240–248

Frenkel JK, Dubey JP, Miller NL (1970) Toxoplasma gondii in cats: fecal stages identified as coccidian oocysts. Science 167:893–896

Hauser WE, Luft BJ, Conley FK, Remington JS (1982) Central nervous system toxoplasmosis in homosexual and heterosexual adults. N Engl J Med 307:498–499

Krogstad DJ, Visvesvara GS, Walls KW, Smith JW (1991) Blood and tissue protozoa. In: Balows A, Hausler WJ, Herrmann KL, Isenberg HD, Shadomy HJ (eds) Manual of clinical microbiology, 5th edn. pp 727–750

McKissick GE, Ratcliffe HL, Koestner A (1968) Enzootic toxoplasmosis in caged squirrel monkeys Saimiri sciureus. Pathol Vet 5:538–560

Nicolle C, Manceaux L (1909) Sur un protozoaire nouveau du Gondi. C R Acad Sci 148:369–372

Ratcliffe HL, Worth CB (1951) Toxoplasmosis of captive wild birds and mammals. Am J Pathol 27:655–667

Remington JS, Cavanaugh EN (1965) Isolation of the encysted form of Toxoplasma gondii from human skeletal muscle and brain. N Engl J Med 273:1308–1310

Seibold HR, Wolf RH (1971) Toxoplasmosis in Aotus trivirgatus and Callicebus moloch. Lab Anim Sci 21:118–120

Werner H, Janitschke K, Köhler H (1969) Observations on marmoset monkeys of the species Saguinus (Oedipomidas) oedipus following oral and intraperitoneal infection by different cyst-forming Toxoplasma strains of varying virulence: I. Communication: clinical, pathological anatomical and parasitological findings. Zentralbl Bakteriol [Orig] 209:553–569

Wilson M, Ware DA, Juranek DD (1990) Serologic aspects of toxoplasmosis. J Am Vet Med Assoc 196:277–281

Zeman DH, Baskin GB (1985) Encephalitozoonosis in squirrel monkeys (Saimiri sciureus). Vet Pathol 22: 24–31

Noma, *Macaca mulatta*

Andrew A. Lackner, Gary C. Armitage, and Morten Schiødt

Synonyms. Cancrum oris; gangrenous stomatitis.

Gross Appearance

Noma is a rapidly progressive gangrenous process that affects the oral cavity, particularly the gingiva, cheeks, and lips, causing bone denudation and extensive facial disfigurement (Fig. 72). The disease process may be unilateral or bilateral and may affect both upper and lower jaws. In the earliest stages the lesions may be limited to a small area of oral mucosal ulceration and edema.

Microscopic Features

The soft tissue lesion consists of an area of mucosal ulceration covered with necrotic debris and a mixed bacterial flora (Fig. 73). Subjacent to the ulcer is a zone of gangrenous necrosis that is sharply demarcated from viable tissue by an irregular band of suppurative inflammation intermixed with bacteria and spirochetes (Fig. 74). The viable tissue is markedly edematous and hypercellular due to an influx of inflammatory cells. Vascular thrombi are often present in the necrotic areas (Fig. 74) and at the junction with viable tissue. The walls of most occluded vessels undergo fibrinoid necrosis. Occasional vessels in viable tissue contain fibrin emboli. Modified gram stains reveal a mixed bacterial flora on the ulcerated surface and at the junction between viable and necrotic tissue. Numerous spirochetes are also recognized in sections stained by the Warthin-Starry technique. Unlike the other bacteria, however, spirochetes can often be found extending several millimeters into viable tissue (Fig. 75).

When the lesion involves bone, there is osteonecrosis, with loss of osteocytes from the lacunae, and osteomyelitis. Sequestration of the affected bone often occurs. In some cases, bone resorption can be seen associated with extensive bacterial colonization of denuded bone (Merrell et al. 1981).

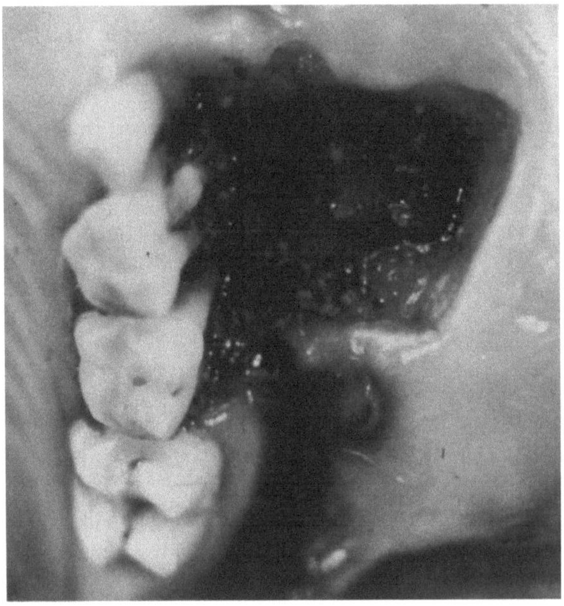

Fig. 72. Teeth and gingiva, *Macaca mulatta* with a relatively mild case of noma. Photo was taken after debridement of necrotic tissue. Note that the necrosis extends to the alveolar and buccal mucosa and that roots of teeth and alveolar bone are visible

Differential Diagnosis

In its severe form, noma is a unique lesion that is difficult to confuse with other oral lesions. In its mildest form, noma should be differentiated from acute necrotizing ulcerative gingivitis. Both are characterized by ulceration of oral mucosa and necrosis; however necrotizing ulcerative gingivitis does not entail denudation of bone and does not extend into the alveolar mucosa.

Fig. 73 (*above*). Noma, *Macaca mulatta*. Necrotic ▶ ulcerated surface of a lesion covered with numerous bacterial colonies. H and E, ×100

Fig. 74 (*middle*). Noma, *Macaca mulatta*. Junction between necrotic and viable tissue marked by an irregular band of suppurative inflammation intermixed with bacteria. A vessel with a thrombus is present in the necrotic tissue. H and E, ×100

Fig. 75 (*below*). Noma, *Macaca mulatta*. Numerous spirochetes in viable tissue several millimeters beyond the necrotic margin. Warthin-Starry stain, ×600

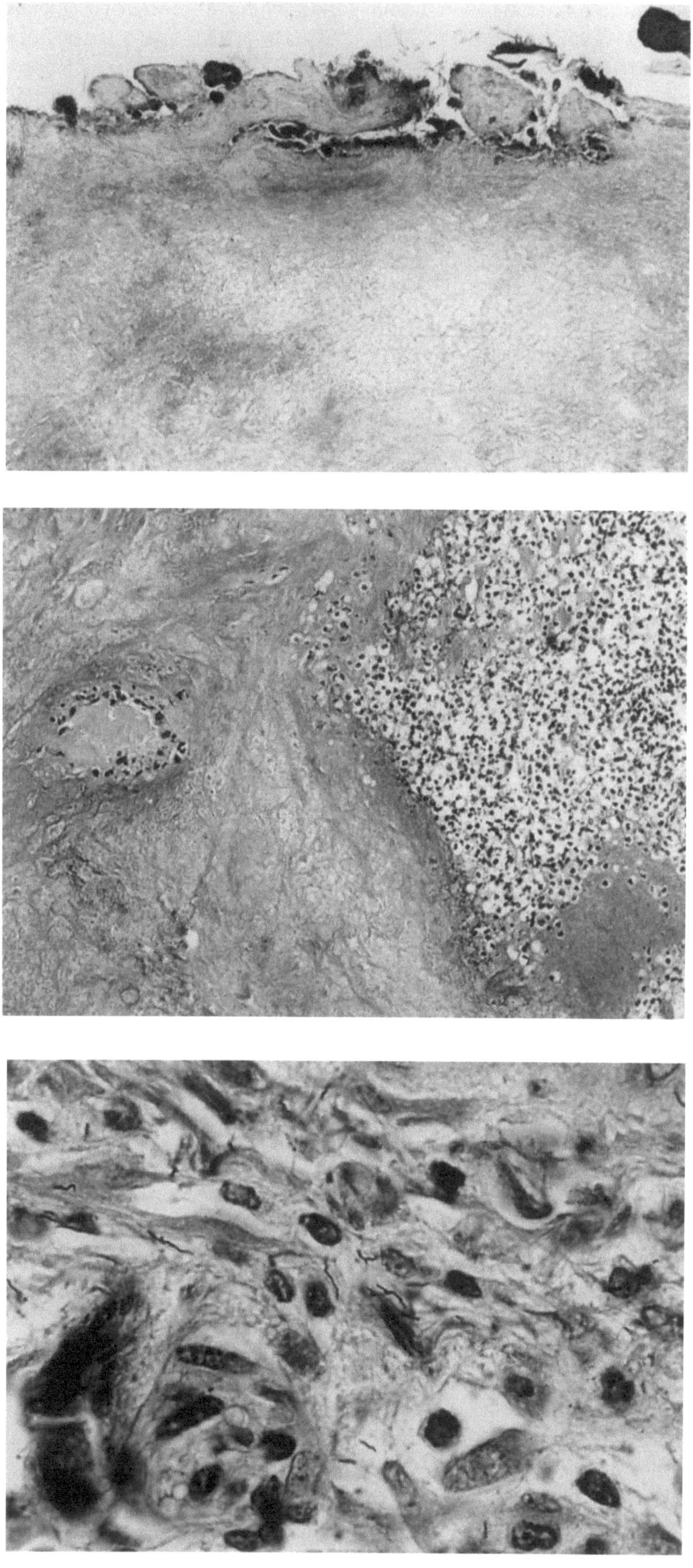

In addition, animals with noma generally have other systemic disorders while animals with this specific gingivitis may be otherwise healthy. The occurrence of noma is highly suggestive of infection with an immunosuppressive type D retrovirus.

Biologic Features

Noma is derived from a Greek work meaning "to devour." As the etymology suggests, noma is a rapidly progressive, disfiguring, and often fatal condition which may occur in macaques of any age. Noma is believed to originate as an ulcerated lesion at the alveolar margin of the gingiva, generally in the region of the premolar and molar teeth (Enwonwu 1972, 1985; Limongelli et al. 1976; Tempest 1966). The ulcerated lesion spreads to other orofacial tissues with extensive destruction of soft and hard tissues. This progression suggests that noma may be the most severe manifestation in a spectrum of lesions beginning with necrotizing gingivitis. Vascular thrombosis in affected tissue is common in noma and may result in necrosis that spreads rapidly (Adams and Bishop 1978). A plethora of anaerobic and aerobic bacteria can be cultured from the lesions including *Bacteroides melaninogenicus*, *Fusobacterium* sp., *Shigella flexneri* serotype 4, *Pseudomonas aeruginosa*, coagulase-positive *Staphylococcus*, *Streptococcus* sp., and *Klebseilla* sp. (Adams and Bishop 1980; Buchanan et al. 1981). Histologically, large numbers of spirochetes are also present at the leading edge of the lesion. The presence of *Shigella flexneri* serotype 4 is of particular interest since this organism is capable of inducing necrotizing gingivitis in normal animals (Armitage et al. 1982). Therefore, *Shigella* may cause the initial necrotizing lesion in which the anaerobic bacteria may grow and produce necrotoxins that lead to progression of the lesion, resulting in noma (Enwonwu 1985).

While noma often develops suddenly, afflicted macaques generally have some preexisting condition that places them at risk. Therefore, noma may be considered secondary to such conditions as malnutrition from experimental diets deficient in one or more nutrients (Chapman and Harris 1941; Topping and Fraser 1939) and, more recently, immune suppression from infection with type D retroviruses (Gardner et al. 1988; Marx et al. 1984; Osborn et al. 1984). Type D retroviruses are widespread in captive macaques and often result in an acquired immune deficiency syndrome that is similar to AIDS in humans (Daniel et al. 1988; Lerche et al. 1987). In our experience, virtually all cases of noma have occurred in macaques infected with type D retrovirus and noma occurs in approximately 10% of rhesus macaques that die from simian AIDS induced by type D retrovirus. In retrospect, many of the previous reports of noma in rhesus macaques were probably manifestations of outbreaks of immune suppression caused by type D retrovirus. Surprisingly, immune suppression in macaques associated with experimental infection with another retrovirus, simian immunodeficiency virus, is rarely associated with noma. This presumably reflects both differences in the mechanisms of immune suppression between these retroviruses and differences in cell tropism (Gardner et al. 1988; Lackner et al. 1989; Maul et al. 1988).

Comparison with Other Species

Noma has only been described in humans and captive nonhuman primates. In addition to rhesus macaques, noma has been described in the Formosan macaque (*Macaca cyclopis*), pig-tailed macaque (*Macaca nemestina*), chimpanzee (*Pan troglodytes*), and the cotton-topped marmoset (*Saguinas oedipus*) (Bourdelle et al. 1931; Brack 1982; Buchanan et al. 1981; Enwonwu 1985). The oral lesions of noma in human and nonhuman primates are similar. In humans, however, noma occurs primarily in immune suppressed children in underdeveloped parts of Africa, Asia, and South America in association with chronic malnutrition, malaria, or other infectious diseases (Enwonwu 1985). Measles is the most common infectious disease associated with the development of noma in Nigerian children (Enwonwu 1985). Noma rarely occurs in adults and has been described in only a few patients with AIDS (Malden 1985; Winkler et al. 1989).

References

Adams RJ, Bishop JL (1978) The comparative pathology of an oral disease resembling noma in six rhesus monkeys. In: Montali RJ, Migaki G (eds) The comparative pathology of zoo animals. Smithsonian Institution, Washington, pp 77–85

Adams RJ, Bishop JL (1980) An oral disease resembling noma in six rhesus monkeys (Macaca mulatta). Lab Animal Sci 30:85–91

Armitage GC, Newbrun E, Hoover CI, Anderson JH (1982) Periodontal disease associated with Shigella flexneri in rhesus monkeys. Clinical, microbiological and histopathologic finding. J Periodont Res 17:131–144

Bourdelle PE, Urbain A, Davesne J (1931) Deuz cas de Gingivo-Stomatite Gangreneuse (noma) due à B. perfringens chez le Chimpanzé. Bull Soc Pathol Exot Fibales 24:787–789

Brack M (1982) Noma in Saguinus oedipus: a report of two cases. Lab Anim 16:361–363

Buchanan W, Sehgal P, Bronson RT, Rodger RF, Horton JE (1981) Noma in a nonhuman primate. Oral Surg Oral Med Oral Pathol 52:19–22

Chapman OD, Harris AE (1941) Oral lesions associated with dietary deficiencies in monkeys. J Infect Dis 69: 7–17

Daniel MD, Letvin NL, Sehgal PK, Schmidt DK, Silva DP, Solomon KR, Hodi FS Jr, Ringler DJ, Hunt RD, King NW, Desrosiers RC (1988) Prevalence of antibodies to 3 retroviruses in a captive colony of macaque monkeys. Int J Cancer 41:601–608

Enwonwu CO (1972) Epidemiological and biochemical studies of necrotizing ulcerative gingivitis and noma (cancrum oris) in Nigerian children. Arch Oral Biol 17:1357–1371

Enwonwu CO (1985) Infectious oral necrosis (cancrum oris) in Nigerian children: a review. Community Dent Oral Epidemiol 13:190–194

Gardner MB, Luciw P, Lerche N, Marx P (1988) Nonhuman primate retrovirus isolates and AIDS. In: Perk K (ed) Immunodeficiency disorders and retrovirus. Vet Sci Comp Med 32:171–226

Lackner AA, Schiødt M, Armitage GC, Moore PF, Munn RJ, Marx PA, Gardner MB, Lowenstine LJ (1989) Mucosal epithelial cells and Langerhans cells are targets for infection by the immunosuppressive type D retrovirus simian AIDS retrovirus serotype 1. J Med Primatol 18:195–207

Lerche NW, Marx PA, Osborn KG, Maul DH, Lowenstine LJ, Bleviss ML, Moody P, Henrickson RV, Gardner MB (1987) Natural history of endemic type D retrovirus infection and acquired immune deficiency syndrome in group-housed rhesus monkeys. J Natl Cancer Inst 79:847–854

Limongelli WA, Clark MS, Williams AC (1976) Noma-like lesion in a patient with chronic lymphocytic leukemia. Review of the literature and report of a case. Oral Surg Oral Med Oral Pathol 41:40–51

Malden N (1985) An interesting case of adult facial gangrene (from Papua, New Guinea). Oral Surg Oral Med Oral Pathol 59:279–281

Marx PA, Maul DH, Osborn KG, Lerche NW, Moody P, Lowenstine LJ, Henrickson RV, Arthur LO, Gilden RV, Gravell M, London WT, Sever JL, Levy JA, Munn RJ, Gardner MB (1984) Simian AIDS: isolation of a type D retrovirus and transmission of the disease. Science 223:1083–1086

Maul DH, Zaiss CP, MacKenzie MR, Shiigi SM, Marx PA, Gardner MB (1988) Simian retrovirus D serogroup 1 has a broad cellular tropism for lymphoid and nonlymphoid cells. J Virol 62:1768–1773

Merrell BR, Joseph SW, CAsazza LJ, Duncan JF (1981) Bacterial bone resorption in noma (gangrenous stomatits). J Oral Pathol 10:173–185

Osborn KG, Prahalada S, Lowenstine LJ, Gardner MB, Maul DH, Henrickson RV (1984) The pathology of an epizootic of acquired immunodeficiency in rhesus macaques. Am J Pathol 114:94–103

Tempest MN (1966) Cancrum oris. Br J Surg 53:949–969

Topping NH, Fraser HF (1939) Mouth lesions associated with dietary deficiencies in monkeys. Public Health Rep 54:416–431

Winkler JR, Murray PA, Hammerle C (1989) Gangrenous stomatitis in AIDS. Lancet ii:108

Herpesviruses of Primates: An Introduction

Ronald D. Hunt

The word "herpes," from Greek meaning to creep or crawl, has been used in medical writing for over 25 centuries. Herodotus, Hippocrates, Celsus, Galen, and other medical writers of ancient times probably included a variety of skin disorders under the name herpes, most of which were probably not herpesvirus infections. *Herpesvirus simplex* and *Herpesvirus varicellae* (*H. zoster*) were, however, likely included among the disorders they described. Although more restrictive descriptions of herpes appeared in the late seventeenth and early eighteenth centuries, it was not until the early nineteenth century that the use of herpes was applied to those maladies we associate with specific herpesviruses today. That herpes were infectious diseases was shown later in the nineteenth century at about the same time that Louis Pasteur and Robert Koch proved the germ theory of disease. *Herpesvirus simplex* was the first herpesvirus to be discovered. *Herpesvirus B*, the first herpesvirus of nonhuman primates to be identified, was isolated by Sabin and Wright in 1934 and for 24 years remained the only known herpesvirus of importance to nonhuman primates.

Today a long list of herpesviruses is known to infect nonhuman primates. Most of these are indigenous to various species of monkeys, a few are viruses of human beings. Certain of these viruses are very important pathogens, one (*Herpesvirus B*) an important zoonotic infection of human beings; but many remain poorly characterized as agents and as causes of disease. Most of the herpesviruses of nonhuman primates are α-herpesviruses. It is these that are associated with the most serious disease when they infect hosts other than the natural host. In the natural host α-herpesviruses are most often associated with a mild disease, when recognized, but the virus remains latent, usually in ganglia, for the life of the individual. β-herpesviruses cause cytomegalic inclusion body disease. These agents are relatively host-specific, infection and disease being confined to the natural host. β-herpesviruses also remain latent, in secretory glands, kidneys, lymphocytes, and other tissues. γ-herpesviruses also have a limited host range and remain latent in either T or B lymphocytes. As a group they are not frequent causes of spontaneous disease in nonhuman primates, but have emerged as very important models for the study of herpesviruses and cancer. Table 4 lists herpesviruses of primates, and those which are of greatest significance are discussed in ensuing chapters in this volume.

References

Ablashi DV, Chopra HC, Armstrong GR (1972) A cytomegalovirus isolated from an owl monkey. Lab Anim Sci 22:190–195

Asher DM, Gibbs CJ Jr, Lang DJ, Gajdusek DC (1974) Persistant shedding of cytomegalovirus in the urine of health rhesus monkeys. Proc Soc Exp Biol Med 145: 794–801

Barahona HH, Meléndez LV, King NW, Daniel MD, Fraser CEO, Preville AC (1973) Herpesvirus aotus Type 2: a new viral agent from owl monkeys (Aotus trivirgatus). J Infect Dis 127:171–178

Black PH, Hartley J, Rowe WP (1963) Isolation of cytomegalovirus from African Green Monkey. Proc Soc Exp Biol Med 112:601–605

Blakely GA, Lourie B, Morton WG, Evans HH, Kaufman AF (1973) A varicella-like disease in macaque monkeys. J Infect Dis 127:617–625

Böcker JF, Tiedemann KH, Bornkamm GW, zur Hausen H (1980) Characterization of an EBV-like virus from African Green Monkey lymphoblasts. Virology 101: 291–295

Clarkson MJ, Thorpe E, McCarthy K (1967) A virus disease of captive vervet monkeys (Cercopithecus aethiops) caused by a new herpesvirus. Arch Gesamte Virusforsch 22:219–234

Corey L, Spear PG (1986) Infections with herpes simplex viruses (parts I and II). N Engl J Med 314:686–691, 749–757

Daniel MD, Meléndez LV, King NW, Fraser CEO, Barahona HH, Hunt RD, García FA (1971) Herpes virus aotus: a latent herpesvirus from owl monkeys (Aotus trivirgatus). Isolation and characterization. Proc Soc Exp Biol Med 138:835–845

Table 4. Herpesvirus of primates

Scientific name	Common name	Host	Disease	Reference
α-Herpesvirus				
Human herpesvirus 1, 2	*H. simplex* (types 1 and 2)	Humans	Usually self-limiting mucocutaneous vesicles	Corey and Spear 1986
		Gibbons	Self-limiting vesicles or encephalitis	Smith et al. 1969
		Owl monkeys, tamarins, marmosets, tree shrews	Generalized fatal infection	Meléndez et al. 1969 Hunt and Melendez, 1969, 1972 McClure et al. 1972
Saimirine herpes 1	*Herpes T* (*Herpes M*, *H. platyrrhinae*)	Squirrel monkey	Usually self-limiting, mucocutaneous vesicles	King et al. 1967
		Owl monkey, marmosets, tamarins	Fatal generalized infection	Hunt and Melendez, 1966, 1972
Cercopithicine herpesvirus 1	*H. simiae* (B virus)	Macaques	Usually self-limiting, mucocutaneous vesicles	Keeble et al. 1958
		Humans	Fatal encephalomyelitis	Sabin and Wright 1934
Cercopithicine herpesvirus 2	SA 8	African green monkey	Myelitis	Malherbe and Harwin 1958 Malherbe et al. 1963
		Baboon	Usually latent	Eichberg et al. 1976
Human herpesvirus 3	*H. varicella*/*H. zoster*	Humans	Chicken pox, shingles	Weller 1953
		Great apes	Chicken pox	Heuschle 1960 Marennikova et al. 1973 McClure and Keeling 1971 White et al. 1972
Cercopithicine herpesvirus 6, 7, 9	Simian varicella (delta, Medical Lake, Liverpool vervet monkey viruses)	African green, patas, *M. nemestrina*, *M. fuscata*, *M. fascicularis*	Chicken pox-like disease	Clarkson et al. 1967 McCarthy et al. 1968 Blakely et al. 1973
Ateline herpesvirus 1	Spider monkey herpesvirus	Spider monkey	Usually latent, may cause generalized fatal infection	Hull et al. 1972
Tupaiaine herpesvirus	Tree shrew herpesvirus Uncertain classification	Tree shrew	None described (suggested to cause lymphoma)	Darai et al. 1982 Darai and Koch 1984 Mirkovic et al. 1970
Cercopithicine herpesvirus 13	*H. cyclopis* Uncertain classification	*M. cyclopis*	None described	Jackman et al. 1977
β-Herpesviruses				
Human herpesvirus 5	Cytomegalovirus	Humans	Cytomegalic inclusion body disease	Smith 1956
Aotine herpesvirus 1, 3, 4	*H. aotus* types 1, 3, 4	Owl monkey	Cytomegalic inclusion body disease/usually latent	Daniel et al. 1971, 1973 Ablashi et al. 1972
Cercopithecine herpesvirus 3	SA-6	Vervet monkeys	Cytomegalic inclusion body disease/usually latent	Goodheart and Plummer 1975
Cercopithecine herpesvirus 4	SA-15	Vervet monkeys	Cytomegalic inclusion body disease/usually latent	Malherbe et al. 1963
Cercopithecine herpesvirus 5	African green CMV	African green monkey	Cytomegalic inclusion body disease/usually latent	Black et al. 1963
Cercopithecine herpesvirus 8	Rhesus monkeys CMV	Rhesus monkey	Cytomegalic inclusion body disease/usually latent	Asher et al. 1969
Callitrichine herpesvirus 1, 2	*H. saguinus* marmoset CMV	Cotton-top tamarin	Cytomegalic inclusion body disease/usually latent	Meléndez et al. 1970; Nigada et al. 1979

Table 4 (*continued*)

Scientific name	Common name	Host	Disease	Reference
Cebine herpesvirus 1	Capuchin herpesvirus (AL-5)	Cebus monkey	Cytomegalic inclusion body disease/usually latent	Lewis et al. 1976
Cebine herpesvirus 2	Capuchin herpesvirus (AP-18)	Cebus monkey	Cytomegalic inclusion body disease/usually latent	Sabin and Wright 1934
Human herpesvirus 6 (HHV-6)	Human B lymphotropic virus (HBLV)	Humans	Exanthem subitum (roseola infantum)	Lopez et al. 1988 Salahuddin et al. 1986
γ-Herpesviruses				
Human herpesvirus 4	Epstein-Barr virus	Humans	Infectious mononucleosis Burkitt's lymphoma Nasopharyngeal carcinoma?	Epstein et al. 1965
		Cotton-top tamarins	Lymphoma	Shope et al. 1973; Miller et al. 1977
		Common marmoset	Mild lymphoproliferative disease	Falk et al. 1976a
Cercopithicine herpesvirus 14	African green EBV-like	African green monkey	None described	Böcker et al. 1980
	Other EBV-related viruses	Macaques (*M. mulatta*, *M. fascicularis*)	Associated with lymphoma	Rangan et al. 1986 Heberling et al. 1982
Saimirine herpesvirus 2	*H. saimiri*	Squirrel monkeys	Latent infection	Meléndez et al. 1968
		Owl monkeys Tamarins Marmosets Spider monkeys Howler monkeys	Lymphoma/leukemia	Hunt et al. 1972, 1979 Rangan et al. 1977 Wolfe et al. 1971
Ateline herpesvirus 2, 3	*H. ateles*	Spider monkeys	Latent infection	Meléndez et al. 1972 Falk et al. 1974
		Owl monkeys	Lymphoma/leukemia	Hunt et al. 1978
Aotine herpesvirus 2	*H. aotus* type 2	Owl monkeys	None described	Barahona et al. 1973
Cercopithicine herpesvirus 12	*H. papio*	Baboon	Associated with lymphoma?	Falk et al. 1976b
Cercopithicine herpesvirus 10, 11	Rhesus leukocyte-associated herpesvirus (strains 1 and 2)	Rhesus monkey	None described	Frank et al. 1973
Pongine herpesvirus 1	Chimpanzee herpes	Chimpanzee	None described	Landon et al. 1968
Pongine herpesvirus 2	Orangutan herpes	Orangutan	None described	Rasheed et al. 1977
Pongine herpesvirus 3	Gorilla herpes	Gorilla	None described	Neubauer et al. 1979

CMV, cytomegalovirus.

Daniel MD, Meléndez LV, King NW, Barahona HH, Fraser CEO, García FA, Silva D (1973) Isolation and characterization of a new virus from owl monkeys: herpesvirus aotus type 3. Am J Phys Anthropol 38: 497–500

Darai G, Koch H-G (1984) Tree shrew herpesvirus: pathogenicity and latency. Curr Top Vet Med Anim Sci 27:91–102

Darai G, Zoller L, Matz B, Flugel RM, Moller P, Hofmann W, Gelderblom H, Delius H (1982) Tupaia herpesviruses: characterization and biological properties. Microbiologica 5:285–298

Eichberg JW, McCullough B, Kalter SS, Thor DE, Rodriquez AR (1976) Clinical, virological, and pathological features of herpesvirus SA 8 infection in conventional and gnotobiotic baboons (Papio cynocephalus). Arch Virol 50:255–270

Epstein MA, Henle G, Achong BG, Barr YM (1965) Morphological and biological studies on virus in cultured lymphoblasts from Burkitt's lymphoma. J Exp Med 121:761–770

Falk LA, Nigida SM, Deinhardt F, Wolfe LG, Cooper RW, Hernandez-Camacho JI (1974) Herpesvirus ateles: properties of an oncogenic herpesvirus isolated

from circulating lymphocytes of spider monkeys (Ateles sp.). Int J Cancer 14:473–482
Falk L, Deinhardt F, Wolfe L, Johnson D, Hilgers J, de-Thé G (1976a) Epstein-Barr virus: experimental infection of Callithrix jacchus marmosets. Int J Cancer 17:785–788
Falk L, Deinhardt F, Nonoyama M, Wolfe LG, Bergholz C, Lapin B, Yakovleva L, Agrba V, Henle G, Henle W (1976b) Properties of a baboon lymphotropic herpesvirus related to Epstein-Barr virus. Int J Cancer 18:798–807
Frank AL, Bissell JA, Rowe DS, Dunnick NR, Mayner RE, Hopps HE, Parkman PD, Meyer HM Jr (1973) Rhesus leucocyte-associated herpesvirus: I. Isolation and characterization of a new herpesvirus recovered from rhesus monkey leukocytes. J Infect Dis 128:618–629
Goodheart C, Plummer G (1975) The densities of herpes viral DNAs. Prog Med Virol 19:324–352
Heberling R, Bieber C, Kalter S (1982) Establishment of a lymphoblastoid cell line from a lymphomatous cynomologous monkey. In: Yohn DS, Balkeslee JR (eds) Advances in comparative leukemia research. Elsevier/North Holland, New York, pp 385–386
Heuschle WP (1960) Varicella (chicken pox) in three young anthropoid apes. J Am Vet Med Assoc 136: 256–257
Hull RN, Dwyer AC, Holmes AW, Nowakowski E, Deinhardt F, Lennette EH, Emmons RW (1972) Recovery and characterization of a new simian virus from a fatally infected spider monkey. J Natl Cancer Inst 49:225–230
Hunt RD, Meléndez LV (1966) Spontaneous Herpes-T infection in the owl monkey (Aotus trivirgatus). Pathol Vet 3:1–26
Hunt RD, Meléndez LV (1969) Herpes virus infections of non-human primates: a review. Lab Anim Care 19: 221–234
Hunt RD, Meléndez LV (1972) Clinical, epidemiologic, and pathologic features of cytocidal and oncogenic herpesviruses in South American monkeys. J Natl Cancer Inst 49:261–271
Hunt RD, Meléndez LV, King NW, García FA (1972) Herpesvirus saimiri: malignant lymphoma in spider monkeys: a new susceptible host. J Med Primatol 1: 114–128
Hunt RD, Barahona H, Daniel MD (1978) Herpesvirus ateles malignant lymphoma in owl monkeys: a new susceptible primate. In: Yohn DS, Bentvelzen P (eds) Advances in comparative leukemia research. Elsevier/ North-Holland, Amsterdam, pp 198–200
Hunt RD, Meléndez LV, King NW, Gilmore CE, Daniel MD, Williamson ME, Jones TC (1979) Morphology of a disease with features of malignant lymphoma in marmosets and owl monkeys innoculated with Herpesvirus saimiri. J Natl Cancer Inst 44:447–465
Jackman DA, King NW, Daniel MD, Sehgal PK, Frazer CEO (1977) H. cyclopis: a new herpesvirus isolated from Macaca cyclopis. Abstracts of the annual meeting of the American Society for Microbiology, abstract S413
Keeble SA, Christofinis GJ, Wood W (1958) Natural virus-B infection in rhesus monkeys. J Pathol Bacteriol 76:189–199
King NW, Hunt RD, Daniel MD, Meléndez LV (1967) Overt herpes-T infection in squirrel monkeys (Saimiri sciureus). Lab Anim Care 17:413–423
Landon JC, Ellis LB, Zeve VH, Fabrizio DPA (1968) Herpes-type virus in cultured leukocytes from Chimpanzees. J Natl Cancer Inst 40:181–192
Lewis MA, Frye LD, Gibbs CJ Jr, Chou SM, Cutchins EC, Gajdusek DC (1976) Isolation and characterization of two new unrelated herpes-like viruses from capuchin monkeys. Infect Immun 14:759–766
Lopez C, Pellett P, Stewart J, Goldsmith C, Sanderlin K, Black J, Warfield D, Feorino P (1988) Characteristics of human herpesvirus-6. J Infect Dis 157:1271–1273
Malherbe H, Harwin R (1958) Neurotropic virus in African monkeys. Lancet 2:530
Malherbe H, Harwin R, Ulrich M (1963) The cytopathic effects of vervet monkey viruses. S Afr Med J 37: 407–411
Marennikova SS, Maltseva NN, Shelukhin EM, Shenkman LS, Korneeva VI (1973) A generalized herpetic infection simulating smallpox in a gorilla. Intervirology 2:280–286
McCarthy K, Thorpe E, Laursen AC, Heymann CS, Beale JA (1968) Exanthematous disease in patas monkeys caused by a herpes virus. Lancet 2:856–857
McClure HM, Keeling ME (1971) Viral diseases noted in the Yerkes Primate Center colony. Lab Anim Sci 21: 1002–1010
McClure HM, Keeling ME, Olberding B, Hunt RD, Meléndez LV (1972) Natural Herpesvirus hominus infection of tree shrews (*Tupaia glis*). Lab Anim Sci 22:517–521
Meléndez LV, Daniel MD, Hunt RD, García FG (1968) An apparently new herpesvirus from primary kidney cultures of the squirrel monkey (Saimiri sciureus). Lab Anim Care 18:374–381
Meléndez LV, España C, Hunt RD, Daniel MD, García FG (1969) Natural Herpes simplex infection in the owl monkey (Aotus trivirgatus). Lab Anim Care 19:38–45
Meléndez LV, Hunt RD, Daniel MD, Trum BF (1970) New world monkeys, herpesviruses, and cancer. In: Balner H, Beveridge WIB (eds) Infections and immunosuppression in subhuman primates. Munksgaard, Copenhagen, pp 111–117
Meléndez LV, Hunt RD, King NW, Barahona HH, Daniel MD, Fraser CEO, García FG (1972) Herpesvirus ateles, a new lymphoma virus of monkeys. Nature New Biol (London) 235:182–184
Miller G, Shope T, Coope D, Waters L, Pagano J, Bornkamm GW, Henle W (1977) Lymphoma in cotton-top marmosets after inoculation with Epstein-Barr virus: tumor incidence, histologic spectrum, antibody responses, demonstration of viral DNA, and characterization of viruses. J Exp Med 145:948–967
Mirkovic R, Voss WR, Benyesh-Melnick M (1970) Characterization of a new herpesvirus indigenous for tree shrews. In: Proceedings of the 10th international congress of microbiology, Mexico City, pp 181–189
Neubauer RH, Rabin H, Strnad BC, Nonoyama M, Nelson-Rees WA (1979) Establishment of a lymphoblastoid cell line and isolation of an Epstein-Barr related virus of gorilla origin. J Virol 31:845–848
Nigada SM, Falk LA, Wolfe LG, Deinhardt F (1979) Isolation of a cytomegalovirus from salivary glands of white-lipped marmosets (Saguinus fuscicollis). Lab Anim Sci 29:53–60
Rangan SRS, Martin LN, Enright FM, Abee CR (1977) Herpesvirus saimiri induced lymphoproliferative disease in howler monkeys. J Natl Cancer Inst 59:165–171
Rangan SRS, Martin LN, Bozelka BE, Wang N, Gormus BJ (1986) Epstein-Barr virus-related herpesvirus from

a rhesus monkey (Macaca mulatta) with malignant lymphoma. Int J Cancer 38:425–432
Rasheed S, Rongey RW, Bruszweski J, Nelson-Ress WA, Rabin H, Neubauer RH, Esra G, Gardner MB (1977) Establishment of a cell line with associated Epstein-Barr-like virus from a leukemic orangutan. Science 198:407–409
Sabin AB, Wright AM (1934) Acute ascending myelitis following a monkey bite with the isolation of a virus capable of producing the disease. J Exp Med 59:115–136
Salahuddin SZ, Ablashi DV, Markham PD, Josephs SF, Sturzenegger S, Kaplan M, Halligan G, Biberfield P, Wong-Staal F, Kramarsky B, Gallo RC (1986) Isolation of a new virus, HBLV, in patients with lymphoproliferative disorders. Science 234:596–601
Shope T, Dechairo D, Miller G (1973) Malignant lymphoma in cotton-top marmosets after inoculation with Epstein-Barr virus. Proc Natl Acad Sci USA 70: 2487–2491
Smith MG (1956) Propagation is tissue cultures of a cytopathogenic virus from human salivary gland virus (SGV) disease. Proc Soc Exp Biol Med 92:424–430
Smith PC, Yuill TM, Buchanan RD, Stanton JS, Chaicumpa V (1969) The gibbon (Hylobates lar); a new primate host for Herpesvirus hominus: I. A natural epizootic in a laboratory colony. J Infect Dis 120:292–297
Weller TH (1953) Serial propagation in vitro of agents producing inclusion bodies derived from varicella and herpes zoster. Proc Soc Exp Biol Med 83:340–346
White RJ, Simmons L, Willson RB (1972) Chickenpox in young anthropoid apes: clinical and laboratory findings. J Am Vet Med Assoc 161:690–692
Wolfe LG, Falk LA, Deinhardt F (1971) Oncogenicity of Herpesvirus saimiri in marmoset monkeys. J Natl Cancer Inst 47:1145–1163

Herpesvirus B Infection

Ronald D. Hunt and Beverly J. Blake

Synonyms. *Herpesvirus simiae*; B virus infection.

Gross Appearance

Ordinary *Herpesvirus B* infection in rhesus monkeys (*Macaca mulatta*) and other macaques is typically an inapparent latent infection (Burnett et al. 1939; Palmer 1987; Hartley 1964, 1966). Gross lesions are generally associated with primary infection, but the lesions are only rarely recorded. This suggests that they are small and do not lead to overt signs or are overlooked due to infrequent and inadequate examination. When present, gross lesions resemble those seen in *H. simplex* infection in humans, (p. 82, this volume), consisting of vesicles which rapidly progress to ulcers (Keeble et al. 1958; Keeble 1960). Lesions are most frequent on the dorsal surface of the tongue and at the mucocutaneous junction of the lips (Fig. 76). Other parts of the oral mucous membranes may also be affected and occasionally the infection occurs as conjunctivitis. Healing without scars is usually complete within 1–2 weeks. Less often, B virus infection causes widespread stomatitis with extension into the pharynx and esophagus. This infection rarely becomes disseminated, which is often fatal. The clinical and gross findings are typically similar in other species of macaques, which include *M. fascicularis, M. fuscata, M. arctoides, M. cyclopis,* and *M. radiata*. Disseminated disease and fatalities have been reported most frequently in *M. radiata* and *M. fascicularis* (Deniel et al. 1975; Espana 1973; Hartley 1964).

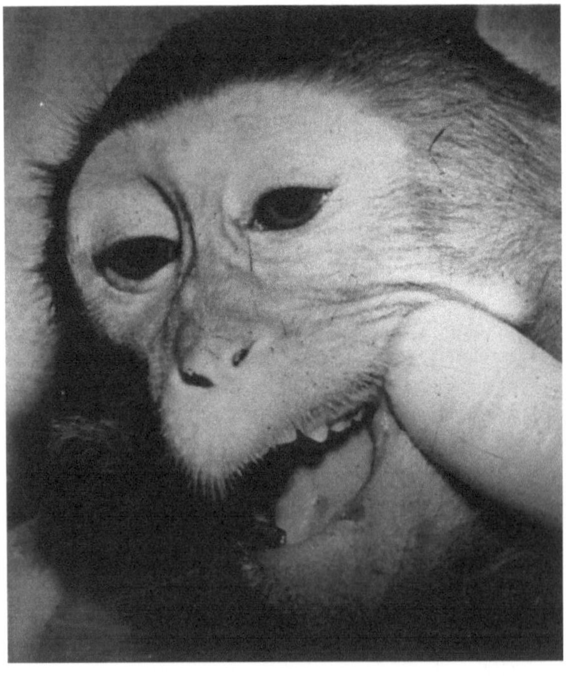

Fig. 76. Rhesus monkey, labial and lingual ulcers caused by *Herpesvirus B*. (Courtesy of Dr. W.E. Cole)

Microscopic Features

Lesions of the oral mucous membranes and skin are characterized by ballooning degeneration of epithelial cells, karyorrhexis and karyolysis of their nuclei, and ultimately their dissolution with the formation of a vesicle. Multinucleated giant cells are admixed in the degenerating epithelium and surrounding vesicles. Intranuclear inclusion bodies are present in epithelial cells and multinucleated giant cells (Fig. 77). These are usually amphoteric, Feulgen-positive, and contain virions. The intranuclear inclusion bodies which are eosinophilic and which lack evidence of virions are generally regarded as residua of viral damage. Inflammatory cells infiltrate the epithelium and are present within the vesicles and the underlying dermis. Necrosis of macrophages and endothelial cells in the dermis often with intranuclear inclusion bodies is a usual feature. With rupture of the vesicle, invasion by bacteria and fungi accentuates the inflammatory reaction. Visceral lesions may develop during the course of clinical disease but rarely lead to systemic signs. The two most frequent alterations are focal hepatic necrosis, with intranuclear inclusion bodies in hepatocytes and Kupffer's cells, and encephalitis. The latter is characterized by neuronal necrosis, gliosis, and perivascular cuffing with lymphocytes. Intranuclear inclusion bodies occur in glial cells and neurons. The lesions are most frequent in the region of the roots of the facial and trigeminal nerves, the descending tract of the trigeminal nerve, and the tractus solitarius in the medulla (Keeble 1960). More severe systemic infection leading to death is associated with necrotizing lesions and inclusion bodies in many organs and tissues including the liver, lung, spleen, lymph nodes, and adrenal cortex (McClure et al. 1973).

Ultrastructure

Typical herpesvirus particles are present in the nucleus of affected cells (Fig. 78).

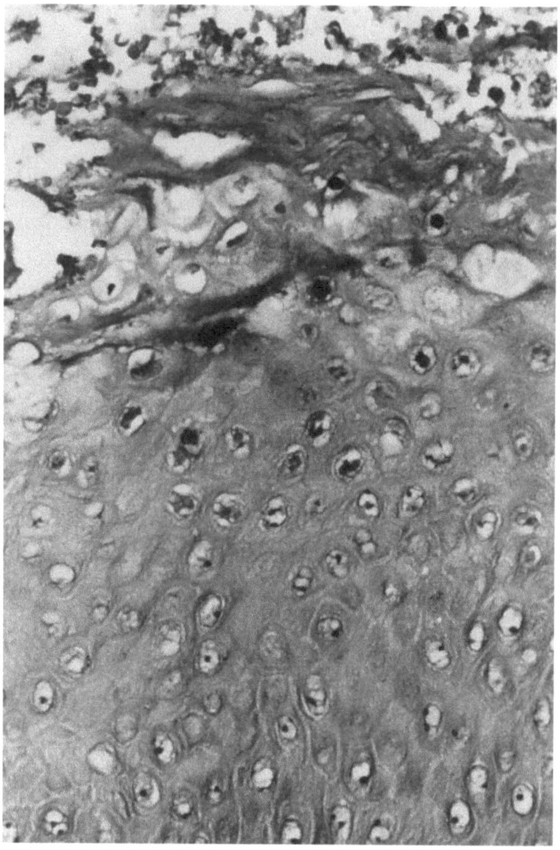

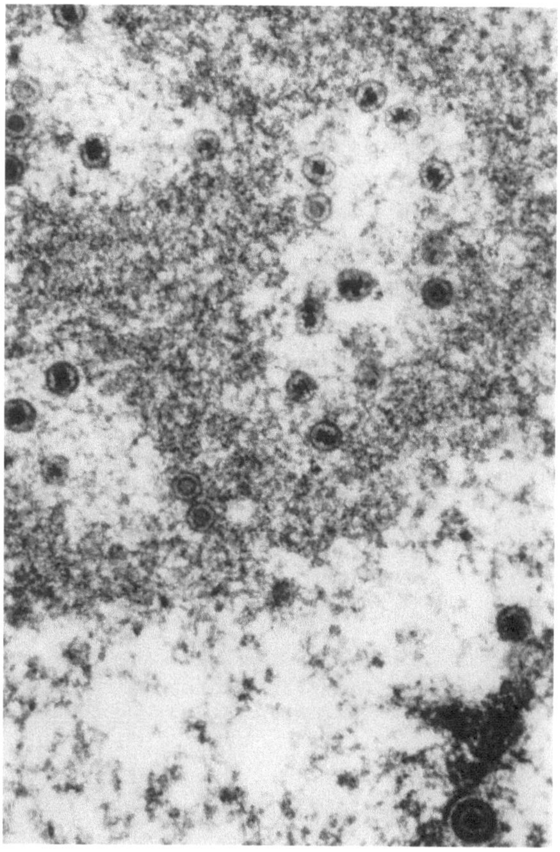

Fig. 77 (*above*). Rhesus monkey, intranuclear inclusion ▶ bodies in keratinocytes in an early lesion caused by *Herpesvirus B*. H and E, ×250

Fig. 78 (*below*). Rhesus monkey, typical *Herpesvirus B* virions within a nucleus. TEM, ×45000

Differential Diagnosis

Herpesvirus B infection should be suspected with any vesicular or ulcerative lesion of the oral cavity, lips, or skin in a macaque monkey. Biopsy of the lesion can confirm a herpesvirus infection by the typical histopathologic lesions. Confirmation of *Herpesvirus B* should be made by virus isolation and identification. Macaques are susceptible to other herpesviruses, but these are not associated with vesicular lesions (see Table 4).

Biologic Features

Natural History and Pathogenesis

Although less thoroughly investigated than *H. simplex* infection in human beings, the available evidence indicates that *Herpesvirus B* infection in macaques follows a similar pathogenesis. Upon exposure of mucosal surfaces the virus replicates in epithelial cells, which may or may not lead to a visible lesion. Released virus then enters nerve endings and is transported intra-axonally, presumably as nucleocapsids, to nerve cells in ganglia, most often the trigeminal, where further replication may occur. Centrifugal spread via sensory nerves and contiguous local extension can expand the mucosal lesions. The virus may extend along nerves to the central nervous system leading to encephalitis and, rarely, viremia which leads to visceral lesions. Once established in ganglia the virus is presumed to remain with the individual for life (Boulter 1975; Vizoso 1975). Recurrence of lesions is less well documented in monkeys than is *H. simplex* infection in humans. There is always the problem of not knowing if a given lesion represents a primary infection of recurrent lesion and most laboratories do not follow animals with lesions for any extended period, but rather sacrifice the animal due to the human health hazard. B virus can be recovered periodically from oral mucosa and occasionally the genital tract of the same monkey in the absence of visible lesions (Zwartouw and Boulter 1984). Reactivation has followed immunosuppression, with recovery of virus from oral and genital swabs (Zwartouw and Boulter 1984). The form the virus takes in neural transport from ganglia to epithelial surfaces is not known. Spontaneous *Herpesvirus B* infection only rarely results in fatal illness in rhesus monkeys, although in *M. fascicularis* and *M. radiata* it may be severe enough to lead to death.

Etiology and Frequency

Herpesvirus B is an α-herpesvirus sharing morphologic, biochemical, and replicative features with other members of this group such as *H. simplex*. *Herpesvirus B* is transmitted horizontally with ease. In colonies of rhesus monkeys the incidence of infection approaches 100% in adult animals (Kessler and Hilliard 1990; Weigler et al. 1990; Zwartouw et al. 1984). The incidence of infection is lowest in young animals and juveniles. In other macaques the incidence is usually lower than in rhesus, but may also be high (Zeitlyonok et al. 1966; Hutt et al. 1981; Kalter and Heberling 1989).

Comparison with Other Species

Herpesvirus B was first isolated in 1932 from the central nervous system of a laboratory worker who had died after being bitten by a rhesus monkey (Sabin and Wright 1934). Over two decades passed before the disease was recognized and defined in rhesus monkeys. In human beings *Herpesvirus B* infection causes encephalomyelitis which is generally fatal (Benson et al. 1989; Breen et al. 1958; Bryan et al. 1975; Cummins 1988; Davidson and Hummeler 1960; Kaplan 1988). Of 22 cases reported in the literature, 15 patients have died and most survivors suffered severe neurologic sequelae. Transmission to human beings usually occurs through exposure to monkey saliva or secretions from bites, scratches, or needle punctures. Other sources are infected tissue cultures. There is one report of human to human transmission (Holmes et al. 1990). Vesicular lesions may develop at the site of viral entry followed by lymphadenitis and subsequent encephalomyelitis. The relative low incidence of disease in humans suggests that susceptibility is not universal or that many asymptomatic cases occur. The possibility that the infection can be latent is supported by the development of B virus infection in a virologist 10 years after the last known exposure to the virus (Fierer et al. 1973). The enzootic relationship of *Herpesvirus B* in macaques is analogous to *H. simplex* infection in humans and *Herpesvirus T* in squirrel monkeys.

Acknowledgement. This work was supported by Public Health Service Award RR00168 from the National Institutes of Health.

References

Benson PM, Malane SL, Banks R, Hicks CB, Hilliard J (1989) B virus (Herpesvirus simiae) and human infection. Arch Dermatol 125:1247–1248

Boulter EA (1975) The isolation of monkey B virus (Herpesvirus simiae) from the trigeminal ganglia of a healthy seropositive rhesus monkey. J Biol Stand 3: 279–280

Breen GE, Lamb SG, Otaki AT (1958) Monkey-bite encephalomyelitis: report of a case – with recovery. BMJ 2:22–23

Bryan BL, Espana CD, Emmons RW, Vijayan N, Hoeprich PD (1975) Recovery from encephalomyelitis caused by Herpesvirus simiae. Arch Intern Med 135: 868–870

Burnett FM, Lush D, Lackson AV (1939) The relationship of herpes and B viruses: immunological and epidemiological considerations. Aust J Exp Biol Med Sci 17:41–51

Cummins LB (1988) Relearning the lessons of Herpes B virus: a report for the monkey handler. Lab Anim 17/3:25–27

Daniel MD, Garcia FG, Melendez LV, Hunt RD, O'Connor J, Silva D (1975) Multiple Herpesvirus simiae isolation from a rhesus monkey which died of cerebral infarction. Lab Anim Sci 25:303–308

Davidson WL, Hummeler K (1960) B virus infection in man. Ann NY Acad Sci 85:970–979

Espana C (1973) Herpesvirus simiae infection in Macaca radiata. Am J Phys Anthropol 38:447–454

Fierer J, Bazeley P, Braude AI (1973) Herpes B virus encephalomyelitis presenting as ophthalmic zoster: a possible latent infection reactivated. Ann Intern Med 79:225–228

Hartley EG (1964) Naturally occurring "B" virus in cynomologus monkeys. Vet Rec 76:555–557

Hartley EG (1966) "B" virus disease in monkey and man. Br Vet J 122:46–50

Holmes GP, Hilliard JK, Klontz KC, Rupert AH, Schindler CM, Parrish E, Griffin DG, Ward GS, Bernstein ND, Bean TW, Ball MR Sr, Brady JA, Wilder MH, Kaplan JE (1990) B virus (Herpesvirus simiae) infection in humans: epidemiologic investigation of a cluster. Ann Intern Med 112:833–839

Hutt R, Guajardo JE, Kalter SS (1981) Detection of antibodies to Herpesvirus simiae and Herpesvirus hominis in nonhuman primates. Lab Anim Sci 31:184–189

Kalter SS, Heberling RL (1989) B virus infection of primates in perspective. Lab Anim 18/3:31–34

Kaplan JE (1988) Herpesvirus simiae (B virus) infection in monkey handlers. J Infect Dis 157:1090

Keeble SA (1960) B virus infection in monkeys. Ann NY Acad Sci 85:960–969

Keeble SA, Christofinis GJ, Wood W (1958) Natural virus-B infection in rhesus monkeys. J Pathol Bacteriol 76:189–199

Kessler MJ, Hilliard JK (1990) Seroprevalence of B virus (Herpesvirus simiae) antibodies in a naturally formed group of rhesus macaques. J Med Primatol 19:155–160

McClure HM, Olberding B, Strozier LM (1973) Disseminated herpesvirus infection in a rhesus monkey (Macaca mulatta). J Med Primatol 2:190–194

Palmer AE (1987) B virus, Herpesvirus simiae: historical perspective. J Med Primatol 16:99–130

Sabin AB, Wright AM (1934) Acute ascending myelitis following a monkey bite, with isolation of a virus capable of reproducing the disease. J Exp Med 59: 115–136

Vizoso AD (1975) Recovery of herpesvirus simiae (B virus) from both primary and latent infections in rhesus monkeys. Br J Exp Pathol 56:485–488

Weigler BJ, Roberts JA, Hird DW, Lerche NW, Hilliard JK (1990) A cross sectional survey for B virus antibody in a colony of group housed rhesus macaques. Lab Anim Sci 40:257–261

Zeitlyonok NA, Chumakova MY, Ralph NM, Goen LS (1966) Distribution area and natural hosts of latent viruses of monkeys. Occurrence of simian vacuolating virus (SV40) and herpesvirus simiae in cynomolgus monkeys in Indonesia. Acta Virol 10:537–541

Zwartouw HT, Boulter EA (1984) Excretion of B virus in monkeys and evidence of genital infection. Lab Anim 18:65–70

Zwartouw HT, MacArthur JA, Boulter EA, Seamer JH, Marston JH, Chamove AS (1984) Transmission of B virus infection between monkeys especially in relation to breeding colonies. Lab Anim 18:125–130

Herpesvirus simplex Infection

Ronald D. Hunt

Synonyms. Herpesvirus hominis infection.

Gross Appearance

Owl Monkeys

The most characteristic gross findings in natural and experimental *Herpesvirus simplex* infection of owl monkeys (*Aotus trivirgatus*) are discrete vesicles, necrotic plaques, and erosions or ulcers on the oral mucous membranes and at the mucocutaneous junction (Fig. 79) of the lips (Meléndez et al. 1969). These are usually gray to red and 1–10 mm in irregular diameter but may affect the entire oral cavity, extending into the pharynx, esophagus, and trachea. The lesions may extend to the adjacent skin and may also appear on the external genitalia. Conjunctivitis is also frequently present (Fig. 80). Internally, gray to red (depending on the amount of blood present in the lesions) foci are present on the capsular and cut surfaces of most organs (i.e., liver, adrenals, spleen, lung, lymph nodes, and occasionally the kidneys) (Fig. 81) (Meléndez et al. 1969; Katzin et al. 1967). Similar gross lesions are seen in naturally infected tree shrews (*Tupaia glis*) (McClure et al. 1972), lemurs (Kemp et al. 1972), and most experimentally infected marmoset and tamarin species (*Callithrix* sp., *Saguinus* sp.) (Hunt and Meléndez 1969; Hunt et al. 1978).

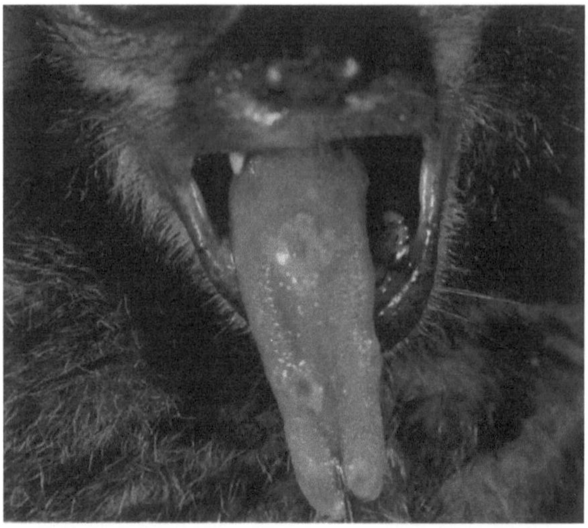

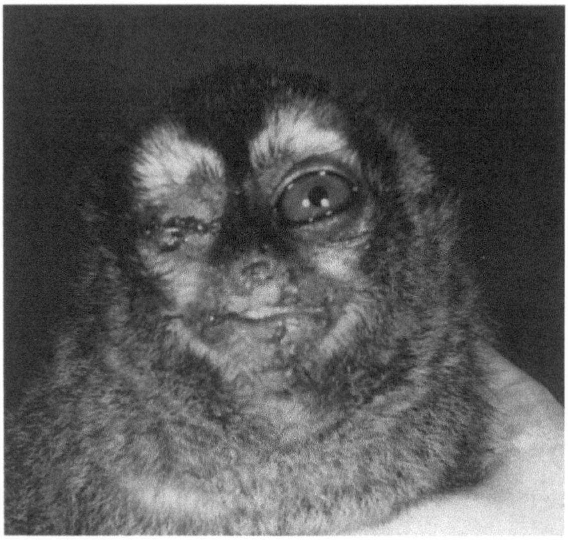

Fig. 79 (*above*). Necrotic plaques on the tongue of an owl monkey which died of *H. simplex* infection

Fig. 80 (*below*). Conjunctivitis and blepharitis caused by infection leading to closure of the eye of an owl monkey

Gibbons, Chimpanzees, and Cebus Monkeys

Natural infection in gibbons (*Hylobates lar*) (Smith et al. 1958, 1969; Emmons and Lennette 1970) and chimpanzees (*Pan* sp.) (McClure et al. 1980) and experimental infection in cebus monkeys (*Cebus* sp.) (Nahmias et al. 1971) with *H. simplex* usually does not progress to a systemic infection: gross lesions are restricted to vesicles; erosions or ulcers in the oral cavity, skin, or external genitalia; and conjunctivitis.

Microscopic Features

Histopathological findings are similar in all species but vary in severity. In the squamous mucous membranes there is increased eosinophilia of affected cells, which become ballooned and then necrotic with karyolytic or karyorrhectic nuclei. Accompanied by intercellular edema the lesions may progress to vesicles, but more usually are seen as necrotic plaques extending from the surface to the submucosa (Fig. 82). Vesicles are more likely to be seen in the skin with its thicker layer of cornified epithelium in contrast to the

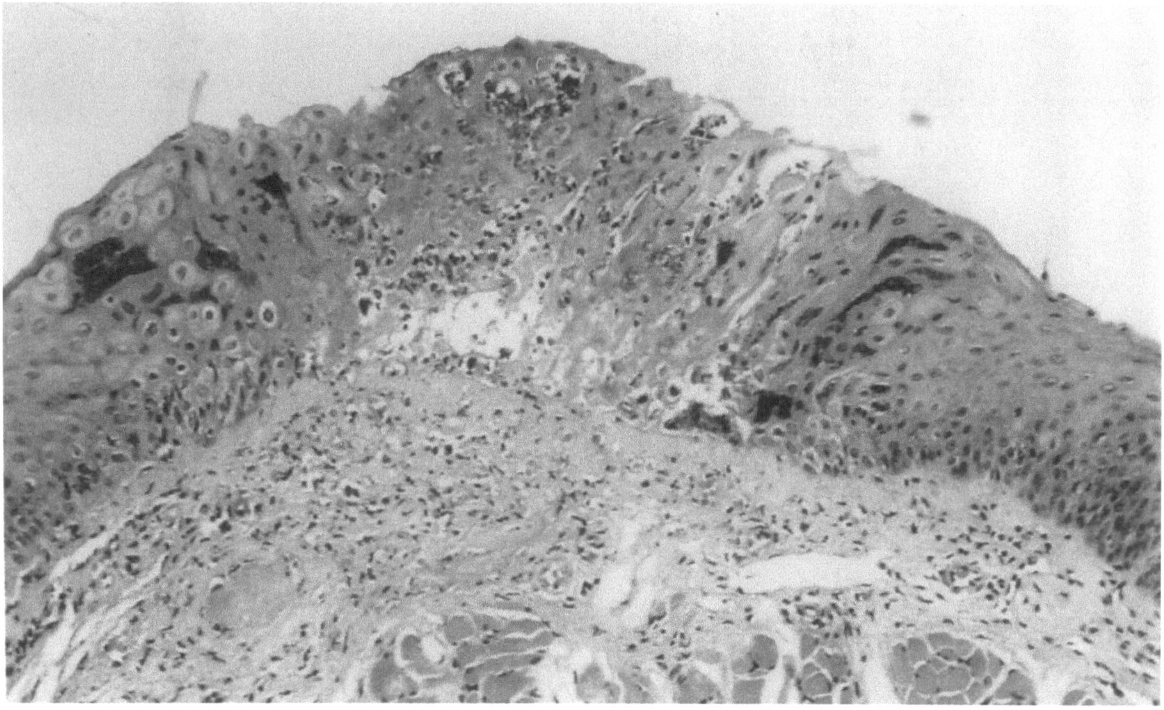

Fig. 81. Necrotic plaque on the tongue caused by *H. simplex* in an owl monkey. Multinucleated syncytial cells are present in the mucosa adjacent to the lesion. H and E, ×100

oral mucous membranes. The necrotic tissue often sloughs, leaving ulcers. Adjacent to the necrotic foci admixed with the ballooned epithelial cells are multinucleated syncytial and epithelial cells, each containing intranuclear inclusion bodies (Fig. 82). These inclusions may either completely fill the nucleus or be surrounded by a clear halo. They are amphoteric to eosinophilic. The mucosa and submucosa are infiltrated with neutrophils and mononuclear cells. Lesions may extend into the submucosa with necrosis of fibroblasts and blood vessels. Inclusion bodies may be present in mononuclear cells and endothelial cells. Necrotic lesions in the stomach through the colon extend into the submucosa and are accompanied by intense inflammatory infiltration. Syncytial cells are rarely seen in these lesions. In the respiratory tract foci of necrosis occur in the epithelium of the trachea, bronchi, and bronchioles and the alveolar walls. In solid tissues such as the liver, adrenals, and lymph nodes, necrotic foci of varying size are randomly distributed and often contain hemorrhage. Intranuclear inclusion bodies are numerous but syncytial cells are rare except in tubular structures such as the bile ducts (Fig. 83).

Encephalitis is sometimes seen in systemically infected owl monkeys and, although not the rule, fatal encephalitis may also develop in gibbons. The most striking feature is necrosis of neurons, particularly of the temporal cerebral cortex and the thalamus. There is increased cytoplasmic eosinophilia, rounding of the neurons, and karyorrhexis of the nuclei. Other neurons contain intranuclear inclusion bodies, which may also be present in astrocytes (Fig. 84). There is minimal perivascular cuffing, but the endothelium of blood vessels may be hypertrophic and hyperplastic.

Differential Diagnosis

H. simplex infection in owl monkeys, marmosets, and tamarins cannot be differentiated from *Herpesvirus T* infection (Hunt and Meléndez 1966, 1969, 1972; Meléndez and Hunt 1966). Both viruses lead to systemic infection with identical gross and microscopic lesions. Encephalitis has not been described in *Herpesvirus T* infection but this would not be a reliable criterion on which to base the differentiation of the two diseases.

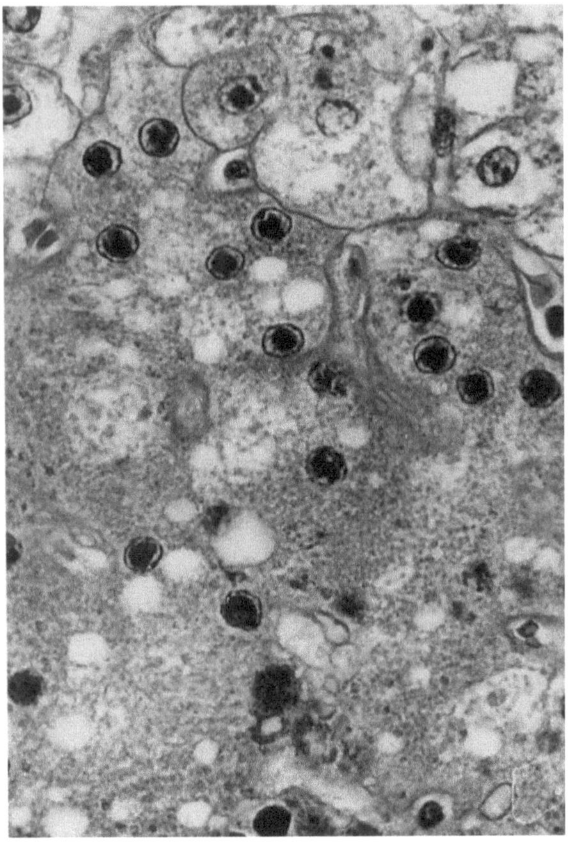

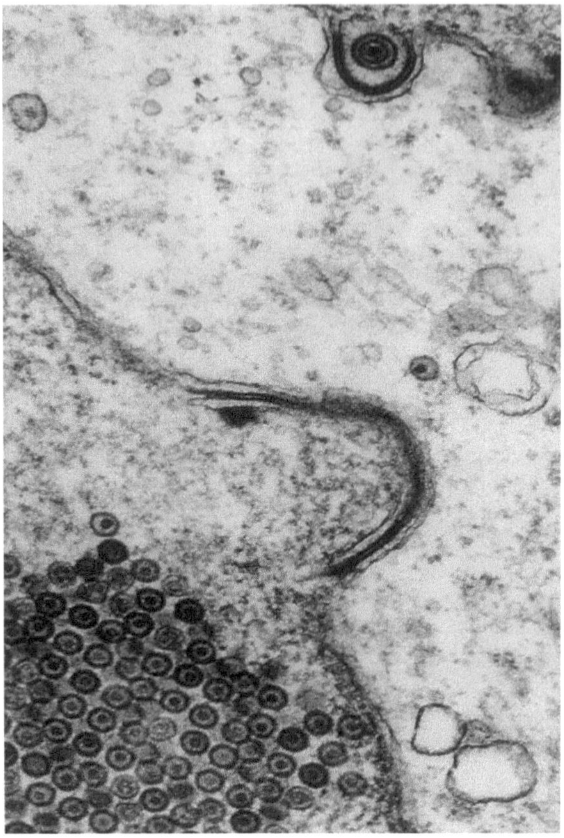

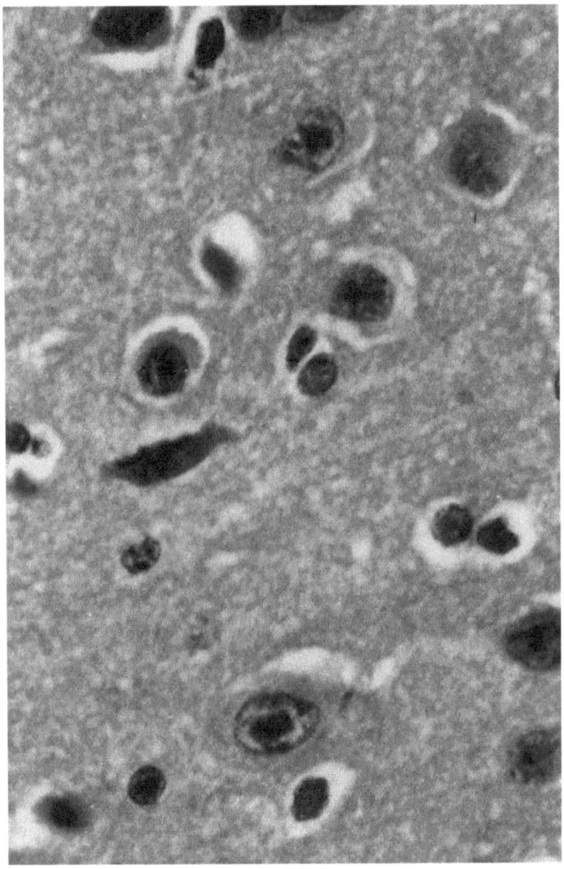

Fig. 82 (*upper left*). Intranuclear inclusion bodies in hepatocytes of an owl monkey which died of *H. simplex* infection. H and E, ×300

Fig. 83 (*lower left*). Intranuclear inclusion bodies in neurons of an owl monkey which died of *H. simplex* infection. H and E, ×350

Fig. 84 (*upper right*). An aggregate of *Herpesvirus simplex* virions in a nucleus of a multinucleated giant cell. A single virion has acquired an envelope from the nuclear membrane. TEM, ×37 500

In gibbons, chimpanzees, and cebus monkeys, in which the lesions are usually confined and self-limiting, the microscopic findings are not duplicated by other viral infections. Although distinct α-herpesviruses indigenous to these species have not been described, they may well exist and morphology alone would not allow differentiation. Chimpanzees, gorillas, and orangutans are susceptible to *H. zoster* (varicella) and it is likely that gibbons are as well (Heuschele 1960; Marennikova et al. 1973; McClure and Keeling 1971; White et al. 1972). The lesions of varicella are generally more widely distributed but very similar in morphology. Squirrel monkeys have not been reported to be susceptible to *H. simplex*, but do carry their own herpesvirus, *Herpesvirus T* (*H. platyrrhini*), in an analogous relationship to *H. simplex* in humans (Daniel et al. 1967; King et al. 1967).

Susceptibility to other cytocidal primate herpesviruses is possible in any species and the lesions could resemble those of *H. simplex*. Definitive diagnosis can be made through virus isolation and identification or by immune staining techniques.

Biologic Features

Natural History and Pathogenesis

H. simplex infection has been recognized clinically and pathologically in humans for centuries (Corey and Spear 1986). Humans are the natural or reservoir host for the virus. Most individuals become infected with the virus as children or adolescents such that by adulthood 90%–95% of the population have serum neutralizing antibodies. Primary infection may go unnoticed or be characterized by localized vesicular lesions most usually in the mouth or at the mucocutaneous junction with the lips. These ordinarily heal completely in about 2 weeks; however, the virus is transported by way of axons and establishes a lifelong latency in ganglia, most usually the trigeminal. Many people, despite the presence of neutralizing antibodies, suffer from periodic recurrence of lesions, secondary *H. simplex*, for much of their lives, often with several episodes each year. Recurrence for no obvious reason or following a variety of stimuli including fever (fever blister), colds (cold sore), menstruation, stress, and fatigue may occur. These lesions appear at approximately the same site as the primary infection and resolve in about 10 days. Neuralgia and hyperesthesia often precede the lesions and may persist for variable periods after healing.

In neonates primary infection can lead to fatal generalized disease with encephalomyelitis (Hunt 1986). More severe disease and encephalomyelitis may also occur in adults, usually in the presence of other disorders, especially those leading to immunosuppression.

Owl monkeys most likely become infected by humans, who can excrete the virus in the absence of visible lesions. Once present in a colony, the disease spreads rapidly with very high morbidity and mortality. It has not been determined if recovered animals remain latently infected to serve as a future source of infection.

Etiology

H. simplex is a member of the family Herpesviridae (Wildy 1986). It is a double-stranded DNA enveloped virus, with an icosahedral capsid containing 162 capsomeres. There are two serotypes, *Herpes simplex* virus type 1 (HSV-1) and *Herpes simplex* virus type 2 (HSV-2). Oral and conjunctival infection and encephalomyelitis in adults are most usually caused by HSV-1, whereas genital infection and neonatal infection are most usually caused by HSV-2. Nonhuman primates are susceptible to both serotypes. No differences between the ensuing diseases have been identified.

Frequency

Although in owl monkeys outbreaks of *H. simplex* infection can be devastating, *H. simplex* infection is not common in any of the primate species.

Prevention

The use of appropriate protective clothing and face masks and preventing individuals with overt *H. simplex* from handling nonhuman primates should greatly reduce risk. A live virus vaccine has proven very effective in owl monkeys (Daniel et al. 1978).

Comparison with Other Species

Several other cytocidal herpesviruses affect nonhuman primates causing similar clinical and pathological diseases. Additionally, a number of cytocidal herpesviruses have been recovered from clinically normal nonhuman primates. A comparison of viruses and species affected is given in Table 4.

Acknowledgement. This work was supported by Public Health Service Award RR00168 from the National Institutes of Health.

References

Corey L, Spear PG (1986) Infections with herpes simplex viruses (parts I and II). N Engl J Med 314:686-691, 749-757

Daniel MD, Karpas A, Meléndez LV, King NW, Hunt RD (1967) Isolation of Herpes-T virus from a spontaneous disease in squirrel monkeys (Saimiri sciureus). Arch Gesamte Virusforsch 22:324-331

Daniel MD, Barahona H, Meléndez LV, Hunt RD, Sehgal P, Marshall B, Ingalls J, Forbes M (1978) Prevention of fatal herpes infections in owl and marmoset monkeys by vaccination. In: Chivers DJ, Ford EHR (eds) Recent advances in primatology: 4. Medicine. Academic, New York, pp 67-69

Emmons RW, Lennette EH (1970) Natural Herpesvirus hominis infection of a gibbon (Hylobates lar). Arch Gesamte Virusforsch 31:215-218

Heuschele WP (1960) Varicella (chicken pox) in three young anthropoid apes. J Am Vet Med Assoc 136: 256-257

Hunt RD (1986) Viral diseases of neonatal and infant nonhuman primates. In: Benirschke K (ed) Primates. The road to self-sustaining populations. Springer, Berlin Heidelberg New York, pp 725-742

Hunt RD, Meléndez LV (1966) Spontaneous Herpes-T infection in the owl monkey (Aotus trivirgatus). Pathol Vet 3:1-26

Hunt RD, Meléndez LV (1969) Herpes virus infections of non-human primates: a review. Lab Anim Care 19: 221-234

Hunt RD, Meléndez LV (1972) Clinical, epidemiologic, and pathologic features of cytocidal and oncogenic herpesviruses in South American monkeys. J Natl Cancer Inst 49:261-271

Hunt RD, Anderson MP, Chalifoux LV (1978) Spontaneous infectious diseases of marmosets. Primates Med 10:239-253

Katzin DS, Connor JD, Wilson LA, Sexton RS (1967) Experimental Herpes simplex infection in the owl monkey. Proc Soc Exp Biol Med 125:391-398

Kemp GE, Losos GL, Causey OR, Emmons RW, Golding RR (1972) Isolation of Herpesvirus hominis from lemurs: a naturally occurring epizootic at a zoological garden in Nigeria. Afr J Med Sci 3:177-185

King NW, Hunt RD, Daniel MD, Meléndez LV (1967) Overt Herpes-T infection in squirrel monkeys (Saimiri sciureus). Lab Anim Care 17:413-423

Marennikova SS, Maltseva NN, Shelukhina EM, Shenkman LS, Korneeva VI (1973) A generalized herpetic infection simulating smallpox in a gorilla. Intervirology 2:280-286

McClure HM, Keeling ME (1971) Viral diseases noted in the Yerkes Primate Center colony. Lab Anim Sci 21: 1002-1010

McClure HM, Keeling ME, Olberding B, Hunt RD, Meléndez LV (1972) Natural Herpesvirus hominis infection of tree shrews (Tupaia glis). Lab Anim Sci 22:517-521

McClure HM, Swenson RB, Kalter SS, Lester TL (1980) Natural genital Herpesvirus hominis infection in chimpanzees (Pan troglodytes and Pan paniscus). Lab Anim Sci 30:895-901

Meléndez LV, Hunt RD (1966) Current views on Herpes T infection in South American monkeys. Vet Med Zootechniques 2:930-936

Meléndez LV, España C, Hunt RD, Daniel MD, Garcia FG (1969) Natural Herpes simplex infection in the owl monkey (Aotus trivirgatus). Lab Anim Care 19:38-45

Nahmias AJ, London WT, Catalano LW, Fuccillo DA, Sever JL, Graham C (1971) Genital Herpesvirus hominis type 2 infection: an experimental model in cebus monkeys. Science 171:297-298

Smith PC, Yuill TM, Buchanan RD (1958) Natural and experimental infection of gibbons with Herpesvirus hominis. Annual progress report of the SEATO medical research laboratory and SEATO clinical research centre, Bangkok, pp 258-261

Smith PC, Yuill TM, Buchanan RD, Stanton JS, Chaicumpa V (1969) The gibbon (Hylobates lar); a new primate host for Herpesvirus hominis: I. A natural epizootic in a laboratory colony. J Infect Dis 120:292-297

White RJ, Simmons L, Wilson RB (1972) Chickenpox in young anthropoid apes: clinical and laboratory findings. J Am Vet Med Assoc 161:690-692

Wildy P (1986) Herpesvirus. Intervirology 25:117-140

Herpesvirus saimiri and *Herpesvirus ateles* Infection

Ronald D. Hunt and Beverly J. Blake

Synonyms. Herpesvirus saimiri lymphoma; *Herpesvirus ateles* lymphoma; herpesvirus-induced lymphoproliferative disease.

Gross Appearance

Experimental injection of *Herpesvirus saimiri* into owl monkeys (*Aotus trivirgatus*), several species of tamarins and marmosets (*Saguinus oedipus, S. fuscicollis, S. nigricollis, S. mystax*, and *Callithrix jaccuhus*), howler monkeys (*Alouatta caraya*), and spider monkeys (*Ateles geoffroyi*) results in lymphoma or lymphocytic leukemia (Cicmanec et al. 1974; Deinhardt 1975; Dienhardt et al. 1974; Fleckenstein and Desrosiers 1982; Hunt et al. 1972a,b, 1976; Hunt 1978; Rangan et al. 1977; Meléndez et al. 1971b, 1972; Wright et al. 1977; Wolfe et al. 1971). Lymphoproliferative disease has been described in cebus monkeys (*C. albifrons*) but it has not been reproduced (Meléndez et al. 1970).

Gross findings may be minimal. On the capsular and cut surfaces of the liver there is a gray and red reticular pattern and, occasionally, small multifocal nodules 1–2 mm in diameter. The spleen is generally enlarged (Fig. 85), often up to ten times normal, and may contain irregular light red to tan nodules 3–5 mm in diameter on the capsular and cut surfaces. Lymph nodes are enlarged, firm, and gray. In young animals the thymus is enlarged and often contains hemorrhages. The kidneys are swollen, soft, and friable and their capsular surface is mottled yellow, gray, tan, and red. The cut surface of the cortex bulges and is dull and mottled. The adrenal glands are almost always enlarged and uniformly tan to yellow with no distinct demarcation between cortex and medulla. Hemorrhage may be presennt. Less frequently, solitary tumors 1–10 cm in irregular diameter may be present; however, these are rare and have been observed in the kidney, orbit, (Fig. 86) peritoneum, and abdominal cavity.

Experimental *H. ateles* infection in *S. oedipus, S. fuscicollis, S. nigricollis, and Aotus trivirgatus* presents the same gross findings. Other species have not been tested.

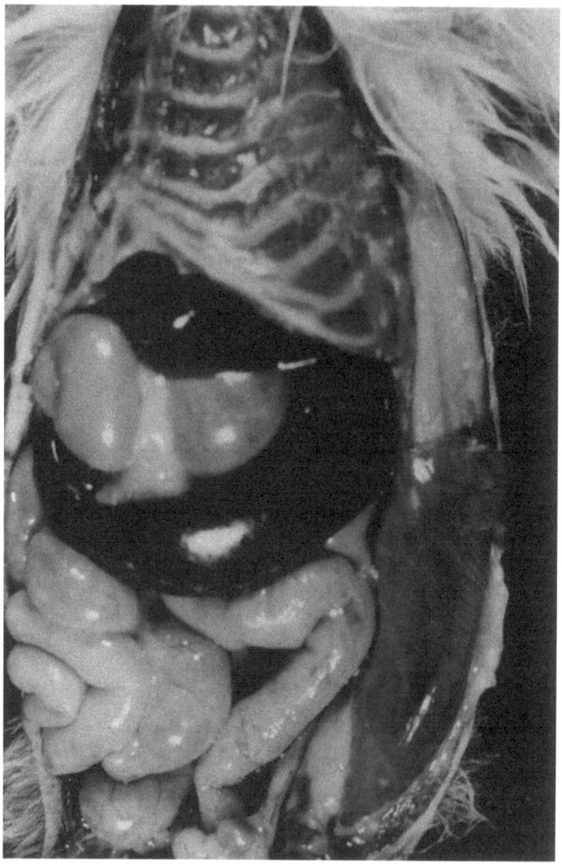

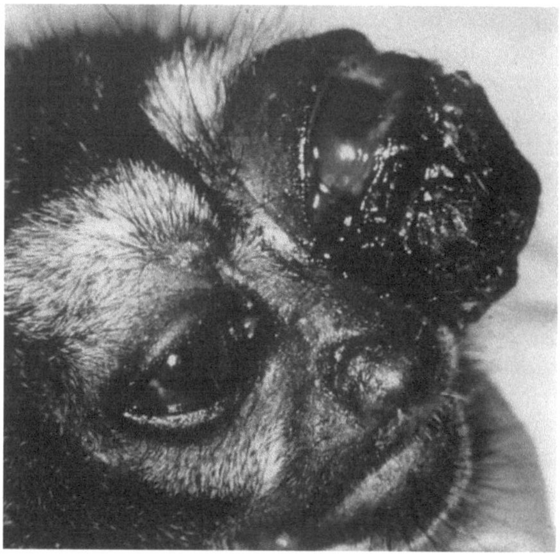

Fig. 85 (*above*). *Herpesvirus saimiri* lymphoma: severe splenomegaly in a cotton-top tarmarin (*Saguinus oedipus*)

Fig. 86 (*below*). *Herpesvirus saimiri* lymphoma: a large orbital lymphoma in an owl monkey (*Aotus trivirgatus*)

Microscopic Features

Cellular infiltration and organ replacement is the basic histopathologic lesion in all affected tissues of all susceptible species (Hunt et al. 1970, 1972c) (Figs. 87–89). The lesions produced by *H. saimiri* and *H. ateles* are essentially identical. As expected from the gross findings, the infiltrates occasionally form discrete nodules but more often exist as sheets of cells intervening between normal structures. The cells are about 12 µm in diameter but vary from 7 to 20 µm and have scant eosinophilic cytoplasm. Nuclei are pleomorphic, most often large, oval, and leptochromatic but sometimes round, indented, or angular. Most cells contain two large distinct nucleoli. Mitotic figures are frequent. Necrotic cells with pyknotic or karyorrhectic nuclei are frequently present. Small lymphocytes commonly are interspersed, and occasional eosinophils and plasma cells are present such that the infiltrate resembles an inflammatory process. In addition to those tissues with gross lesions, cellular infiltrates may be found in almost any tissue or organ including the bone marrow and central nervous system. In the liver, the infiltrate is most extensive in the periportal tissue, with finger-like extensions into the hepatic lobules (Fig. 87). Hepatocytes adjacent to the invasion front are necrotic (Fig. 88). In the kidney, the infiltrate is restricted to the cortex, which may be almost entirely replaced. Although variable, the cytoarchitecture of lymph nodes can be entirely effaced with extension of the infiltrate through the capsule into perinodal tissues. There is necrosis of malpighian follicles in the spleen, congestion of sinusoids, and variable degrees of replacement of cords of Bilroth and sinusoids by infiltrating cells. There is cellular infiltration of both the adrenal cortex and medulla. In other tissues, such as the testes, salivary gland, prostate, lung, bone marrow, meninges and choroid plexus of the brain, mesentery, lamina propria of the gastrointestinal tract, the infiltrate is generally less extensive but invariably present.

Leukemia may or may not be present. It has been most consistently observed in owl monkeys in whom it occurs in about half of the animals inoculated with *H. saimiri*. Total peripheral lymphocyte counts range from about 30000 to 175000/dl, with many prolymphocytes and lymphoblasts in circulation (Cicmanec et al. 1974; Hunt et al. 1976). The invading cells and leukemic cells are T lymphocytes (Fleckenstein and Desrosiers 1982).

Ultrastructure

The cells comprising the infiltrates do not have unique ultrastructural characteristics. Most cells are round to oval, but there is extreme pleomorphism resulting from apparent ameboid invasiveness. Their nuclei are round to oval and occasionally indented. Dense-staining heterochromatin tends to be clumped near the nuclear membrane or adjacent to the nucleolus. Moderately dense granules of euchromatin are widely dispersed throughout the nucleus. Nucleoli are large, dense, and near the nuclear membrane. The cytoplasm of the majority of cells contains numerous free and polyribosomes, large swollen mitochondria, and sparse rough endoplasmic reticulum. The Golgi apparatus is poorly developed. These cells are considered lymphoblasts. Other cells are present with fewer free ribosomes but a large complement of polyribosomes and a more developed rough endoplasmic reticulum. These have been interpreted as reticulum cells or primitive stem cells (Fig. 90). No viral particles are evident in tissues but are seen in cultures of tissues in vitro (Fig. 91) (King and Meléndez 1972).

Differential Diagnosis

As an experimental disease, differential diagnosis is usually not of concern. Recovery of either virus from spontaneous lymphomas or leukemia in owl monkeys, tamarins, or other species in which the viruses cause experimental disease is reasonably good evidence for a causal relationship, as these species do not carry these viruses as latent infections. In squirrel monkeys or spider monkeys a causal relationship is much more difficult to establish since, as the natural hosts for the viruses, a high percentage of

Fig. 87 (*upper left*). *Herpesvirus saimiri* lymphoma: ▶ periportal infiltration of lymphoblasts in the liver of a cotton-top tamarin (*Saguinus oedipus*). H and E, ×150

Fig. 88 (*upper right*). *Herpesvirus saimiri* lymphoma: hepatic infiltrate of large cells resembling lymphoblasts and reticulum cells in a cotton-top tamarin (*Sanguinus oedipus*). H and E, ×400

Fig. 89 (*below*). Effacement of a lymph node with a pleomorphic infiltrate in a cotton-top tamarin inoculated with *H. saimiri*. Most of the cells are immature lymphoblasts but more mature ells are also present. H and E, ×400

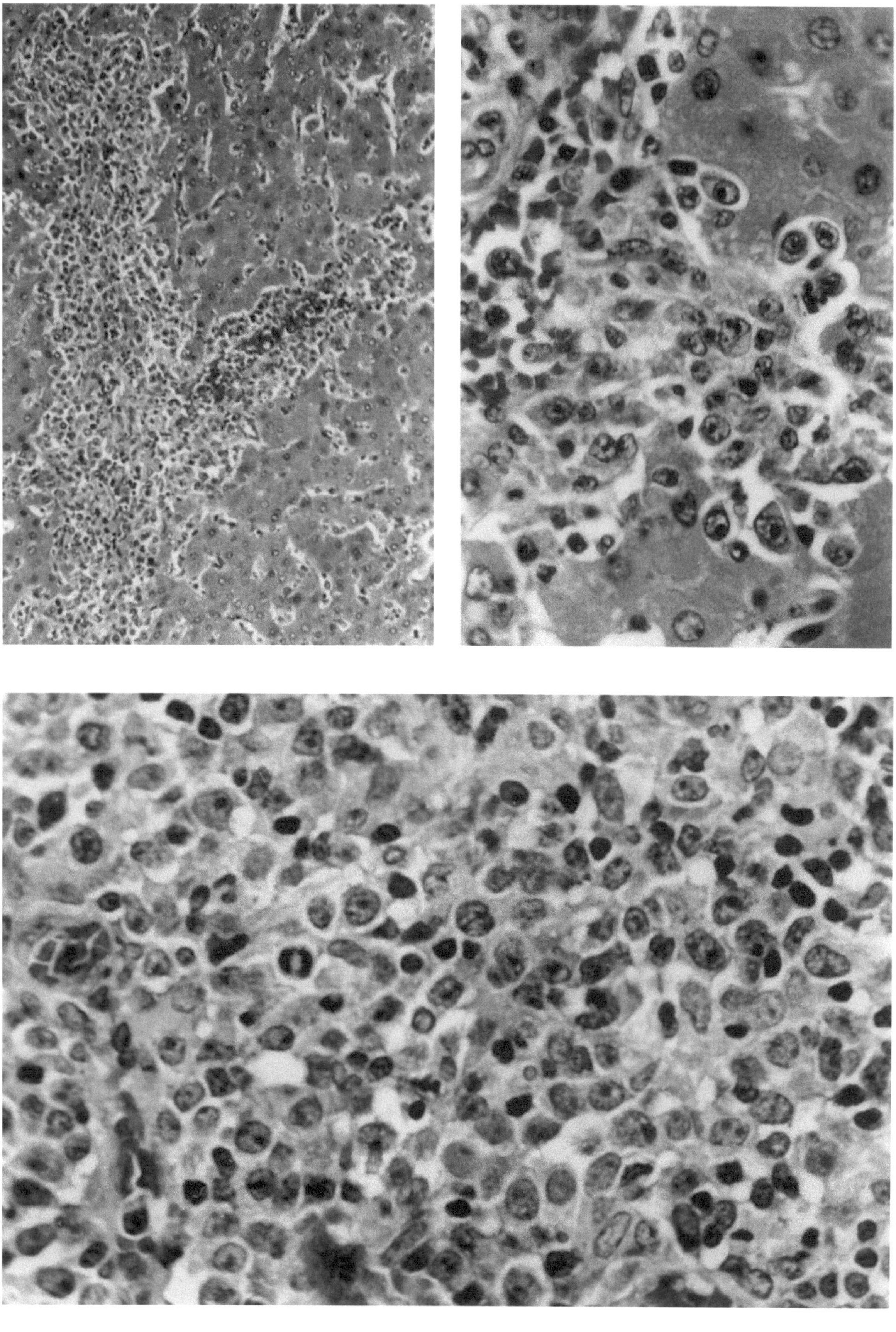

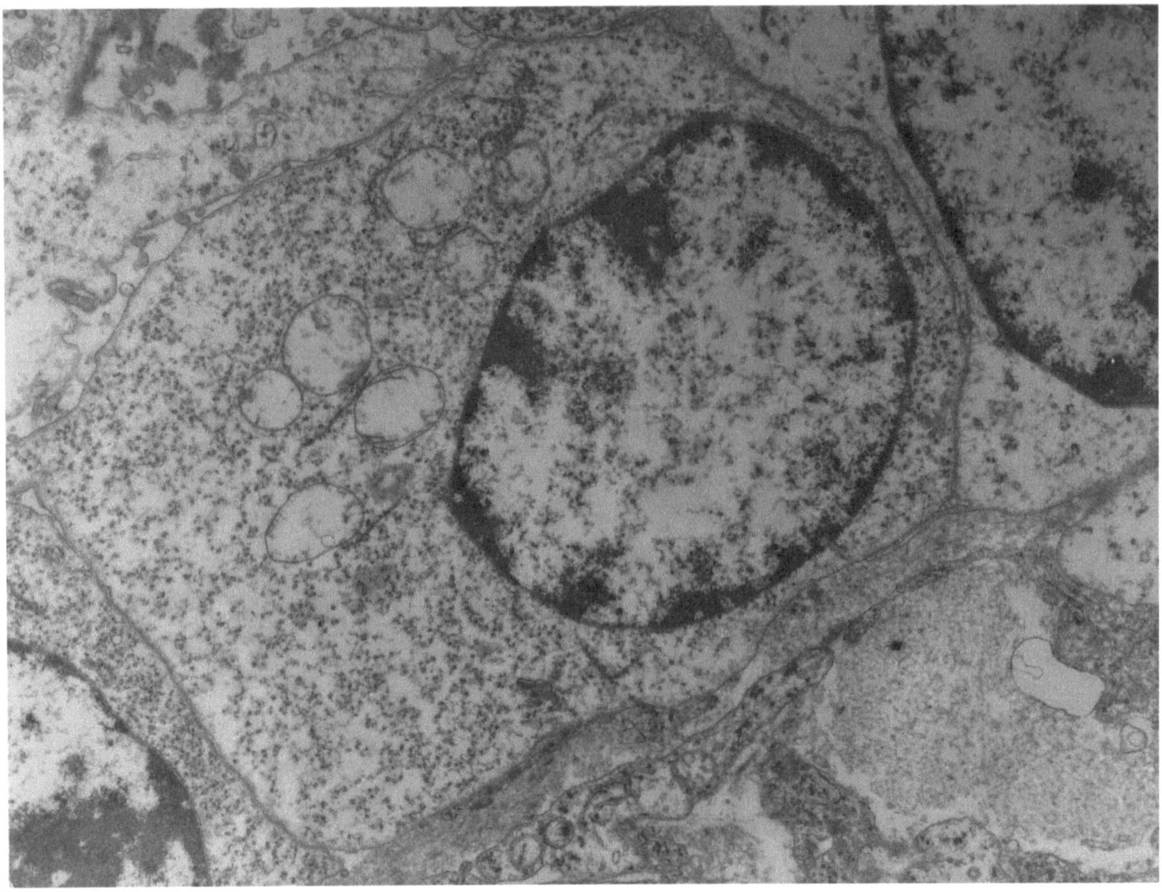

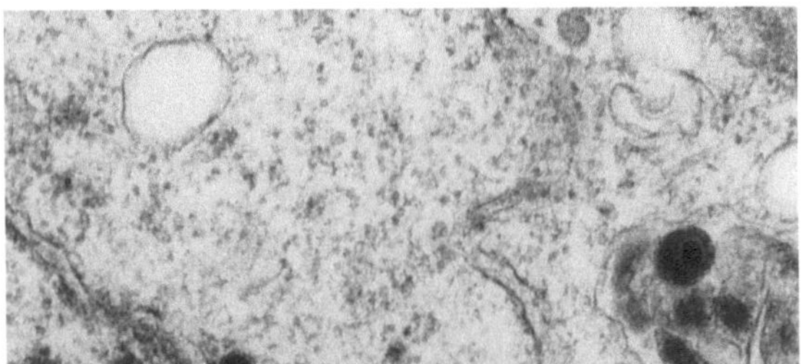

Fig. 90 (*above*). *Herpesvirus saimiri* lymphoma: reticulum cell in the liver of a cotton-top tamarin (*Saguinus oedipus*). Note absence of distinctive morphologic features and no evidence of virions. TEM, ×5130

Fig. 91 (*left*). *Herpesvirus saimiri* in owl monkey kidney tissue culture in which infection is productive. Immature virions are evident in the nucleus and a single viron is seen in the process of acquiring its envelope from the nuclear membrane. TEM, ×37500 normal animals carry the virus as a latent infection of lymphocytes.

Biologic Features

Natural History and Pathogenesis

H. saimiri is indigenous to squirrel monkeys and *H. ateles* is indigenous to spider monkeys. A high percentage of each species is naturally infected, the incidence increasing with age. No disease has been associated with infection in these species. The virus is latent in T lymphocytes and can be recovered from blood and, in the case of squirrel monkeys, from oropharyngeal secretions. Transmission is horizontal (Barahona et al. 1975; Meléndez et al. 1971a,c, 1972; Falk et al. 1973; Deinhardt et al. 1974; Hunt 1978; Fleckenstein and Desrosiers 1982). One example of *H. saimiri* lymphoma in an owl monkey, presumably from contact with a squirrel monkey, has been reported (Hunt et al. 1973).

Following experimental infection both viruses establish themselves in T lymphocytes which are transformed, leading to lymphoma and/or leukemia (Hunt et al. 1970). In *Saguinus* sp. most infected animals develop disease and succumb in a relatively short interval regardless of age or size of the inoculum, whereas the incidence and survival time are more variable in other susceptible monkeys. Survival time for various species is summarized in Table 5. Interestingly, rabbits are susceptible to both viruses (Daniel et al. 1974, 1977; Hunt et al. 1975).

Etiology

H. saimiri and *H. ateles* are members of the family Herpesviridae and as such are double-stranded DNA, enveloped viruses with an icosahedral capsid containing 162 capsomeres. In tissue culture with either virus there is a productive infection leading to typical morphological features of a herpesvirus cytopathology. This includes syncytial cell formation, intranuclear inclusion bodies, and ultimately cell destruction. There is a spectrum of permissive cells for both viruses. A specific region of the genome has been reported to be required for oncogenicity (Desrosiers et al. 1985, 1986, 1987; Medveczky et al. 1984).

Table 5. Incidence and survival time in various species injected with *Herpesvirus saimiri* and *H. ateles*

Species	Survival time (days)	
	H. saimiri	*H. ateles*
Saguinus oedipus	18–126	14–40
Saguinus fuscicollis, nigricollis, mystax	19–116	26–28
Callithrix jacchus	22–42	36–104
Aotus trivirgatus	33–330	21–28
Ateles geoffroyi	129–179	ND
Cebus albifrons	18–20	ND
Alouatta caraya	36–52	ND
Saimiri sciureus	ND	ND
Rabbits	17–196	20–100

ND, no disease.

Frequency

Nearly all squirrel monkeys beyond 1.5–2 years of age possess antibodies to *H. saimiri* (Falk et al. 1973). The natural history of *H. ateles* has not been studied as extensively, but it has been reported that the incidence in spider monkeys exceeds 50% (Meléndez et al. 1971a,b,c; Falk et al. 1974).

Comparison with Other Species

As lymphotropic herpesviruses, *H. saimiri* and *H. ateles* share many biologic features with other viruses in this group, termed γ-herpesviruses. These include Epstein-Barr virus, Marek's disease virus, and *H. sylvilagus*, each of which is associated with lymphoproliferative diseases. Epstein-Barr virus is indigenous to humans, most of whom become infected and remain persistently infected for life (de-The 1982). This virus infects B lymphocytes, is excreted in the saliva, and transmitted horizontally. Up to 20% of asymptomatic individuals secrete virus in their saliva. Although most individuals do not develop overt disease the virus is the cause of infectious mononucleosis and a very strong causal relationship exists between Burkitt's lymphoma and nasopharyngeal carcinoma. Infectious mononucleosis is usually a nonfatal disease of adolescents and young adults. Proliferation of B lymphocytes occurs in this disease but much of the lymphadenopathy and cellular infiltration represents an immune response to infected cells by T lymphocytes.

Burkitt's lymphoma is a B cell lymphoma that in Africa classically is seen in children as a neoplasm of the jaw. Outside of Africa, Burkitt's lymphoma also principally affects children and young adults but is usually seen as an abdominal tumor. Lymph nodes are rarely affected. Microscopically, the tumor is characterized by a sea of large lymphoblasts with oval to round nuclei containing two or more nucleoli, a high mitotic index, and a large number of necrotic cells. The latter attract numerous macrophages giving rise to the term "starry sky" pattern. This may be seen also in other non-Burkitt type lymphomas. Nasopharyngeal carcinoma in human patients is a highly malignant, anaplastic, squamous cell carcinoma.

Cotton-top tarmarins are susceptible to Epstein-Barr virus and may develop lymphoma (Miller et al. 1977). Fatal lymphoproliferative disease has been described in *Callithrix jacchus* and there is evidence that owl monkeys are also susceptible (Epstein et al. 1973; Falk et al. 1976).

Marek's disease virus causes T cell lymphoma in chickens. *H. sylvilagus* is associated with lymphomas in wild cottontail rabbits. This infection has also been likened to infectious mononucleosis. Other herpesviruses which are members of the gamma or lymphotropic group have been recovered from other species of nonhuman primates including gorillas, orangutans, chimpanzees, baboons, and rhesus monkeys. These viruses have not been shown to cause disease but the rhesus monkey virus and baboon virus have been associated with lymphoma in these species.

Acknowledgement. This work was supported by Public Health Service Award RR00168 from the National Institutes of Health.

References

Barahona H, Meléndez LV, Hunt RD, Forbes M, Fraser CEO, Daniel MD (1975) Experimental horizontal transmission of Herpesvirus saimiri from squirrel monkeys to an owl monkey. J Infect Dis 132:694–697

Cicmanec JL, Loeb WF, Valerio MG (1974) Lymphoma in owl monkeys (Aotus trivirgatus) inoculated with Herpesvirus saimiri: clinical, hematologic and pathologic findings. J Med Primatol 3:8–17

Daniel MD, Meléndez LV, Hunt RD, King NW, Anver M, Fraser CEO, Barahona H, Baggs RB (1974) Herpesvirus saimiri: VII. Induction of malignant lymphoma in New Zealand white rabbits. J Natl Cancer Inst 53:1803–1807

Daniel MD, Hunt RD, Fleckenstein B, King NW, Ingalls JK (1977) The oncogenicity of Herpesvirus ateles in rabbits. In: Abstracts of the 3rd international symposium on oncogenesis and herpesviruses, Cambridge, p 213

Deinhardt F (1975) Virus induced lymphoproliferative disease in non-human primates. Br J Cancer [Suppl II]: 140–146

Deinhardt FW, Falk LA, Wolfe LG (1974) Simian herpesviruses and neoplasia. Adv Cancer Res 19:167–205

Desrosiers RC, Bakker A, Kamine J, Falk LA, Hunt RD, King NW (1985) A region of the Herpesvirus saimiri genome required for oncogenicity. Science 228: 184–187

Desrosiers RC, Silva DP, Waldron LM, Letvin NL (1986) Nononcogenic delation mutants of herpesvirus saimiri are defective for in vitro immortalization. J Virol 57: 701–705

Desrosiers RC, Murthy S, Trimble J (1987) A region of the Herpesvirus saimiri genome required for oncogenicity. In: Gallow RC, Haseltine W, Klein G, zur Hausen H (eds) Viruses and human cancer. Liss, New York, pp 309–317 (UCLA symposia on molecular and cellular biology, new series, vol 43)

De-The G (1982) Epidemiology of Epstein-Barr virus and associated diseases in man. In: Roizman B (ed) The herpesviruses, vol 1. Plenum, New York, pp 25–103

Epstein MA, Hunt RD, Rabin H (1973) Pilot experiments with EB virus in owl monkeys (Aotus trivirgatus): I. Reticuloproliferative disease in an inoculated animal. Int J Cancer 12:309–318

Falk LA, Nigida S, Deinhardt F, Cooper RW, Hernandez-Camacho JI (1973) Oral excretion of Herpesvirus saimiri in captive squirrel monkeys and incidence of infection of feral squirrel monkeys. J Natl Cancer Inst 51:1987–1989

Falk LA, Nigida S, Deinhardt F, Wolfe LG, Cooper RW, Hernandez-Camacho JI (1974) Herpesvirus ateles: properties of an oncogenic herpesvirus isolated from circulating lymphocytes of spider monkeys (Ateles sp). Int J Cancer 14:473–482

Falk LA, Dienhardt F, Wolfe LG, Johnson D, Hilgers J, de-The G (1976) Epstein-Barr virus: experimental infection of Callithrix jacchus marmosets. Int J Cancer 17:785–788

Fleckenstein B, Desrosiers RC (1982) Herpesvirus saimiri In: Roizman B (ed) The herpesviruses, vol 1. Plenum, New York, pp 253–332

Hunt RD (1978) Herpesvirus oncogenesis in New World monkeys: history of herpesvirus oncogenesis and overview of Herpesvirus saimiri. In: Chivers DJ, Ford EHR (eds) Recent advances of primatology. Academic, London, pp 117–124

Hunt RD, Melendez LV, King NW, Gilmore CE, Daniel MD, Williamson ME, Jones TC (1970) Morphology of a disease with features of malignant lymphoma in marmosets and owl monkeys inoculated with Herpesvirus saimiri. J Natl Cancer Inst 44:447–465

Hunt RD, Melendez LV, King NW, Garcia FG (1972a) Herpesvirus saimiri malignant lymphoma in spider monkeys: a new susceptible host. J Med Primatol 1: 114–128

Hunt RD, Melendez LV, Trum BF (1972b) Herpesvirus ateles lymphoma in cotton-topped marmosets. The virus, the disease, and evidence for horizontal transmission. In: Goldsmith EI, Moor-Jankowski J (eds) Medical primatology, part III. Karger, Basel, pp 159–162

Hunt RD, Melendez LV, Garcia FG, Trum BF (1972c) Pathologic features of Herpesvirus ateles lymphoma in cotton-topped marmosets (Saquinus oedipus). J Natl Cancer Inst 49:1631-1639

Hunt RD, Garcia FG, Barahona HH, King NW, Fraser CEO, Melendez LV (1973) Spontaneous Herpesvirus saimiri lymphoma in an own monkey. J Infect Dis 127:723-725

Hunt RD, Daniel MD, Baggs RB, Blake BJ, Silva D, DuBose D, Melendez LV (1975) Clinicopathologic characterization of Herpesvirus saimiri malignant lymphoma in New Zealand white rabbits. J Natl Cancer Inst 54:1401-1412

Hunt RD, Blake BJ, Daniel MD (1976) Herpesvirus saimiri lymphoma in owl monkeys (Aotus trivirgatus): susceptibility, latent period, hematologic picture and course. Theriogenology 6:139-151

King NW, Meléndez LV (1972) The ultrastructure of herpesvirus saimiri-induced lymphoma in cotton-topped marmosets (Saguinus oedipus). Lab Invest 26:682-693

Medveczky P, Szomolanyi E, Desrosiers RC, Mulder C (1984) Classification of herpesvirus saimiri into three groups based on extreme variation in a DNA region required for oncogenicity. J Virol 52:938-944

Meléndez LV, Hunt RD, Daniel MD, Fraser CEO, Garcia FG, Williamson ME (1970) Lethal reticuloproliferative disease induced in Cebus albifrons monkeys by Herpesvirus saimiri. Int J Cancer 6:431-435

Meléndez LV, Daniel MD, Barahona HH, Fraser CEO, Hunt RD, Garcia FG (1971a) New herpesviruses from South American monkeys. Preliminary report. Lab Anim Sci 21:1050-1054

Meléndez LV, Hunt RD, Daniel MD, Blake B, Garcia FG (1971b) Acute lymphocytic leukemia in own monkeys inoculated with Herpesvirus saimiri. Science 171:1161-1163

Meléndez LV, Hunt RD, King NW, Barahona HH, Daniel MD, Fraser CEO, Garcia FG (1971c) Herpesvirus ateles, a new lymphoma virus of monkeys. Nature New Biol (London) 235:182-184

Meléndez LV, Hunt RD, Daniel MD, Fraser CEO, Barhona HH, Garcia FG, King NW (1972) Lymphoma viruses of monkeys; Herpesvirus saimiri and herpesvirus ateles, the first oncogenic herpes-viruses of primates - a review. In: Biggs PM, de-The G, Payne LN, (eds) Oncogenesis and herpesviruses. IARC, Lyon, pp 451-461

Miller G, Shope T, Coope D, Waters L, Pagano J, Bornkamm GW, Henle W (1977) Lymphoma in a cotton-top marmosets after inoculation with Epstein-Barr virus: tumor incidence histologic spectrum antibody response, demonstration of vial DNA and characteristics of viruses. J Exp Med 145:948-967

Rangan SRS, Martin LN, Enright FM, Abee CR (1977) Herpesvirus saimiri-induced lymphoproliferative disease in howler monkeys. J Natl Cancer Inst 59: 165-171

Wolfe LG, Falk LA, Deinhardt F (1971) Oncogenicity of Herpesvirus saimiri in marmoset monkeys. J Natl Cancer Inst 47:1145-1162

Wright J, Falk LA, Wolfe LG, Ogden J, Deinhardt F (1977) Susceptibility of common marmosets (Callithrix jacchus) to oncogenic and attenuated strains of Herpesvirus saimiri. J Natl Cancer Inst 59:1475-1478

Simian Varicella

E. Donald Roberts

Synonyms. Delta herpes, Liverpool vervet virus, patas herpesvirus, Medical Lake macaque virus.

Gross Appearance

Simian varicella is an acute fatal disease of African green monkeys (*Cercopithecus aethiops*), patas monkeys (*Erythrocebus patas*), and pigtailed macaques (*Macaca nemestrina*) and is characterized by epidermal rash (Fig. 92), cutaneous vesicle formation, hemorrhagic ulceration throughout the digestive tract (Fig. 93), multifocal hemorrhagic necrosis of the liver, spleen, and lymph nodes, and focal necrosis of various endocrine organs. The lesion of the lung (Fig. 94) consists of severe pulmonary edema with large areas of necrosis and hemorrhage of the alveolar septae producing an "infarct type" necrotic lesion (Loomis et al. 1981; McClure et al. 1973; Roberts et al. 1984a).

Microscopic Appearance

Simian varicella affects most organ systems. Stratified squamous epithelium of the skin, tongue, and esophagus have focal areas of epidermal hyperplasia, ballooning degeneration, and blood-filled vesicle formation, many of which ulcerate. A characteristic eosinophilic intranuclear (herpes-type) (Fig. 95) inclusion of epidermal cells is evident. Most changes in viscera and lymphoid organs include necrosis (Fig. 96) and hemorrhage with intranuclear inclusions in many cells. Hepatic necrosis (Fig. 97) and hemorrhage are consistent visceral lesions producing large focal to coalescing areas

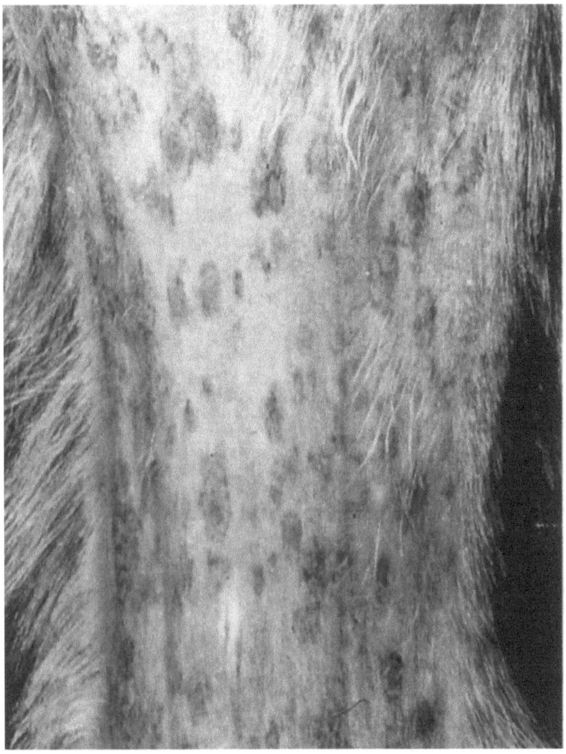

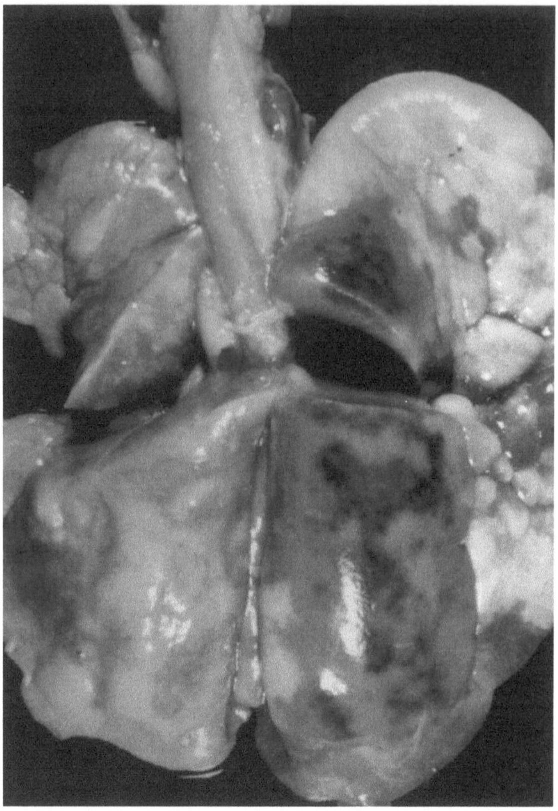

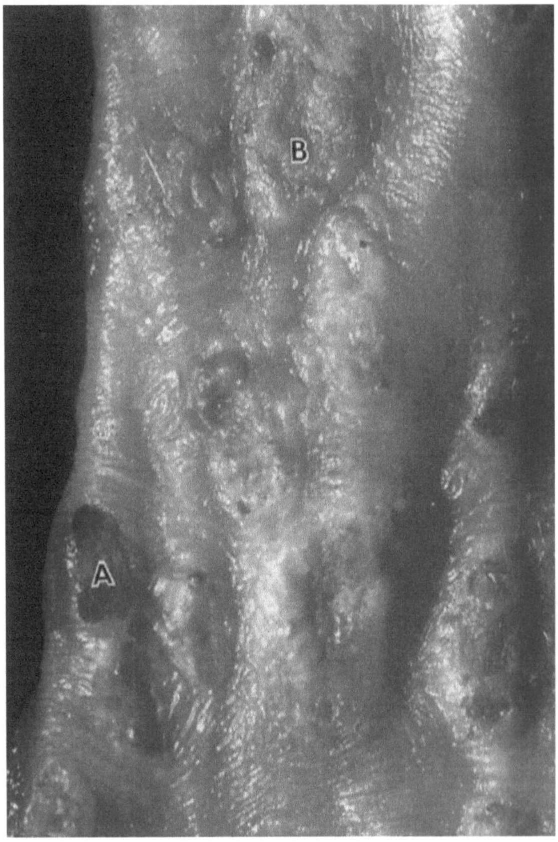

Fig. 92 (*upper left*). Abdomen of an African green monkey experimentally infected with delta herpesvirus, 10 days postinoculation. The erythematous plaques demonstrated may develop into blood-filled vesicles which result in cutaneous ulcers

Fig. 93 (*lower left*). Esophagus from an African green monkey experimentally infected with delta herpesvirus (10 days postinoculation). Lesions vary from hemorrhagic ulceration (*A*) to a partly healed ulcerative lesion (*B*)

Fig. 94 (*upper right*). Lung from an African green monkey experimentally infected with delta herpesvirus (11 days postinoculation). Note sharply demarcated lesions of the posterior and intermediate lobes of the lung

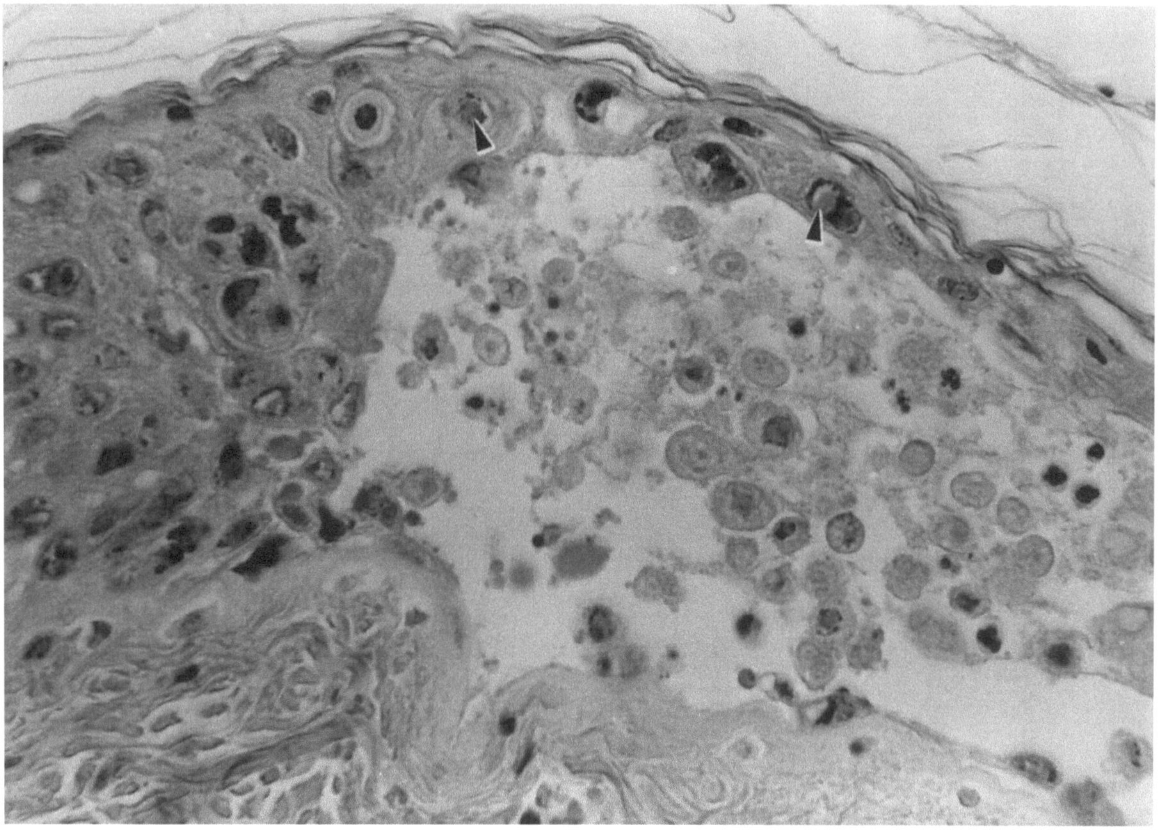

Fig. 95. Skin from African green monkey 11 days after being infected with delta herpesvirus. Note the vesicle formation with intranuclear inclusions of the adjacent epithelial cells (*arrowheads*). H and E, ×250

of necrosis with intranuclear inclusions evident in the adjacent hepatocytes. A generalized vascular involvement occurs with intranuclear inclusions formed in endothelial and smooth muscle nuclei. Vascular damage and leakage are particularly prominent in the lung where necrosis and hemorrhage of the alveolar septae occur (Fig. 98). Intranuclear inclusions are present in the alveolar walls of affected lung.

Ultrastructure

The ultrastructural lesion of the skin consists of vacuolation of epithelial cells with herpesvirus virions in nuclear, cytoplasmic, and extracellular spaces (Lehner et al. 1984). The changes affect both epithelial cells of the airway and endothelial cells of the alveolar septae (Fig. 99). Edema of the interalveolar septae with swelling of the capillary endothelial cells are evident with numerous virions in endothelial and epithelial cell nuclei (Roberts et al. 1984b).

Differential Diagnosis

Exanthematous disease resembling varicella among primates of Old World origin may be caused by one of at least three closely related herpesviruses. High morbidity and mortality rates are observed among African green (*Cercopithecus aethiops*), patas (*Erythrocebus patas*), and macaca species. Though severity of the lesions and their distribution may vary from animal to animal, all three viruses produce the same tissue changes. Characterization and identification of the virus is required for definitive diagnosis even though the presence of a herpes-type inclusion body with multiple organ involvement is diagnostic for the broad disease designation of simian varicella. In young macaques, B virus infection should be considered in a differential diagnosis.

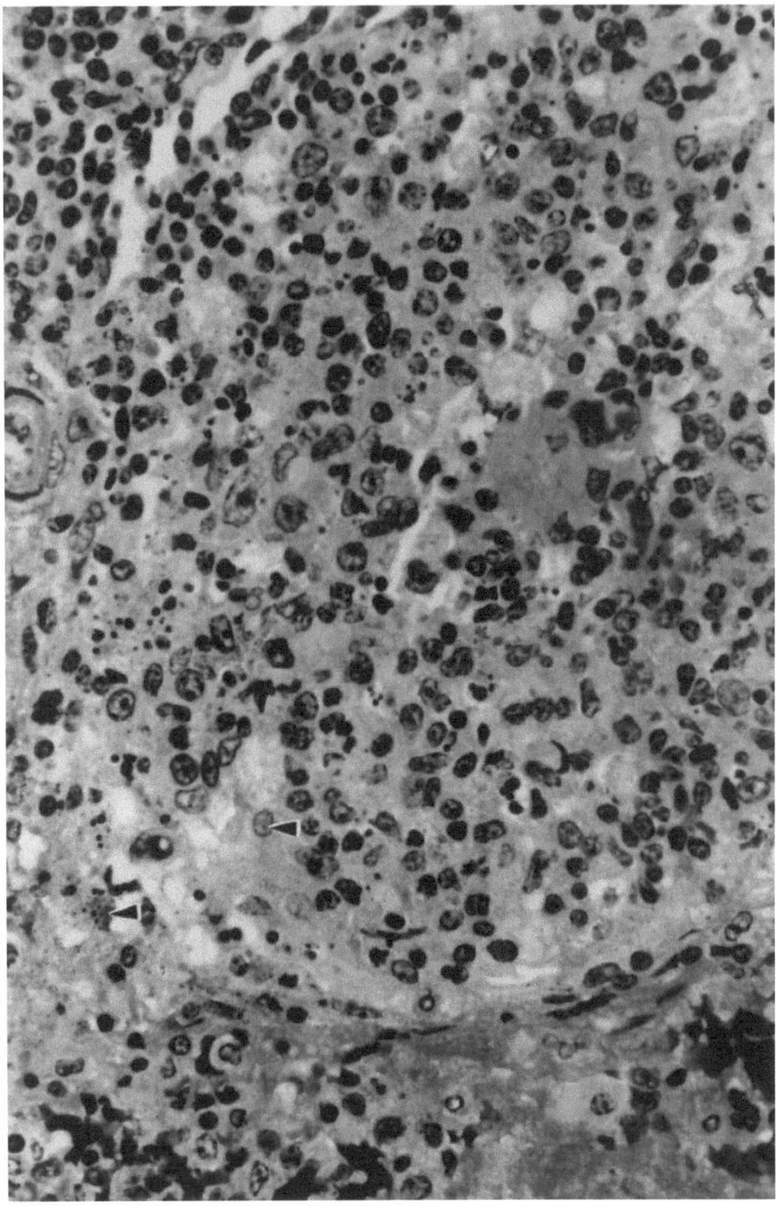

Fig. 96. Lymph node of an African green monkey infected 11 days earlier with delta herpesvirus. Note the necrosis of the germinal center, evidenced by cellular debris. Also note the intranuclear herpes-type inclusions (*arrowheads*). H and E, ×250

Biologic Features

Simian varicella has been reviewed (Hunt and Meléndez 1969; Daniel et al. 1972; Hull 1973; Barahona et al. 1974; Padovan and Cantrell 1986). Outbreaks of an often fatal exanthematous herpesvirus infection have been reported. The first outbreak occurred in 1966 at the Liverpool School of Tropical Medicine among African green monkeys and the agent was identified as Liverpool vervet virus (Clarkson et al. 1967). One year later in Great Britain a second herpesvirus epizootic with many fatalities occurred among patas monkeys and the causative agent was called the patas herpesvirus (McCarthy et al. 1968). Outbreaks occurred at the Delta Primate Research Center in patas caused by the δ-herpesvirus in 1968 (Felsenfeld and Schmidt 1975; Migaki et al. 1971; Wolf et al. 1974). Three separate outbreaks occurred in 1969 and 1970 in pig-tailed macaques, Japanese macaques, cynomolgus, and possibly Formosan

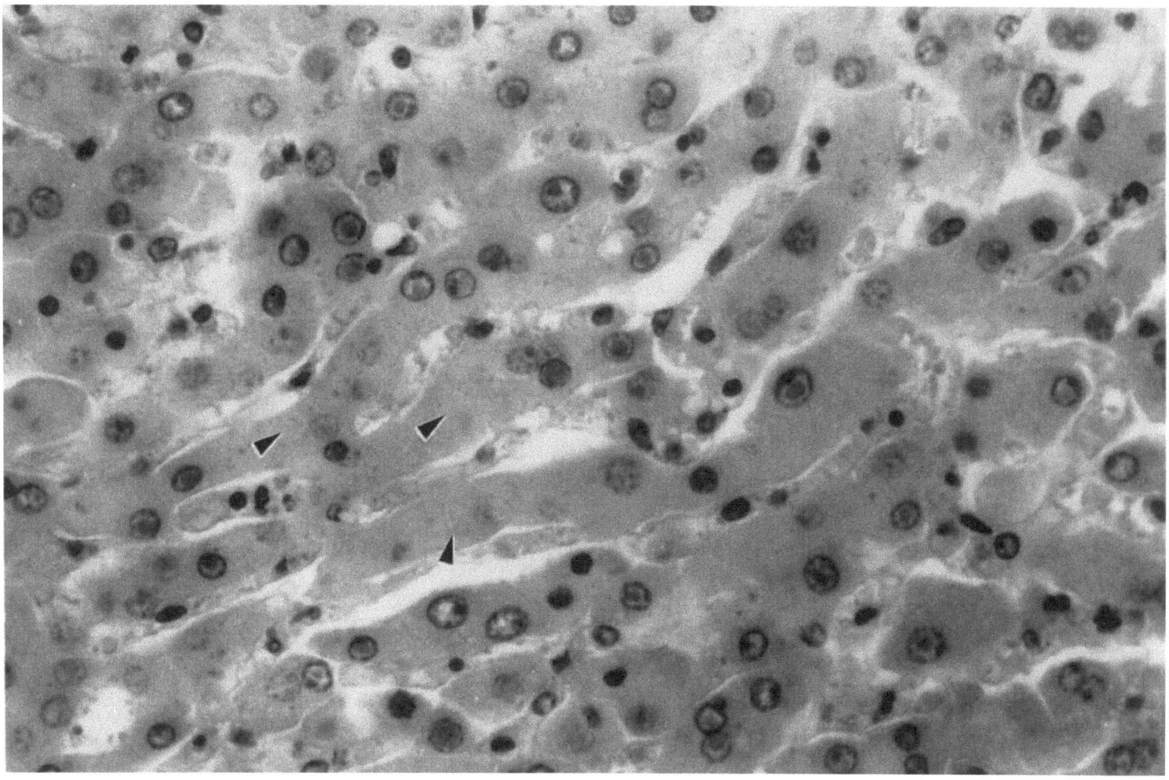

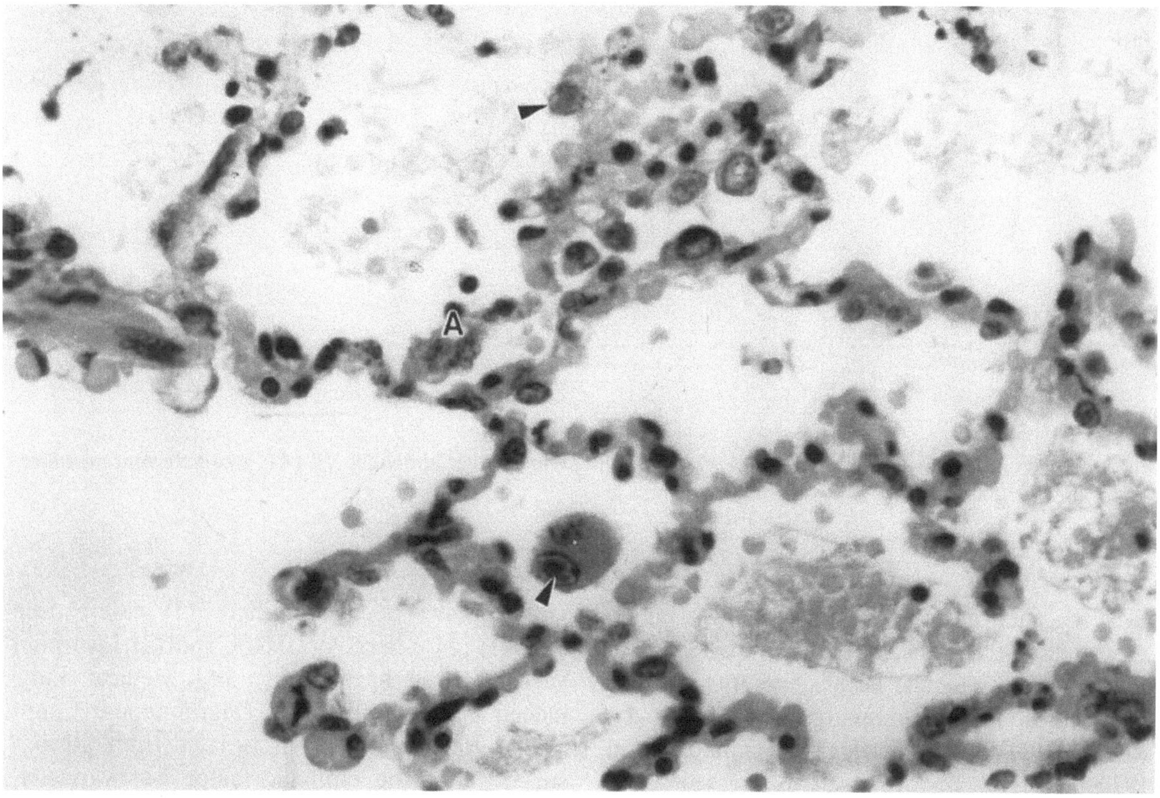

Fig. 97 (*above*). Liver of an African green monkey 10 days after experimental infection with delta herpesvirus. Necrosis of hepatic cells (*arrowheads*) and numerous intranuclear inclusions are evident. H and E, ×400

Fig. 98 (*below*). Lung of an African green monkey 10 days following infection with delta herpesvirus. Note alveolar hemorrhage, fibrin necrosis of alveolar wall (*A*), and intranuclear inclusion (*arrowhead*). H and E, ×400

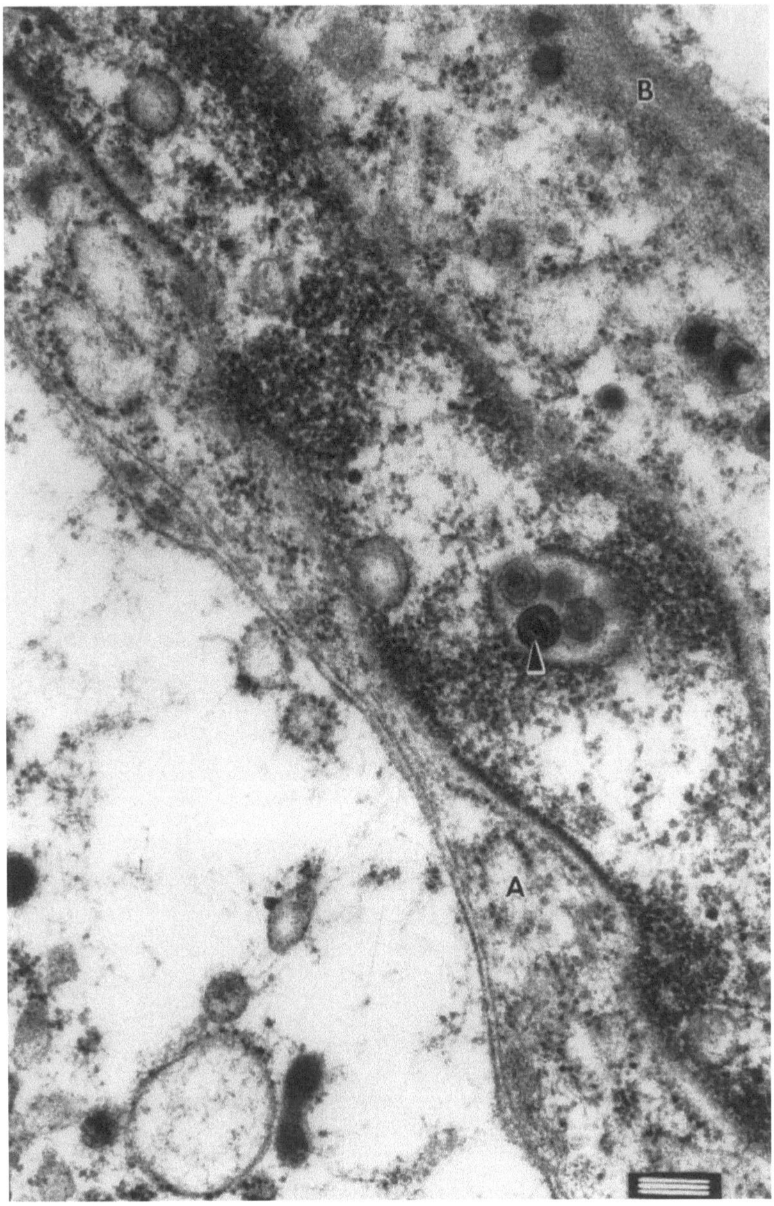

Fig. 99. Delta herpesvirus viral particles (*arrowheads*) within an alveolar epithelial cell (*A*); note basement membrane (*B*); *white bar*, 200 nm

rock macaque monkeys. This agent was named the Medical Lake macaque virus (Blakely et al. 1973). A mild vesicular herpesvirus infection was reported in infant chimpanzees (McClure and Keeling 1971) and in gorillas (Marennikoza et al. 1973; Myers et al. 1987) a virus, named gorilla herpesvirus, was recovered. Outbreaks in African green monkeys have been reported (Arvin et al. 1983; Gard and London 1983; Lehner et al. 1984). Serologic comparison of the various non-human primate isolates suggests that three distinct viruses have been identified: (1) chimpanzee (CZHI) (2) δ-herpes (DHV), and (3) Liverpool vervet, patas herpesvirus, and Medical Lake macaque group of viruses (Harbour and Caunt 1979). Recent studies of isolates from gorillas suggest that the etiologic factor was varicella-zoster virus (Harbour and Caunt 1979; Myers et al. 1987). Immunologically, the other non-human primate varicella-type agents are closely related to varicella-zoster virus of humans (Ayers 1971; Alle et al. 1974; Felsenfeld and Schmidt

1975). The various spontaneous outbreaks suggest viral latency, with stress, such as shipping, triggering an acute outbreak. Recovery from experimental infection may be followed by another acute manifestation when the animals are stressed.

Comparison with Other Species

Simian varicella is a very severe, often fatal disease of cutaneous, digestive, respiratory, and lymphoid organs. The usual course of human varicella (chicken pox) is confined to a focal, disseminated, cutaneous rash. Fatal varicella-zoster disease of humans resembles very closely the disease in nonhuman primates, although the simian infection is generally more severe and of shorter duration than the human disease. Immunosuppressed individuals are particularly prone to develop the fatal form of varicella-zoster. The involvement of the respiratory system in African green monkeys has been compared to the pulmonary disease in fatal varicella-zoster, using transmission and electron microscopy (Roberts et al. 1984b). Endothelial cell shrinkage was prominent with edema and necrosis of the interalveolar septae. Numerous intranuclear and cytoplasmic viral particles were observed in the areas of necrosis. Thrombosis was not an important component of the lesion in the lung. The pathogenesis of the clinical and pathologic features of simian varicella indicates a close parallel with fatal human varicella-zoster. Generalized herpes diseases in other species such as the dog and horse do not produce a cutaneous rash. Thus the simian varicella more closely resembles varicella-zoster. Delta herpesvirus has also been used to screen new antiviral drugs (Soike et al. 1980, 1981) and as a viral disease model for evaluating tumor necrosis factor (Soike et al. 1989).

References

Allen WP, Felsenfeld AD, Wolf RH, Smetana HF (1974) Recent studies on the isolation and characterization of delta herpesvirus. Lab Anim Sci 24:222–228

Arvin AM, Martin DP, Gard EA (1983) Interferon prophylaxis against simian varicella in Erytrocebus patas monkeys. J Infect Dis 147:149–154

Ayers JP (1971) Studies of the delta herpesvirus isolated from the patas monkey (Erythrocebus patas). Lab Anim Sci 21:685–695

Barahona H, Meléndez LV, Melnick JL (1974) A compendium of herpesviruses isolated from nonhuman primates. Intervirology 3:175–192

Blakely GA, Lourie B, Morton WG, Evans HH, Kaufman AF (1973) A varicella-like disease in macaque monkeys. J Infect Dis 127:617–625

Clarkson MJ, Thorpe E, McCarthy K (1967) A virus disease of captive vervet monkeys (Cercopithecus aethiops) caused by a new herpesvirus. Arch Gesamte Virusforsch 22:219–234

Daniel MD, Melendez LV, Hunt RD, Trum BF (1972) Pathology of simian primates: II. Infections and parasitic diseases. The herpesvirus group. In: Fiennes RN (ed) Pathology of simian primates. Karger, Basel, pp 592–611

Felsenfeld AD, Schmidt NJ (1975) Immunological relationship between delta herpesvirus of patas monkeys and varicella-zoster virus of humans. Infect Immun 12:261–266

Felsenfeld AD, Schmidt NJ (1979) Varicella-zoster virus immunized patas monkeys against simian varicella-like disease. J Gen Virol 42:171–178

Gard EA, London WT (1983) Clinical history and viral characterization of delta herpesvirus infection in a patas monkey colony. Proceedings of the symposium on the use of non-human primates in exotic viral and immunological disease. Monographs in Primatology, vol 7. University of Texas, San Antonio, pp 211–212

Harbour DA, Caunt AE (1979) The serological relationship of varicella-zoster virus to other primate herpesviruses. J Gen Virol 45:469–477

Hull RN (1973) The simian herpesviruses. In: Kaplan AS (ed) The herpesviruses. Academic, New York, pp 389–425

Hunt RD, Meléndez LV (1969) Herpes virus infections of non-human primates: A review. Lab Anim Care 19:221–234

Lehner NDM, Bullock BC, Jones ND (1984) Simian varicella infection in the African green monkey (Cercopithecus aethiops). Lab Anim Sci 34:281–285

Loomis MR, O'Neill T, Bush M, Montali RJ (1981) Fatal herpesvirus infection in patas monkeys and a black and white colobus monkey. J Am Vet Med Assoc 179:1236–1239

Marennikoza SS, Maltseva NN, Skelukhima EM, Shenkman LS, Kornieva VI (1973) A generalized herpetic infection simulating smallpox in a gorilla. Intervirology 2:280–286

McCarthy K, Thorpe E, Laursen AC, Heymann CS, Beale AJ (1968) Exanthematous disease in patas monkeys caused by a herpes virus. Lancet 2:856–857

McClure HM, Keeling ME (1971) Viral diseases noted in the Yerkes primate center colony. Lab Anim Sci 21:1002–1010

McClure HM, Olberding B, Strozier LM (1973) Disseminated herpesvirus infection in a rhesus monkey (Macaca mulatta). J Med Primatol 2:190–194

Migaki G, Seibold HR, Wolf RH (1971) Pathologic conditions in the patas monkey. J Am Vet Med Assoc 159:549–556

Myers MG, Kramer LW, Stanberry LR (1987) Varicella in a gorilla. J Med Virol 23:317–322

Padovan D, Cantrell CA (1986) Varicella-like herpesvirus infections of nonhuman primates. Lab Anim Sci 36: 7–13

Roberts ED, Baskin GB, Soike K, Gibson SV (1984a) Pathologic changes of experimental simian varicella

(delta herpesvirus) infection in African green monkeys. Am J Vet Res 45:523–530
Roberts ED, Baskin GB, Soike K, Meiners N (1984b) Transmission and scanning electron microscopy of experimental pulmonary simian varicella (delta herpsevirus) infection in African green monkeys (Cercopithecus aethiops). J Comp Pathol 94:323–328
Soike KF, Felsenfeld AD, Gibson S, Gerone PJ (1980) Ineffectiveness of adenine, arabinoside and adenine arabinoside 5′-monophosphate in simian varicella infection. Antimicrob Agents Chemother 18:142–147
Soike KF, Felsenfeld AD, Gerone PJ (1981) Acyclovir treatment of experimental simian varicella infection of monkeys. Antimicrob Agents Chemother 20:291–297
Soike KF, Czarniecki CW, Baskin G, Blanchard J, Liggitt D (1989) Enhancement of simian varicella virus infection in African green monkeys by recombinant human tumor necrosis factor (alpha). J Infect Dis 159:331–335
Wolf RH, Smetana HF, Allen WP, Felsenfeld AD (1974) Pathology and clinic history of delta herpesvirus in patas monkeys. Lab Anim Sci 24:218–221

Herpesvirus platyrrhinae Infection

Ronald D. Hunt and Beverly J. Blake

Synonym. Herpesvirus T infection.

Gross and Microscopic Features

The pathological and biological features of *Herpesvirus T* infection are remarkably similar to *Herpesvirus simplex* infection as described on p. 82, this volume. The virus, however, is distinct and only New World monkeys are naturally affected (Holmes et al. 1964; Melnick et al. 1964). *Herpesvirus T* is carried by squirrel monkeys (*Saimiri sciureus*) (Meléndez et al. 1966) in which serum antibodies are very frequent but clinical disease only rarely recorded. Documented lesions in squirrel monkeys consist of vesicles and ulcers of the oral mucous membranes and lips and have a very similar gross and microscopic appearance to *H. simplex* lesions in human beings (Figs. 100–105) (Daniel et al. 1967; King et al. 1967). Circulating serum neutralizing antibodies to *Herpesvirus T* have been detected also in cinnamon ringtail monkeys (*Cebus albifrons*) and spider monkeys (*Ateles* spp.), which might also be natural reservoir hosts but no clinical disease has been described in these two species (Holmes et al. 1966).

Natural or experimental *Herpesvirus T* infection in owl monkeys (*Aotus trivirgatus*) and several species of marmosets and tamarins is a very severe systemic infection comparable to *Herpesvirus simplex* infection in owl monkeys described on p. 82 this volume (Hunt and Meléndez 1966, 1969). There are oral, labial and dermal lesions characterized by vesicles, erosions and ulcers and variable-sized foci of necrosis in most any organ and tissue of the body. Microscopically the dis-

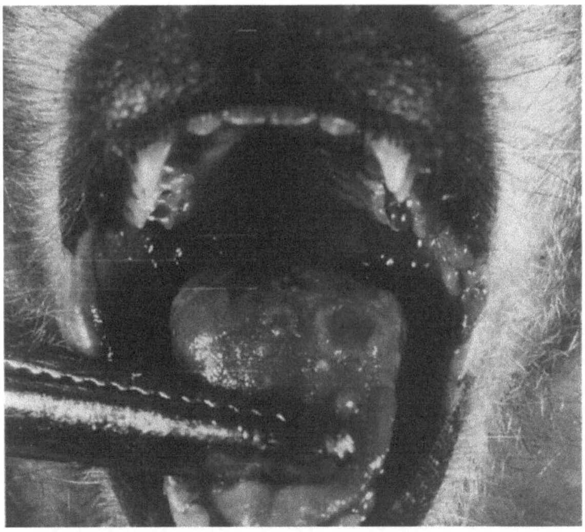

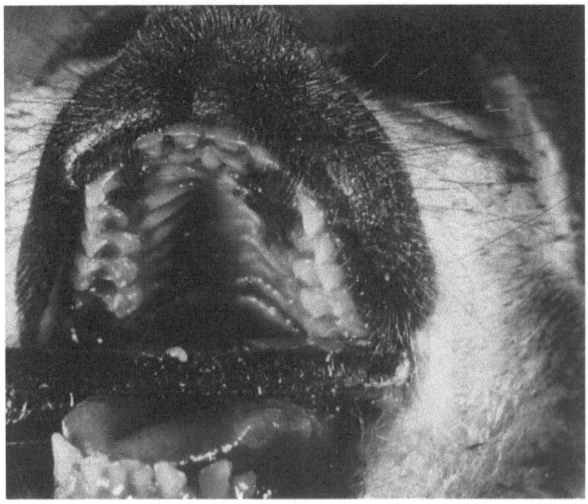

Fig. 100 (*above*). Multiple ulcers and plaques on the tongue of a squirrel monkey infected by *Herpesvirus T*

Fig. 101 (*below*). Ulcers on the hard palate caused by *Herpesvirus T* in a squirrel monkey

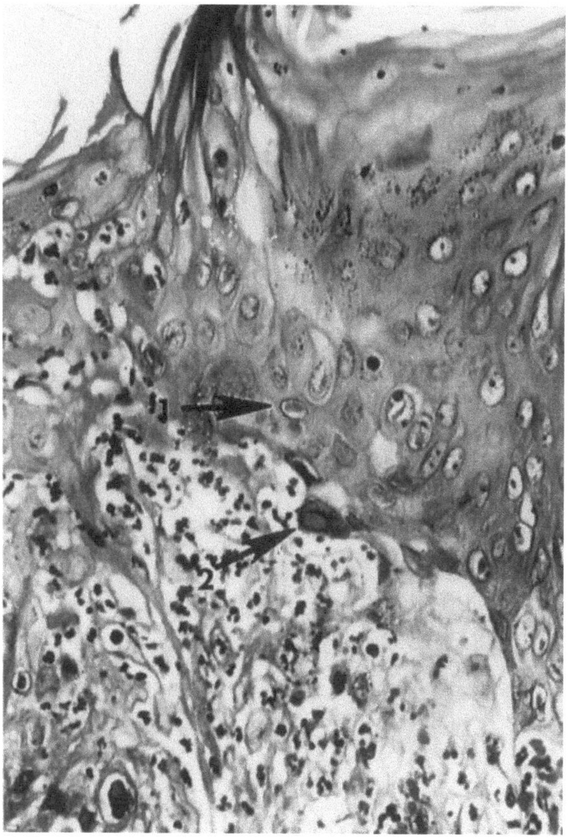

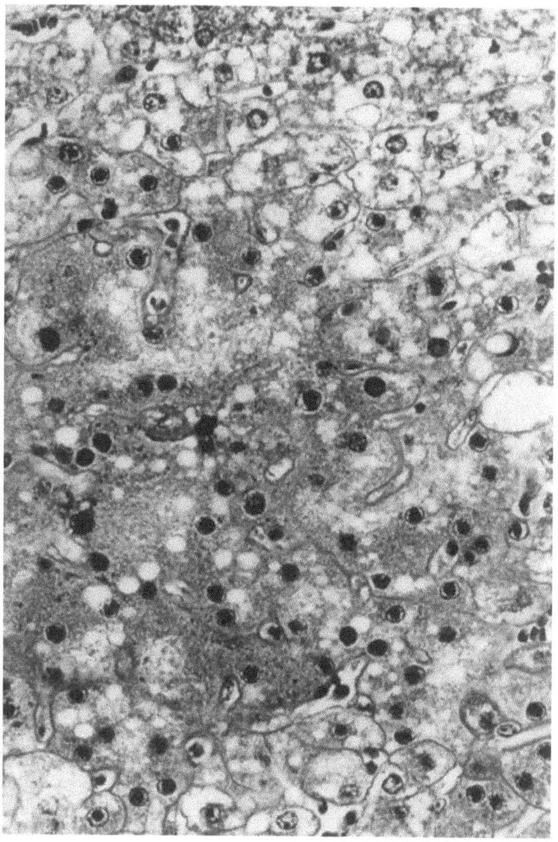

Fig. 102 (*upper left*). A vesicle infiltrated with neutrophils on the tongue of a squirrel monkey infected by *Herpesvirus T*. Note inclusion body surrounded by a clear halo (*1*) and another which completely fills the nucleus of a syncytial cell (*2*). H and E, ×200

Fig. 103 (*below*). Necrotic lesion of the tongue of an owl monkey which died of *Herpesvirus T* infection. H and E, ×80

Fig. 104 (*upper right*). Multiple intranuclear inclusion bodies in hepatocytes of an owl monkey which died of *Herpesvirus T*. H and E, ×200

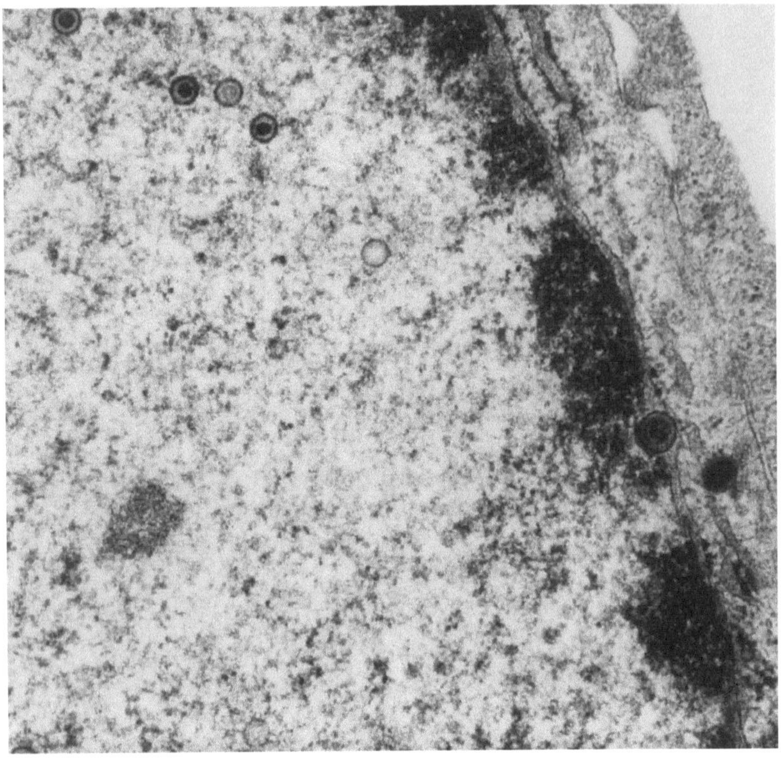

Fig. 105. Intranuclear naked nucleocapsids of *Herpesvirus T*, one of which lacks a core. TEM, ×45 000

tinctive features of the lesions are multinucleated giant cells and intranuclear inclusion bodies. In contrast to *H. simplex* infection, encephalitis is not common and when present not extensive.

Differential Diagnosis

In squirrel monkeys the lesions of *Herpesvirus T* are not known to be duplicated by any other herpesvirus. Squirrel monkeys do, however, carry other herpesviruses such as *H. saimiri* which is excreted in the oral cavity. Neutralization tests or other techniques are required to differentiate the two. Both cause similar cytopathological changes in vitro.

In owl monkeys, tamarins and marmosets the lesions of *Herpesvirus T* are essentially identical to those caused by *H. simplex* and cannot be differentiated without the use of immunohistochemical or other special staining techniques or by virus isolation and identification.

Biologic Features

Natural History and Pathogenesis

All available evidence indicates that the relationship of *Herpes T* virus to squirrel monkeys is comparable to *H. simplex* virus in human beings. Squirrel monkeys probably become infected at an early age with the virus establishing a latent infection in ganglia. Little information is available on the recurrence of clinically recognizable lesions after primary infection, and recurrence has not been documented. However, virus has been recovered from oral secretions from clinically healthy adult squirrel monkeys. It could be anticipated that in neonatal squirrel monkeys the virus might cause fatal systemic infection comparable to *H. simplex* in human infants or SA8 in baboons, but this has not been recognized.

Owl monkeys, marmosets and tamarins become infected by exposure to squirrel monkeys carrying the virus. In these species the infection is nearly always fatal and when it occurs in a colony there is exceptionally high morbidity.

Etiology

Herpesvirus T is a distinct member of the family Herpesviridae with characteristics similar to those described elsewhere for *H. simplex* (p. 82).

Frequency

Based upon serologic surveys, *Herpes T* infection is common in squirrel monkeys. Outbreaks of the disease in colonies of owl monkeys, marmosets and tamarins have not been frequent.

Prevention

In order to prevent cross species infection, squirrel monkeys and those species in which the infection is more severe should not be housed together. A live virus vaccine has proven highly effective in owel monkeys (Daniel et al. 1967). The efficacy of the vaccine in tamarins and marmosets is less certain. Although it offers protection, the vaccinated animal may develop myelitis months after vaccination.

Comparison with Other Species

The relationship of *Herpesvirus T* infection in squirrel monkeys is analogous to that of *Herpesvirus simplex* in human beings or *Herpesvirus B* in macaques. The severe disease seen in owl monkeys, marmosets and tamarins is analogous to the disease seen in these species caused by *H. simplex* or the severe disease caused by *Herpes B* in human beings.

Acknowledgement. This work was supported by Public Health Service Award RR00168 from the National Institutes of Health.

References

Daniel MD, Karpas A, Meléndez LV, King NW, Hunt RD (1967) Isolation of Herpes T virus from a spontaneous disease in squirrel monkeys (Saimiri sciureus). Arch Gesamte Virusforsch 22:324–331

Holmes AW, Caldwell RG, Dedmon RE, Deinhardt F (1964) Isolation and characterization of a new herpes virus. J Immunol 92:602–610

Holmes AW, Devine JA, Nowakowski E, Deinhardt F (1966) The epidemiology of a herpes virus infection of New World monkeys. J Immunol 96:668–671

Hunt RD, Meléndez LV (1966) Spontaneous Herpes T infection in the owl monkey (Aotus trivirgatus). Pathol Vet 3:1–26

Hunt RD, Meléndez LV (1969) Herpes virus infection of non-human primates: a review. Lab Anim Care 19:221–234

King NW, Hunt RD, Daniel MD, Meléndez LV (1967) Overt Herpes T infection in squirrel monkeys (Saimiri sciureus). Lab Anim Care 17:413–423

Meléndez LV, Hunt RD, Garcia FG, Trum BF (1966) A latent Herpes T infection in Saimiri sciureus (squirrel monkey). In: Fiennes RN, Fiennes TW (eds) Recent developments in comparative medicine. Academic, London, pp 393–397

Melnick JL, Midulla M, Wimberly I, Barrera-Oro JG, Levy BM (1964) A new member of the herpesvirus group isolated from South American marmosets. J Immunol 92:596–601

OTHER VIRAL DISEASES

Encephalomyocarditis Virus Infection, Nonhuman Primates

Gary B. Baskin

Synonyms. Columbia-SK viruses, cardiovirus infection.

Gross Appearance

In animals that die in the acute phase of infection, there is usually an increased amount of clear pericardial fluid that may contain fibrin. All parts of the myocardium may contain diffuse or patchy pale areas that are often friable. These areas frequently appear as pale or yellowish streaks oriented parallel to the muscle fibers (Fig. 106). Petechial hemorrhages sometimes occur in the epicardium. In some animals there is pulmonary edema. The myocardium of animals that survive the acute phase often has irregular white firm areas of scarring. Gross changes are not usually detectable in the brain, pancreas, or other organs (Blanchard et al. 1987).

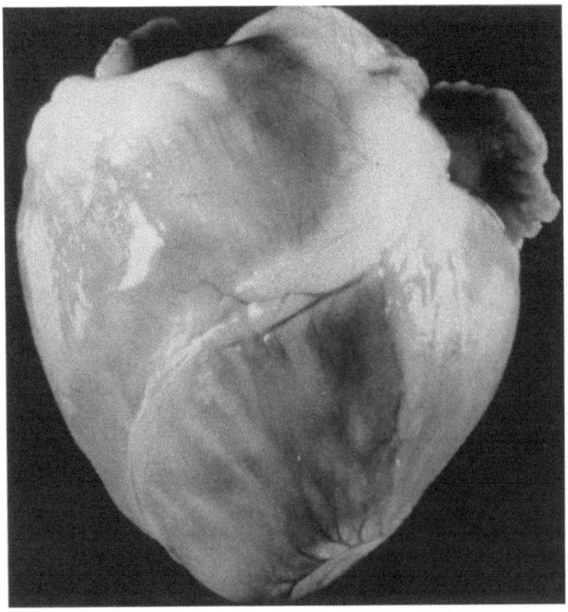

Fig. 106. Myocarditis caused by encephalomyocarditis virus in an African green monkey. Note linear pale areas of myocardial necrosis and inflammation

Microscopic Features

Extensive irregular areas of myocardium are degenerated and necrotic. Degenerated fibers are pale and have lost their striations. Necrotic fibers are fragmented and hyalinized, may contain granular basophilic material, and are sometimes infiltrated with neutrophils and macrophages. Fibers are often separated by interstitial edema and sometimes small amounts of hemorrhage. There is a mild to severe infiltration of lymphocytes and macrophages, along with lesser numbers of neutrophils and eosinophils, depending upon the stage at which the heart is examined (Fig. 107). In animals that survive the acute phase, there is often subsequent focal myocardial scarring with lymphocytic infiltration and hypertrophy of myocardial fibers (Fig. 108). Lesions in the brain, common in laboratory rodents, are rare in primates and consist of lymphocytic infiltration of the meninges and

Fig. 107 (*upper left*). Acute myocarditis caused by encephalomyocarditis virus in an African green monkey. Note fiber degeneration and inflammatory cell infiltration. H and E, ×250 ▶

Fig. 108 (*lower left*). Encephalomyocarditis virus infection, heart, African green monkey. Lymphocytic infiltration, scarring, and hypertrophy or myocardial fibrils. H and E, ×250

Fig. 109 (*upper right*). Encephalomyocarditis virus infection, pancreas, squirrel monkey. Note necrosis of pancreatic acinii. The islet of Langerhans (*I*) is unaffected. H and E, ×400

Fig. 110 (*lower right*). Encephalomyocarditis virus infection, pancreas, squirrel monkey. Note acinar atrophy and interstitial lymphocytic infiltration. H and E, ×400

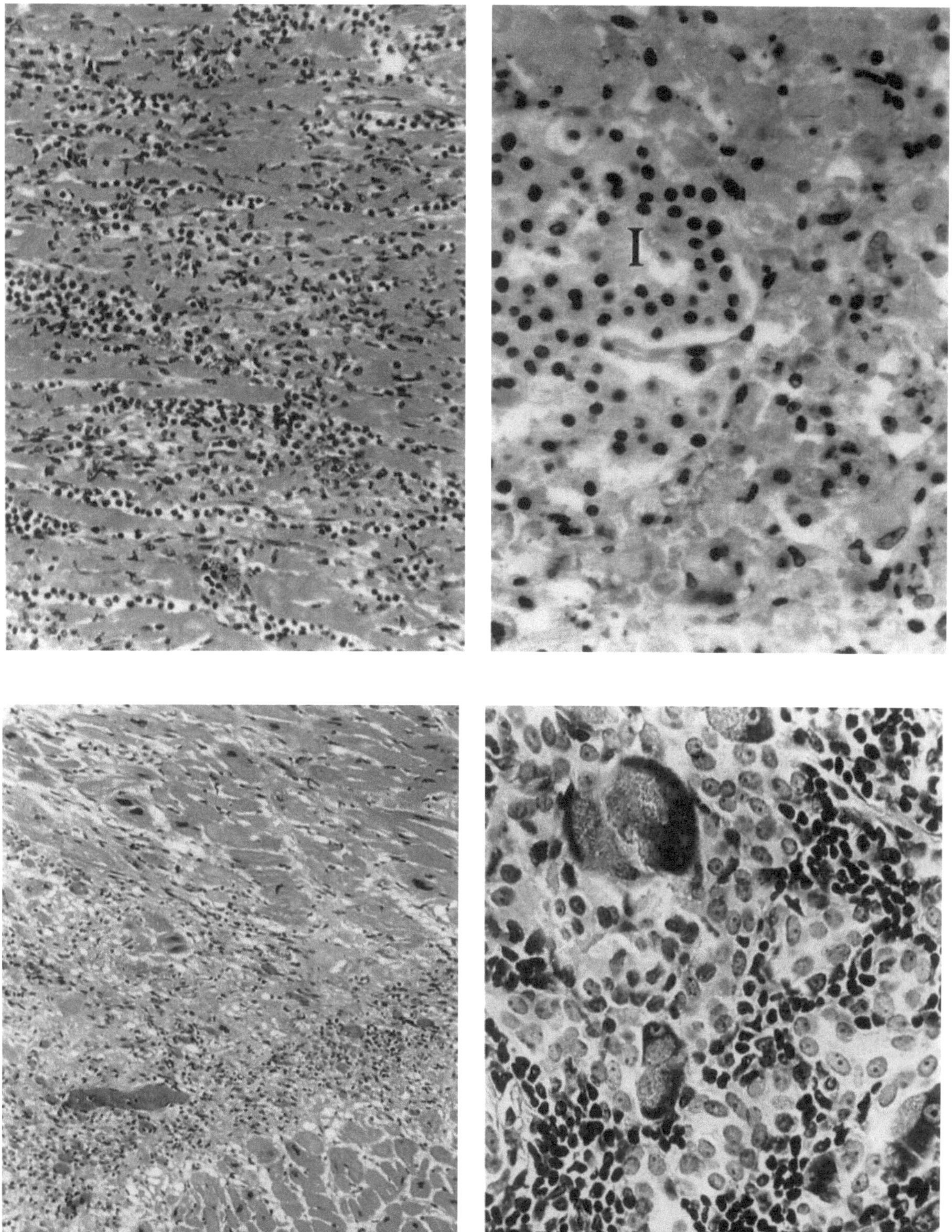
I

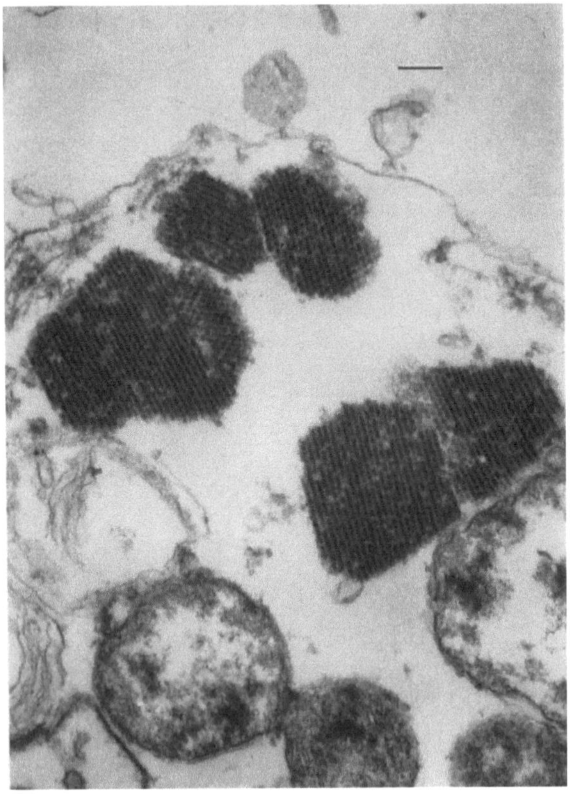

Fig. 111. Crystalline arrays of encephalomyocarditis virus in cell culture; *bar*, 200 nm. ×25 000

multifocal perivascular accumulations of neutrophils and lymphocytes in the cerebral cortex. There may be focal areas of neuronal necrosis with neuronophagia and infiltration of neutrophils and macrophages.

Early lesions in the pancreas consist of severe acinar necrosis with minimal to severe neutrophilic infiltration (Fig. 109). Later lesions are characterized by acinar atrophy and interstitial lymphocytic infiltration (Fig. 110). The islets of Langerhans are spared.

Monkeys may have mild focal necrosis with infiltration of neutrophils and macrophages in skeletal muscles. There may also be centrilobular hepatic degeneration, pulmonary edema, and generalized congestion associated with congestive heart failure (Blanchard et al. 1987).

Ultrastructure

Mature viral particles are icosahedral, nonenveloped, and measure 22–30 nm in diameter. Virus is assembled in the cytoplasm but may resemble ribosomes. Crystalline arrays of particles may form (Fig. 111).

Differential Diagnosis

Encephalomyocarditis virus infections must be differentiated from other causes of viral myocarditis, particularly Coxsackie B virus. This must be done by viral isolation and serological identification, as neither virus induces characteristic inclusion bodies and both are identical by electron microscopy. *Toxoplasma gondii* causes myocarditis in New World monkeys, particularly squirrel monkeys, and sporadically in other species. *Trypanosoma cruzi* causes myocarditis in monkeys from South America and others housed in the southern United States. Organisms should be visible in histologic sections in the latter instances.

Biologic Features

Encephalomyocarditis viruses belong to the cardiovirus group of picornaviruses which also includes Columbia-SK, M.M., and Mengo virus (Murnane 1981). They have been isolated from human beings, insects, mammals, and birds worldwide (Tesh and Wallace 1978), although natural disease has been documented only in nonhuman primates, pigs, and elephants. Natural and experimental infections have been reported in several species of nonhuman primates (Blanchard et al. 1987; Citino et al. 1988; Dick et al. 1948; Gaskin et al. 1980; Helwig and Schmidt 1945; Kilham et al. 1956; Kissling et al. 1956; Levin et al. 1986; Roca-Garcia and Sanmartin-Barberi 1957; Stebbins et al. 1983; Weck et al. 1983; Wells et al. 1989). The lesions produced in experimentally inoculated animals vary with the strain of virus and with the species infected. Some isolates produce predominately myocardial lesions, while others induce lesions in the brain or pancreas. One isolate causes severe myocarditis in African green monkeys and severe pancreatitis in squirrel monkeys (Blanchard et al. 1987). Rhesus monkeys (*Macaca mulatta*) are fairly resistant to the disease, while owl monkeys (*Aotus trivirgatus*), marmosets, African green monkeys (*Cercopithecus aethiops*), lemurs, tamarins, colobus monkeys, chimpanzees (*Pan troglodytes*), baboons, and siamangs are highly susceptible (Blanchard et al. 1987; Gaskin et al.

1980; Levin et al. 1986; Murnane 1981; Roca-Garcia and Sanmartin-Barberi 1957; Wells et al. 1989). Tissue destruction is probably caused by the immune response to the virus, rather than as a direct viral effect. Inflammation persists longer than the presence of culturable virus. Asymptomatically infected rodents, particularly rats, are presumed by many to be the natural reservoir for this infection, although the details of natural transmission are unclear and this contention is disputed (Tesh and Wallace 1978). Animals could become infected by consuming feed or water contaminated with the feces or urine of rodents and by consuming the rodent themselves. Although encephalomyocarditis has been isolated from mosquitoes and ticks, there is not epidemiological or experimental evidence to suggest that the virus is spread by arthropod vectors in nature. Rodent control is probably the most important factor in preventing infection in captive primates. The naturally occurring cases reported in the United States have all been in southern states. Most affected primates are found dead without prior clinical signs. Animals that survive often have cardiac conduction abnormalities and are poor risks for anesthesia.

Comparison with Other Species

Although encephalomyocarditis viruses naturally infect a wide variety of mammals and birds, naturally occurring epizootics have been described only in nonhuman primates, pigs, elephants, and possibly lions. Most natural infections in rodents are probably asymptomatic, although laboratory infections may produce severe lesions. There is serologic evidence of widespread infection in some human populations, but infection has only rarely been linked to neurological signs which completely resolve. Most human infections are probably asymptomatic (Tesh 1978). Myocardial lesions predominate in most species of nonhuman primates, as in swine and elephants, while many strains are highly neurotropic in experimentally inoculated mice.

References

Blanchard JL, Soike K, Baskin GB (1987) Encephalomyocarditis virus infection in African green and squirrel monkeys: comparison of pathologic effects. Lab Anim Sci 37:635–639

Citino SB, Homer BL, Gaskin JM, Wickman DJ (1988) Fatal encephalomyocarditis virus infection in a Sumatran orangutan (Pongo pygmaeus abelii). J Zoo Anim Med 19:214–218

Dick GWA, Smithburn KC, Haddow AJ (1948) Mengo encephalomyelitis virus: isolation and immunological properties. Br J Exp Pathol 29:547–558

Gaskin JM, Jorge MA, Simpson CF, Lewis AL, Olson JH, Schobert EE, Wollenman EP, Marlowe C, Curtis MM (1980) The tragedy of encephalomyocarditis virus infection in zoological parks of Florida. Am Assoc Zoo Vet Ann Proc:1–7

Helwig FC, Schmidt EDH (1945) A filter-passing agent producing interstitial myocarditis in anthropoid apes and small animals. Science 102:31–33

Kilham L, Mason P, Davies JPN (1956) Host-virus relations in encephalomyocarditis (EMC) virus infections: II. Myocarditis in mongooses. Am J Trop Med Hyg 5:655–663

Kissling RE, Vanella JM, Schaeffer M (1956) Recent isolations of encephalomyocarditis virus. Proc Soc Exp Biol Med 91:148–150

Levin R, Yaari A, Gitelman J, Bino T, Rosenberg H, Kohn A (1986) EMC virus infection in baboons as a model for studies on antiviral substances. Antiviral Res 6:277–283

Murnane TG (1981) Encephalomyocarditis. In: Beran GW (ed) CRC handbook series in zoonoses: B. Viral zoonoses, vol 2. CRC Press, Boca Raton, pp 137–147

Roca-Garcia M, Sanmartin-Barberi C (1957) The isolation of encephalomyocarditis virus from Aotus monkeys. Am J Trop Med Hyg 6:840–852

Stebbing N, Weck PK, Fenno JT, Estell DA, Rinderknecht E (1983) Antiviral effects of bacteria-derived human leukocyte interferons against encephalomyocarditis virus infection of squirrel monkeys. Brief report. Arch Virol 76:365–372

Tesh RB (1978) The prevalence of encephalomyocarditis virus neutralizing antibodies among various human populations. Am J Trop Med Hyg 27:144–149

Tesh RB, Wallace GD (1978) Observations on the natural history of encephalomyocarditis virus. Am J Trop Med Hyg 27:133–143

Weck PK, Harkins RN, Stebbins N (1983) Antiviral effects of human fibroblast interferon from Escherichia coli against encephalomyocarditis virus infection of squirrel monkeys. J Gen Virol 64/2:415–419

Wells SK, Gutter AE, Soike KF, Baskin GB (1989) Encephalomyocarditis virus: epizootic in a zoological collection. J Zoo Wildl Med 20:291–296

Measles Virus Infection, Nonhuman Primates

Linda J. Lowenstine

Synonyms. Rubeola, systemic paramyxovirus infection, morbillivirus infection, giant cell interstitial pneumonia.

Gross Lesions

The lesions of naturally occurring and experimental measles in nonhuman primates have been well described in the literature (Potkay et al. 1966; Scott and Keymer 1975; Hall et al. 1971; Sota and Deauville 1964; Manning et al. 1968; Blake and Trask 1921b; Lorenz and Albrecht 1980; Albrecht et al. 1980; Montrey et al. 1980; Levy and Mirkovic 1971; Taniguchi et al. 1954; Sergiev et al. 1960; Kamahora and Nii 1965). Facial edema and erythema accompanied by chemosis are common features on external examination (Fig. 112). In most species there is also an erythematous maculopapular rash on the ventral abdomen and inner thighs that often has a silvery scaly surface as it begins to fade (Fig. 113). Vesicles are rare, but pustules are occasionally present due to secondary bacterial infection. There may be one or more 1 mm diameter, raised, white spots (Koplik spots), sometimes rimmed in red and with bluish centers, on the buccal, labial, or glossal mucosa; these usually precede the generalized rash and, if present, are pathognomonic for measles. Rash may be absent in some species or outbreaks, e.g., *Colobus guereza* (Hime et al. 1975) and *Saguinus mystax* (Albrecht et al. 1980). Death may precede the development of the rash or may occur after the rash has faded. Rash may be difficult to appreciate in darkly pigmented species, such as gorillas.

Pneumonia and diarrhea with dehydration secondary to enteric involvement are the most common causes of death. Gross lesions in the respiratory tract may be inapparent or the lungs may have distinct lobular consolidation secondary to bronchiolitis (Fig. 114), or diffusely scattered patchy areas of consolidation ranging from tan to hemorrhagic. More extensive lesions may be noted if secondary bacterial pneumonia occurs. Gross changes in the gastrointestinal track are infrequent, but congestion and hemorrhage have

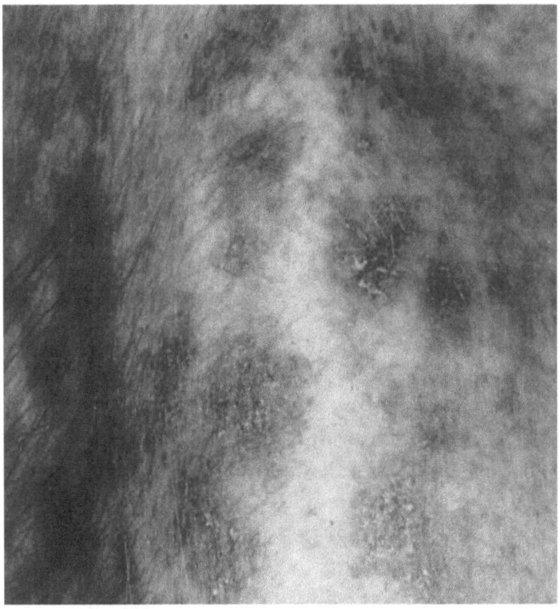

Fig. 112 (*above*). Facial exanthema and conjunctivitis in an adult rhesus macaque naturally infected with measles virus. (Courtesy of Primate Medicine, California Regional Primate Research Center)

Fig. 113 (*below*). Abdominal rash in a rhesus macaque naturally infected with measles virus. Mild exfoliation causes the scaly appearance. (Courtesy of California Regional Primate Research Center)

been noted in some species such as marmosets (Fraser et al. 1978).

Enlargement of peripheral or visceral lymph nodes and spleen may sometimes be present.

Microscopic Appearance

The skin lesions are often more subtle histologically than they are grossly. The erythema is due to congested small vessels with swollen endothelium in the papillary dermis. Mitotic figures may be seen in endothelium. Mild superficial and follicular parakeratosis, accompanied by multinucleated epithelial syncytia, may also be seen (Figs. 115, 116). These syncytia only rarely have intranuclear or intracytoplasmic inclusions and are most frequent in the follicular epithelium. Very rarely one may see ballooning degeneration of epidermis with intranuclear and intracytoplasmic inclusions (L.J. Lowenstine, unpublished). Inflammation is variable and often reflects secondary bacterial infection.

Lesions in the lung are centered on small bronchi and bronchioles, often with necrosis of the epithelium and filling of the lumina by mucus and cell debris (Fig. 117). This may lead to atelectasis of the lung distal to the obstruction. Neutrophils are present in varying numbers depending on the presence of bacterial invaders. The lesion extends into the interstitium and alveoli from the respiratory bronchioles and alveolar ducts. Acute alveolar damage is reflected in the exudation of protein-rich edema fluid, increased number of alveolar macrophages, and sloughed cellular debris.

The characteristic, but not inevitable, feature of measles pneumonia is the presence of multinucleated giant cells, the so called giant cell interstitial pneumonia. Syncytia may be present early in the course of infection in tracheal or bronchial epithelium and later in bronchioles; air sac epithelium may also become infected (Fig. 118). In the parenchyma, the giant cells are found in the presence of hyperplasia of type-2 pneumocytes, and the syncytia appear to be

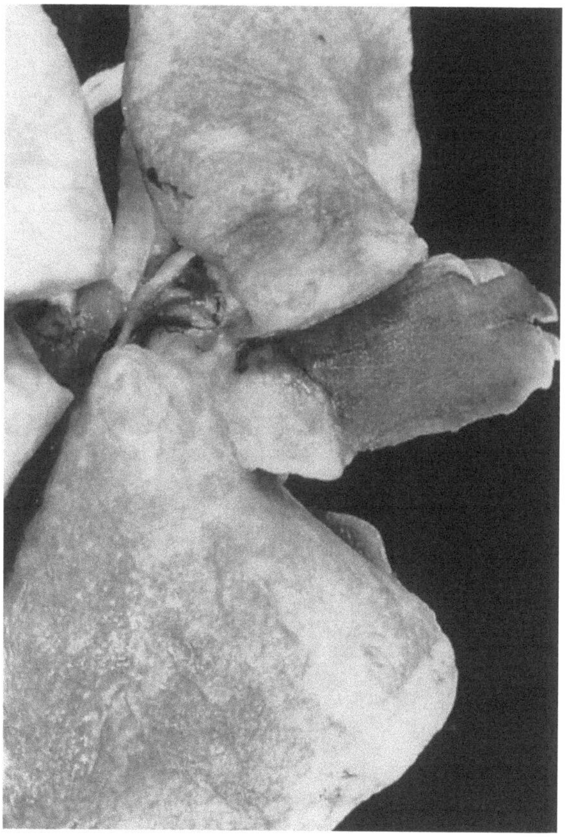

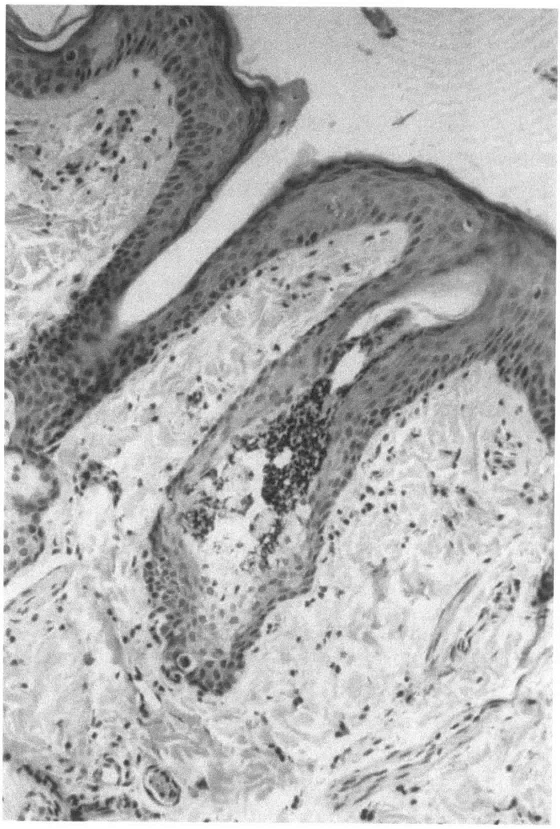

Fig. 114 (*above*). Lobar atelectasis secondary to bronchiolitis and plugging with mucus and desquamated cells, *Macaca mulatta*; lung naturally infected with measles ▶

Fig. 115 (*below*). Measles, *Macaca mulatta*. Skin with mild edema of the papillary dermis, desquamative folliculitis, and a syncytial cell at the base of the follicle. H and E, ×120

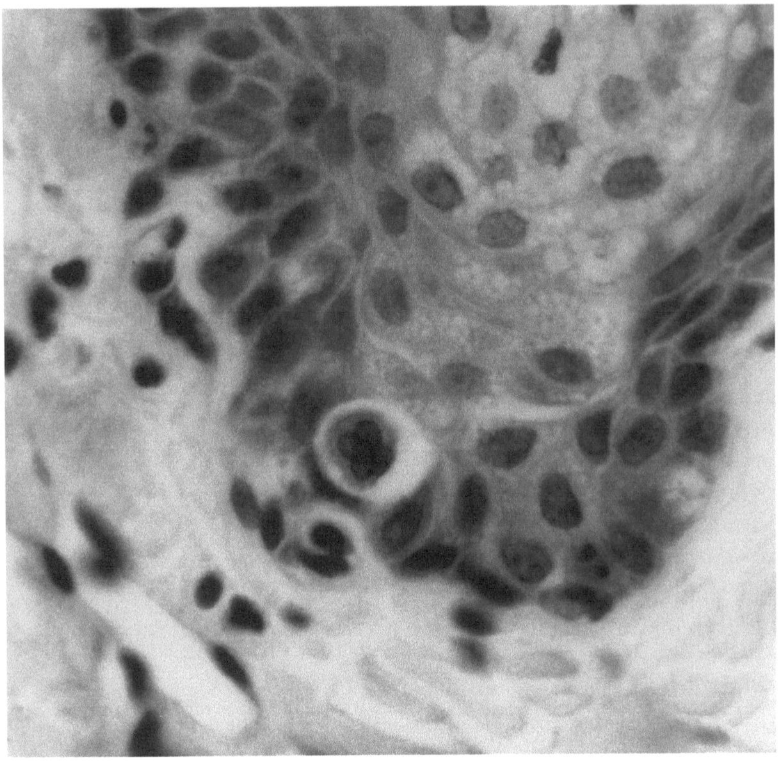

Fig. 116. Measles, *Macaca mulatta*, same monkey as Fig. 115. Syncytial cell at the base of the follicle. H and E, ×640

derived from pneumocytes or possibly macrophages (Fig. 119). Inclusion bodies, both intranuclear and intracytoplasmic, may be seen in these syncytia and less frequently in other cells. At the time of death the characteristic giant cells and inclusions may no longer be present.

Syncytia are also a feature of the gastroenteric form of the disease which is often seen in New World primates (see paramyxovirus enteritis). Enteric involvement in other nonhuman primates and humans may be more frequent than was originally appreciated. Syncytia are present in mucosal epithelium, in the lamina propria, and/or in gut-associated lymphoid tissue. The entire length of the gastrointestinal track may be involved. Hemorrhagic enteritis with epithelial necrosis has been described in New World species and colobus monkeys (Fraser et al. 1978; Scott and Keymer 1975). Death due to diarrhea and dehydration may occur after syncytia and inclusions are no longer present (L.J. Lowenstine, unpublished) and, as in children, may be due to secondary infections with enteropathogenic organisms (El Behairy et al. 1986; Pongrithsukda et al. 1986).

Measles virus affects the lymphoid system as well. Syncytia may be seen in lymph nodes, gut, and bronchial-associated lymphoid tissues. These syncytia are formed from two different cell types, lymphocytes (Warthin-Finkeldy cells) and mononuclear phagocytic cells, as has been described in human cases of measles. Necrosis of the lymphocytes is seen in gut-associated lymphoid tissue and in lymph nodes.

Syncytial giant cells with intranuclear and intracytoplasmic inclusions have been seen in the uterus of rhesus which aborted and died during a measles outbreak (Renne et al. 1973). Abortions and stillbirths are frequent during measles epizootics in macaques (Hertig et al. 1971; Roberts et al. 1988). Syncytia and inclusions may also be found in other organs such as bladder (MacArthur et al. 1982) or pancreatic ducts and biliary tree (L.J. Lowenstine, unpublished observations), and use of immunohistochemistry can demonstrate wide distribution of viral antigen in respiratory, gastrointestinal, lymphoid tissues, salivary gland, pancreas, liver, kidney, and many other tissues (Lorenz and Albrecht 1980).

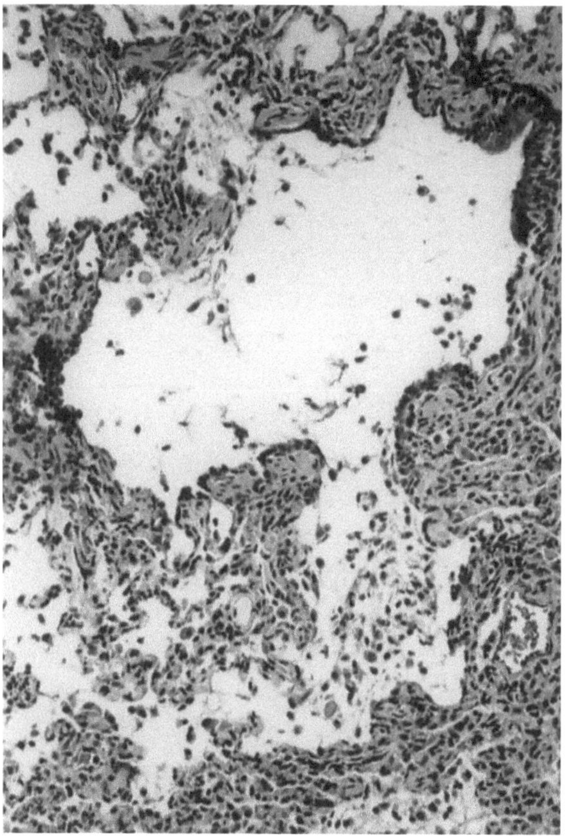

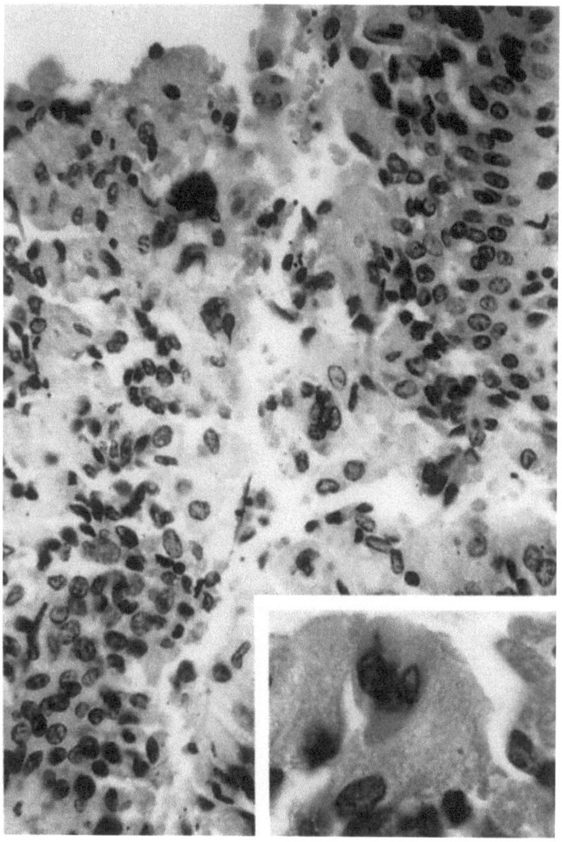

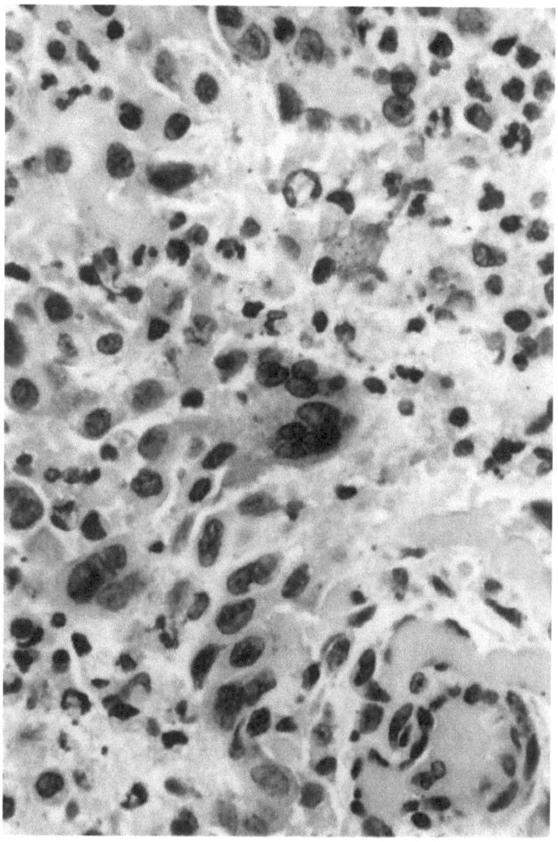

Fig. 117 (*upper left*). Measles, *Macaca mulatta*. Necrosis, desquamation, and plugging of bronchioles in the rhesus lung from Fig. 114 Syncytia were not seen. H and E, ×120

Fig. 118 (*upper right*). Adult male mountain gorilla) (*G. gorilla beringi*). Syncytial cells in the air sac in which paramyxovirus nucleocapsids were demonstrated by electron microscopy. H and E, ×300

Fig. 119 (*lower right*)

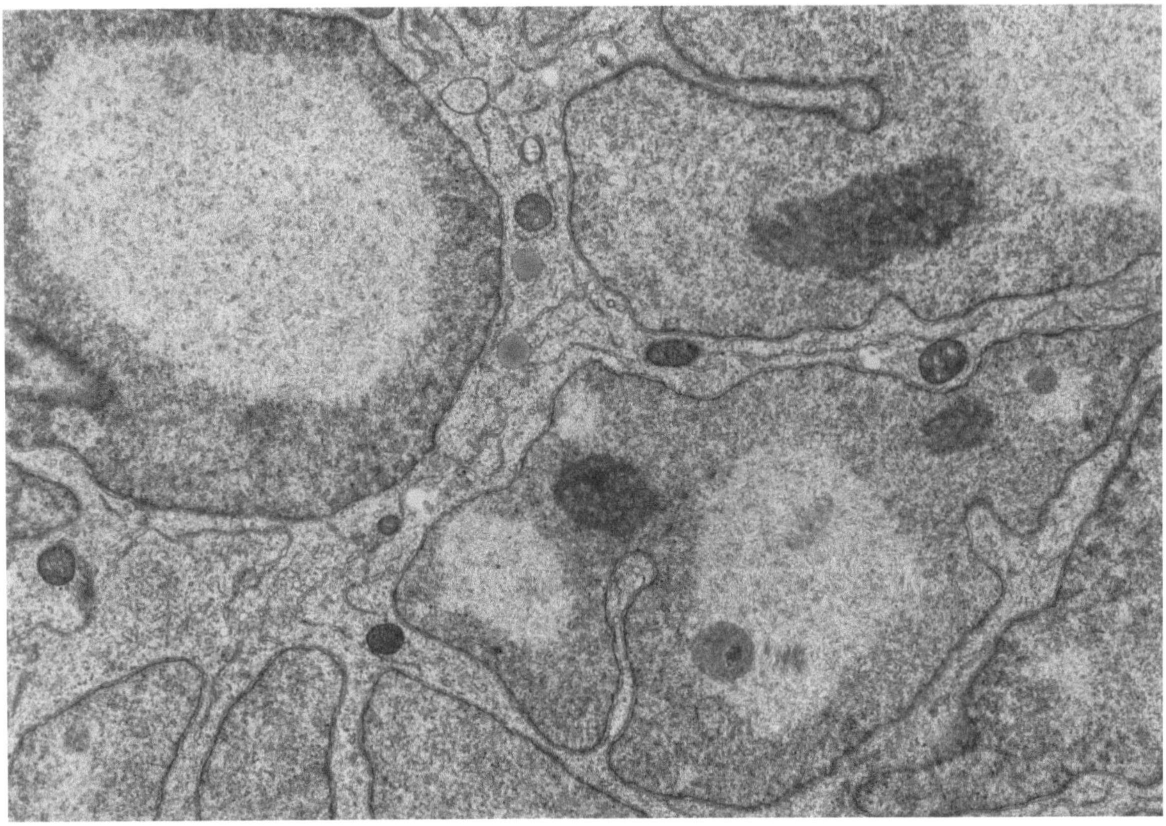

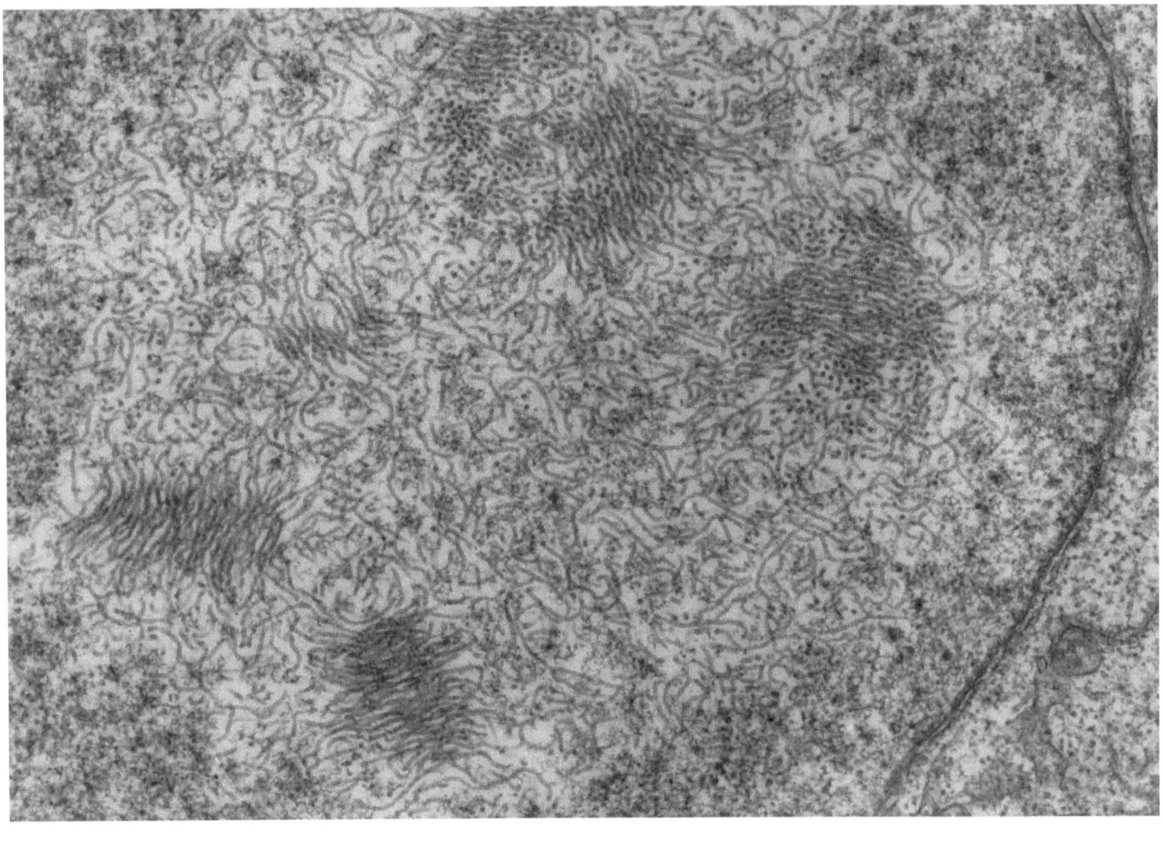

Ultrastructure

Measles virus has the usual tangled, tubular, nucleocapsids of paramyxoviruses. These are seen in the nuclei and cytoplasm of infected cells in areas corresponding to the inclusion bodies (Figs. 120, 121). Areas of thickened cell membrane overlying nucleocapsid material represent virion assembly regions prior to budding. Mature virions are infrequently seen in transmission electron micrographs (TEM) of necropsy specimens but have been identified in negatively stained scrapings of oropharynx from people with measles. The virus envelope is fragile and rupture of particles may account for the apparent absence of mature virions. Specific ultrastructural cellular alterations other than inclusions or virions in various stages of assembly have not been documented.

Differential Diagnosis

The main differential diagnostic problem in nonhuman primates with bronchointerstitial pneumonia includes other paramyxoviruses such as PMV 3 (Dick and Dick 1974) or PMV 1 (Sendai virus), as described in marmosets (Flecknell et al. 1983), or influenza viruses, as described in gibbons (Johnsen et al. 1971). If syncytial cells are found, respiratory syncytial virus (chimpanzee coryza agent or RSV) infection must be considered (Morris et al. 1956). This virus however is seldom associated with fatal disease and is usually confined to the upper airways. Simian virus 5 is a primate paramyxovirus infection which must be considered in the differential, but lesions produced by this virus are poorly documented. The simian immunodeficiency virus, an immunosuppressive lentivirus, is also associated with a giant cell interstitial pneumonia, but there is no necrotizing bronchiolitis or inclusion body (see p. 11). Other viral causes of pneumonia include adenoviruses, Herpesvirus B in bonnet monkeys, SA 8 herpesvirus (Cercopithicine herpesvirus 2) infection in baboons, cytomegalovirus, and varicella-like herpesviruses which can be differentiated from measles, based on the type of inclusions and the distribution of the pneumonias (Brack et al. 1985; Roberts et al. 1984; Espana 1973; Boyce et al. 1978). Enteric lesions accompanied by syncytial cells are strongly suggestive of measles infection. However, in macaques infected with simian immunodeficiency virus, syncytial cells without inclusions can be found in the lamina propria of the gastrointestinal tract (see p. 14). Encephalitic forms of the infection must be differentiated from other nonsuppurative encephalidities such as herpesvirus (*Herpesvirus simiae* or B virus in bonnet macaques) or enterovirus (e.g., encephalomyocarditis virus) encephalitis in squirrel monkeys (Blanchard et al. 1987; Weigler 1992), or even mild cases of progressive multifocal leukoencephalopathy due to the papovavirus SV40 in which giant reactive astrocytes are not present (Gribble 1975). Measles, *Herpes B* and progressive multifocal leukoencephalopathy all produce intranuclear inclusions, but only in measles are the inclusions also intracytoplasmic. Canine distemper virus, which is also classified in the genus *Morbillivirus*, can experimentally cause encephalitis in squirrel monkeys but is unlikely to cause a differential dilemma as a natural infection (Nagata et al. 1990). Serology, virus isolation or immunocytochemistry may be necessary to arrive at a definitive diagnosis.

◀ **Fig. 120** (*above*). Intranuclear inclusion bodies in a multinucleated syncytial cell formed in tissue culture (Raji cell) by cocultivation with peripheral blood mononuclear cells from a rhesus with naturally occuring measles infection. Lead citrate and uranyl acetate, TEM, ×9200

Fig. 121 (*below*). Nucleocapsids with smooth surfaces are characteristic of the intranuclear inclusions seen in cells infected with measles virus. Lead citrate and uranyl acetate, TEM, ×33 000. (Courtesy of R.J. Munn, Department of Pathology, School of Medicine, University of California-Davis)

Biologic Features

Natural History

Measles is pandemic in the human population and becomes important as a pathogen in nonhuman primates when they come in contact with humans in their natural environment or, more frequently, in captivity (Black 1989; Meyer 1962). Measles in humans is most severe in countries in which the populace experiences poor nutrition and sanitation, overcrowding and intercurrent diseases; the case fatality rate may be in excess of 20% (Aaby et al. 1985). Although usually thought of as a respiratory problem, diarrhea is an important cause of morbidity and mortality in measles infection in children in third world countries

(Pongrithsukda et al. 1986; Axton 1975; El Behairy 1986; Black 1989). The virus is spread through aerosolization and is highly contagious.

The epidemiology of measles in humans has been well described. In unvaccinated populations measles occurs as cyclic outbreaks; epidemics occur every 2–5 years and last for 3–4 months when there is a susceptible cohort of naive preschool children. The epidemiology of measles infection in nonhuman primates has been less well studied but outbreaks in free living and laboratory primates are often coincident with human outbreaks. (Roberts et al. 1988; Phillips et al. 1975).

Measles in nonhuman primates varies from an inapparent infection, recognized only by seroconversion, to very mild disease, with high morbidity and low mortality, to one with devastating mortality. It is apt to appear in association with the shipment of animals and when new monkeys are introduced into an established colony. One outbreak in marmosets resulted in 326 deaths (Levy and Mirkovic 1971). In a shipment of ten colobus the mortality was 100% (Scott and Keymer 1975), and in silver leafmonkeys (*Presbytis cristatus*) mortality was about 25% (Montrey et al. 1980). In one epizootic in recently imported juvenile rhesus there was no mortality in a shipment of 100 animals, out of which 58 developed a rash, while in another outbreak (Roberts et al. 1988; McChesney et al. 1989) in long-term captive rhesus and cynomolgus macaques, there was a mortality rate of 27% with most deaths being the result of diarrhea and dehydration.

Pathogenesis

The pathogenesis of measles in human and nonhuman primates is similar (Kempe and Fulginiti 1965; Sergiev et al. 1960; Blake and Trask 1921a; Black 1989). The virus is highly contagious, spread by aerosolization, and enters through the respiratory or conjunctival mucosa. In rhesus monkeys experimentally inoculated via the respiratory membranes the incubation period is 6–10 days (mean and median of 7 days), while in naturally infected children the incubation period is 10–12 days (Blake and Trask 1921b; Black 1989). During the initial part of incubation period, there is proliferation of the virus in regional lymphoreticular tissue (Carter and ter Meulen 1990). Lymphoid hyperplasia and necrosis have been demonstrated during this period along with multinucleated cells in germinal centers (Warthin-Finkeldy cells) (Sergiev et al. 1960) and associated leukopenia. Primary viremia results in spread of the virus to the entire lymphoreticular system and the upper respiratory epithelium. This is followed by a second period of viremia in which virus is spread throughout the body, presumably by infected circulating lymphoid cells. Resulting necrosis of affected epithelial surfaces ushers in the prodromal stage of 1–5 days in monkeys and 2–4 days in children before the development of rash or exanthema. Signs during the prodrome include conjunctivitis, sneezing and cough, and the development of Koplik spots or exanthema of the oral cavity. Koplik spots are not always present in nonhuman primates, or may be overlooked. The virus has broad tissue tropism and during the prodrome virus is shed in tears, urine, and nasal secretions. The hemogram is characterized by leukopenia involving neutrophils, eosinophils, and lymphocytes. The characteristic rash usually begins on day 14 after infection in humans and day 5–10 in rhesus monkeys and baboons. Individuals may die before the onset of the rash as a result of respiratory and gastrointestinal injury or immune suppression. Antibodies are detected at the time of onset of rash. Rapid clearance of detectable virus ensues; however, the lifelong persistence of antibodies suggest the persistence of antigen. Virus can be recovered years after initial infection in individuals developing subacute sclerosing panencephalitis.

Protection against measles virus is dependent upon both humoral and cellular immunity. Immunity after natural infection is complete and neutralizing antibodies persist for the life of the individual (McChesney and Oldstone 1989). Cytolytic complement-dependent antibodies are important in eliminating measles virus during acute infection (Hicks and Albrecht 1976) in rhesus monkeys. Other components of the immune system such as NK cells may also be important (Hicks et al. 1977).

Measles virus has been shown to cause immunosuppression in humans and nonhuman primates (McChesney and Oldstone 1989; McChesney et al. 1989). The virus infects mononuclear cells in the peripheral blood including B and T lymphocytes, alters T cell-mediated functions such as cutaneous reactions to tuberculin testing, and suppresses immunoglobulin synthesis. Secondary bacterial infections, including mycoplasmosis, are the usual manifestation of this deficiency of immunoglobulin in skin, lung, and gastrointestinal tract (El Behairy et al. 1986; McChesney et al. 1989) and may be principally responsible for measles fatalities (Remfry 1976; Montrey et al. 1980). Rarely, other opportunistic infections are observed such as cytomegalovirus sialoadenitis in rhesus (L.J. Lowenstine, unpublished) and cryptosporidiosis in children (DeMol et al. 1984).

The respiratory and gastrointestinal lesions, lymphopenia, and immune suppression are attributed to direct cell necrosis in these organs and tissues and to secondary invaders. The encephalitic complications of the disease are thought to be the result of immune-mediated damage rather than direct viral neuropathogenicity, although viral inclusions are present in neurons and glial cells. The pathogenesis of subacute sclerosing panencephalitis has not been completely elucidated but is probably also immune-mediated. The rash is thought to be the result of immune complex-induced endothelial damage.

Etiology

Along with canine and phocine distemper, rinderpest and kata viruses, measles virus is a member of the *Morbillivirus* genus of the family Paramyxoviridae (Carter and Meulen 1990; DeLay et al. 1965; Cosby et al. 1988). Morphologically, the mature enveloped virion is pleomorphic with a floppy fragile membrane with spikes and a diameter of 120–250 nm. The viral genome is single-stranded negative sense RNA which encodes for six virus-specific structural proteins. Three or four are within the core. These include: (1) N (nucleocapsid) protein; (2) L (large) protein, which is probably the RNA-dependent RNA polymerase; (3) the P (phospho-)protein which may also be needed for transcription; (4) the M (matrix) protein which is needed for virus assembly; (5) the envelope protein HA (hemagglutinin); and (6) the F (fusion) protein. The latter two proteins make up the envelope spikes seen on electron microscopy. An additional nonstructural C protein has been identified in the intranuclear and cytoplasmic inclusions of infected cells. The order of the genes is N, P/C, M, F, HA, L. The measles virus is unique among the *Morbilliviruses* in having a hemagglutinin; it agglutinates nonhuman primate but not human red blood cells. HA protein is responsible for this; it is the major immunogenic moiety and induces neutralizing antibodies. Hemolysis is mediated by the F protein, which is also responsible for the formation of syncytial giant cells that characterize the cytopathic effect both in vitro and in vivo. It also is immunogenic, and antibodies to it prevent fusion and cell to cell spread of the virus.

The virus attaches to host cell receptors via the HA protein spikes. The F protein then causes fusion of viral envelope and host cell membrane. The viral contents are discharged into the cytoplasm where RNA polymerase activation allows for mRNA transcription. Association of the viral-encoded mRNA with host cell ribosomes leads to translation of viral proteins. Nucleocapsid material accumulates within the cell and is seen as inclusion bodies in both the nucleus and cytoplasm. These inclusions consist of spaghetti-like tangled aggregations of linear nucleocapsids (Figs. 120, 121). The envelope proteins are bound to rough endoplasmic reticulum, processed through the Golgi apparatus, and inserted diffusely in the cell membrane. Replication of virions is probably triggered by L, N, and P protein concentrations within the cell. Viral (negative sense) RNA coupled with N protein form the nucleocapsid which then binds to the P protein. M protein combines with nucleocapsids and plasma membrane and draws the two together, causing localized concentration of the virus glycoproteins in the cell membrane. These complexes of nucleocapsids and aggregated glycoprotein bud and are pinched off to form complete enveloped virions. Lack of M protein results in the inability to form completed virions and is probably involved in the pathogenesis of subacute sclerosing panencephalitis. During viral replication large amounts of viral glycoproteins in the host cell membrane can cause hemadsorption and F protein-induced fusion with adjacent cells. Fusion is cytotoxic and if fusion is prevented there is more virus produced within the longer lived single cells.

There is only one serotype of measles virus, and the viral proteins are amazingly homogeneous among various isolates by both electrophoretic and genomic comparisons. Variation in severity of disease therefore is thought to be primarily the result of as yet poorly understood host factors. However, different laboratory strains of virus have different virulence; JM and Edmonston strains caused 100% and 25% mortality, respectively, in experimentally inoculated marmosets (Albrecht et al. 1980).

Frequency

Numerous serological surveys have been published concerning the incidence of measles in free-ranging and captive nonhuman primates, primarily macaques (Meyer et al. 1962; MacArthur et al. 1979; Kalter et al. 1967; Shah and Southwick 1965; Bhatt et al. 1966; Mei et al. 1990; Suzuki et al. 1981; Yamanouchi et al. 1969). The data indicate that measles is not a natural disease of macaques; that it is acquired through contact with humans; and that once in nonhuman primates it can become endemic or epizootic (Meyer et al. 1962; Shishido 1966; MacArthur et al. 1979). Serum antibodies were not present in 170 bonnet monkeys (*M. radiata*) and 195 Hanuman langurs (*Presbytis entellus*) that were forest-dwelling in India but were found in 12/47 (25%) of free-living urban rhesus (Bhatt et al. 1966). At capture none of 116 rhesus were positive, but by 6–8 weeks after capture nearly 100% had immunologic evidence of infection (Meyer et al. 1962). In a more recent serosurvey of Chinese rhesus in China, 8/25 (32%) of newly caught, 7/20 (35%) of caged, and 30/33 (91%) of group-housed animals were positive (Mei et al. 1990). Seropositivity in newly imported cynomolgus monkeys varies from 0% to 74%, depending on site of origin and shipment group (Yamanouchi et al. 1969; Suzuki et al. 1981). Another report of viral testing in captive primates revealed a marked variation in measles seropositivity in different groups of primates: 233/499 (49%) macaques (rhesus and cynomologous macaques), 0/111 African green monkeys, 9/341 (20%) chimpanzees and gorillas (Kalter and Heberling 1990). Captive African green monkeys had been previously shown to be susceptible (Meyer et al. 1962). Protective titers to measles virus wanes in captivity and offspring born in captivity are fully susceptible to measles, making them vulnerable to infection from humans or newly imported primates (MacArthur et al. 1979; Roberts et al. 1988; Phillips et al. 1975).

Comparison with Other Species

Measles in nonhuman primates closely mimics the infection in humans. Since measles worldwide is still one of the principal causes of childhood mortality, with an estimated 1.5×10^6 deaths per year (Katz 1985), nonhuman primates have served as invaluable animal models. It also has some similarities to canine and phocine distemper virus infections and rinderpest and kata in ruminants.

Encephalitis is a rare consequence of measles infection in both human and nonhuman primates. Three different neurologic syndromes have been described in humans (Carter and ter Meulen 1989). Acute postinfectious measles encephalitis occurs in about 1/1000–5000 human cases and there is one report of "paramyxovirus encephalitis," probably due to measles, in macaques (Steel et al. 1982). Perivascular cuffing with large lymphocytes is present primarily in gray matter. Spherical, eosinophilic, intranuclear and intracytoplasmic inclusions may be seen in cortical astrocytes, neurons, and astrocytes in basal ganglia and cerebellar Purkinje cells. Subacute measles encephalitis has been recognized recently in immunosuppressed children but has not been documented in naturally infected nonhuman primates. Subacute sclerosing panencephalitis is an infrequent sequel to measles infection in humans and has been reported in two adult baboons (Kim et al. 1970). Lesions include marked perivascular cuffing, demyelination and edema of white matter, gliosis, neuronophagia, and occasionally intranuclear inclusion bodies. Similar lesions have been induced experimentally in rhesus monkeys, inoculated intracerebrally with a hamster brain-adapted measles virus (Albrecht et al. 1972).

References

Aaby P, Coovadia H, Bukh J, Lisse IM, Smits AJ, Wesley A, Kiepiela P (1985) Severe measles: a reappraisal of the role of nutrition, overcrowding and virus dose. Med Hypotheses 18:93–112

Albrecht P, Shabo AL, Burns GR, Tauraso NM (1972) Experimental measles encephalitis in normal and cyclophosphamide-treated rhesus monkeys. J Infect Dis 126:154–161

Albrecht P, Lorenz D, Klutch MJ, Vickers JH, Ennis FA (1980) Fatal measles infection in marmosets. Pathogenesis and prophylaxis. Infect Immun 27:969–978

Axton JHM (1975) Measles: a protein-losing enteropathy. Br Med J 3:79–80

Bhatt PN, Brandt CD, Weiss RA, Fox JP, Shaffer MF (1966) Viral infections of monkeys in their natural habitat in southern India: II. Serological evidence of viral infection. Am J Trop Med Hyg 15:561–566

Black FL (1989) Measles. In: Evans AS (ed) Viral infections of humans, 3rd edn. Plenum, London, pp 451–469

Blake FG, Trask JD Jr (1921a) Studies on measles: I. Susceptibility of monkeys to the virus of measles. J Exp Med 33:385–412

Blake FG, Trask JD Jr (1921b) Studies on measles: II. Symptomatology and pathology in monkeys experimentally infected. J Exp Med 33:413–422

Blanchard JL, Soike K, Baskin GB (1987) Encephalomyocarditis virus infection in African green and squirrel monkeys: comparison of pathologic effects. Lab Anim Sci 37:635–639

Boyce JT, Giddens WE, Valerio M (1978) Simian adenoviral pneumonia. Am J Pathol 91:259–276

Brack M, Eichberg JW, Heberling RL, Kalter SS (1985) Experimental herpes neonatalis in SA 8-infected baboons (Papio cynocephalus). Lab Anim 19:125–131

Carter MJ, ter Meulen V (1990) Measles. In: Zuckerman AJ, Banatvala JE, Pattison JR (eds) Principles and practice of clinical virology, 2nd edn. Wiley, Chichester, pp 313–337

Cosby SL, McQuaid S, Duffy N, Lyons C, Rima BK, Allan GM, McCullough SJ, Kennedy S, Smyth JA, McNeilly F, Draig C, Orgell C (1988) Characterization of a seal morbillivirus. Nature 336:115–116

DeLay PD, Stone SS, Karzon DT, Katz S, Enders J (1965) Clinical and immune response of alien hosts to inoculation with measles, rinderpest and canine distemper viruses. Am J Vet Res 26:1359–1373.

DeMol P, Mukashema S, Bogaerts J, Hemelhof W, Butzler JP (1984) Cryptosporidium related to measles diarrhoea in Rwanda. Lancet 8393:42–43

Dick EC, Dick CR (1974) Natural and experimental infections of nonhuman primates with respiratory viruses. Lab Anim Sci 24:177–181

El Behairy F, Hablas R, El Diwany K, Eissa AM, Ezzat M, Hassan F (1986) Clinico-bacteriological study of diarrhoea complicating measles. J Trop Pediatr 32: 44–46

Espana C (1973) Herpesvirus simiae infection in Macaca radiata. Am J Phys Anthropol 38:447–454

Flecknell PA, Parry R, Needham JR, Ridley RM, Baker HF, Bowes P (1983) Respiratory disease associated with parainfluenza type I (Sendai) virus in a colony of marmosets (Callithrix jacchus). Lab Anim 17:111–113

Fraser CEO, Chalifoux L, Sehgal P, Hunt RD, King NW (1978) A paramyxovirus causing fatal gastroenterocolitis in marmoset monkeys. Primates Med 10:261–270

Gribble DH, Haden CC, Schwartz LW, Henrickson RV (1975) Spontaneous progressive multifocal leukoencephalopathy (PML) in macaques. Nature 254:602–604

Hall WC, Kovatch RM, Herman PH, Fox JG (1971) Pathology of measles in rhesus monkeys. Vet Pathol 8:307–319

Hertig AT, King NW, MacKey (1971) Spontaneous abortion in wild-caught rhesus monkeys, Macaca mulatta. Lab Anim Sci 21:510–519

Hicks JT, Albrecht P (1976) Cytolytic, complement-dependent antibodies to measles virus in rhesus monkeys after administration of liver or killed virus. J Infect Dis 133:648–654

Hicks JT, Sullivan JL, Albrecht P (1977) Immune responses during measles infection in immunosuppressed rhesus monkeys. J Immunol 119:1452–1456

Hime JM, Keymer IF, Baxter CJ (1975) Measles in recently imported colobus monkeys (Colobus guereza). Vet Rec 97:392

Johnsen DO, Wooding WL, Tanticharoenyos P, Karnjanaprakorn C (1971) An epizootic of A2/Hong Kong/68 influenza in gibbons. J Infect Dis 123:365–370

Kalter SS, Heberling RL (1990) Viral battery testing in nonhuman primate colony management. Lab Anim Sci 40:21–23

Kalter SS, Ratner J, Kalter GV, Rodriguez AR, Kim CS (1967) A survey of primate sera for antibodies to viruses of human and simian origin. Am J Epidemiol 86:552–568

Kamahora J, Nii S (1965) Pathological and immunological studies of monkeys infected with measles virus. Archiv Gesamte Virusforsch 16:161–167

Katz S (1985) Measles – forgotten but not gone. N Engl J Med 313:577–578

Kempe CH, Fulginiti (1965) The pathogenesis of measles virus infection. Archiv Gesamte Virusforsch 16:103–128

Kim CS, Kriewaldt FH, Hagino N, Kalter SS (1970) Subacute sclerosing panencephalitis-like syndrome in the adult baboon (*Papio sp*). J Am Vet Med Assoc 157:730–735

Levy BM, Mirkovic RR (1971) An epizootic of measles in a marmoset colony. Lab Anim Sci 21:33–39

Lorenz D, Albrecht P (1980) Susceptibility of tamarins (Saguinus) to measles virus. Lab Anim Sci 30:661–665

MacArthur JA, Mann PG, Oreffo V, Scott GBD (1979) Measles in monkeys: an epidemiological study. J Hyg 83:207–212

MacArthur JA, Mann PG, Oreffo V, Scott GBD (1982) The effect of measles vaccination in rhesus monkeys (Macaca mulatta). Lab Anim 16:267–270

Manning PJ, Banks KL, Lehner NDM (1968) Naturally occurring giant cell pneumonia in the rhesus monkey (Macaca mulatta). J Am Vet Med Assoc 153:899–904

McChesney MB, Oldstone MBA (1989) Virus-induced immunosuppression: infections with measles virus and human immunodeficiency virus. Adv Immunol 45:335–380

McChesney MB, Fujinami RS, Lerche NW, Marx PA, Oldstone MBA (1989) Virus-induced immunosuppression: infection of peripheral blood mononuclear cells and suppression of immunoglobulin synthesis during natural measles virus infection of rhesus monkeys. J Infect Dis 159:757–760

Mei A, Meng Y, Gouxiang L, Shufan C, Jingyn Z, Ping S (1990) Virological survey of rhesus monkeys in China. Lab Anim Sci 40:29–32

Meyer HM, Brooks BE, Douglas RD, Rogers NG (1962) Ecology of measles in monkeys. Am J Dis Child 103: 307–313

Montrey RD, Huxsoll DL, Hildebrandt PK, Booth BW, Arimbalam S (1980) An epizootic of measles in captive silvered leaf-monkeys (Presbytis cristatus) in Malaysia. Lab Anim Sci 30:694–697

Morris JA, Blount RE Jr, Savage RE (1956) Recovery of cytopathogenic agent from chimpanzees with coryza. Proc Soc Exp Biol Med 92:544–549

Nagata T, Ochikubo F, Yoshikawa Y, Yamanouchi K (1990) Encephalitis induced by a canine distemper virus in squirrel monkeys. J Med Primatol 19:137–149

Phillips IR, Colman G, Clarke M (1975) Measles in monkeys. Vet Rec 97:436

Pongrithsukda V, Phonboon K, Manunpichu K (1986) Measles-associated diarrhoea in northeastern Thailand. S E Asian J Trop Med Public Health 17:43–47

Potkay S, Ganaway JR, Rogers NG, Kinard R (1966) An epizootic of measles in a colony of rhesus monkeys (Macaca mulatta). Am J Vet Res 27:331–334

Remfry J (1976) A measles epizootic with 5 deaths in newly imported rhesus monkeys (Macaca mulatta). Lab Anim 10:49–57

Renne RRA, McLaughlin R, Jenson AB (1973) Measles virus-associated endometritis, cervicitis and abortion in a rhesus monkey. J Am Vet Med Assoc 163:639–641
Roberts ED, Baskin GB, Soike K, Meiners N (1984) Pathologic changes of experimental simian varicella (delta herpesvirus) infection in African green monkeys (Cercopithecus aethiops). Am J Vet Res 45:523–530
Roberts JA, Lerche NW, Anderson JH, Markovits JE, Maul DH (1988) Epizootic measles at the California primate research center. Lab Anim Sci 38:491 (abstr)
Scott GBD, Keymer IF (1975) The pathology of measles in Abyssinian colobus monkeys (Colobus guerza): a description of an outbreak. J Pathol 117:229–233
Sergiev PG, Ryazantseva NE, Shroit IG (1960) The dynamics of pathological processes in experimental measles in monkeys. Acta Virol 4:265–273
Shah KV, Southwick CH (1965) Prevalence of antibodies to certain viruses in sera of free-living rhesus and of captive monkeys. Ind J Med Res 53:488–500
Shishido A (1966) Natural infection of measles virus in laboratory monkeys. Jpn J Med Sci Biol 19:221–222
Sota PJ, Deauville GA (1964) Spontaneous simian giant-cell pneumonia with coexistent B virus infection. Am J Vet Res 25:793–804
Steel MD, Giddens WE Jr, Valerio M, Sumi SM, Stetzer ER (1982) Spontaneous paramyxoviral encephalitis in nonhuman primates (Macaca mulatta and M. nemestrina). Vet Pathol 19:132–139
Suzuki M, Sasagawa A, Inayoshi T, Nakamura F, Honjo S (1981) Serological survey for SV5, measles and herpes simplex infections in newly imported cynomolgus monkeys. Jpn J Med Sci Biol 34:69–80
Taniguchi T, Kamahora J, Kato S, Hagiwara K (1954) Pathology in monkeys experimentally infected with measles virus. Med J Osaka Univ 5:367–396
Weigler BJ (1992) Biology of B virus in macaque and human hosts: a review. Clin Infect Dis 14:555–567
Yamanouchi K, Fukuda A, Kobune F, Hikita M, Shishido A (1969) Serologic survey with the sera of monkeys in regard to their natural infection with measles virus. Jpn J Med Sci Biol 22:117–121

Simian Hemorrhagic Fever

Philip M. Zack

Synonyms. Hemorrhagic fever in monkeys.

Clinicopathologic Features

Infection with simian hemorrhagic fever virus causes an acute febrile disease dominated by generalized impairment of hemostasis and coagulation (Abildgaard et al. 1975). Measurable fibrin degradation products and elevated body temperature are the only abnormal parameters observed during the initial stage of infection, which may last several days. Subsequently, there is a transient period in which the concentrations of several serum enzymes begin to increase, including lactate dehydrogenase, aspartate aminotransferase, alanine aminotransferase, and creatine kinase. The terminal stage of the clinical syndrome is characterized by fever, anorexia, facial edema, depression, photophobia, ataxia, cyanosis, diarrhea, dehydration, epistaxis, ecchymotic and petechial hemorrhages, and abortions. A typical clinical pathology profile of a macaque in this stage of the disease is presented in Table 6. Hemostatic abnormalities indicate disseminated intravascular coagulation. Thrombocytopenia may be severe with platelet numbers reaching nadirs of less than 30 000/μl. Additionally, there is progressive total loss of platelet aggregation in response to adenosine diphosphate or collagen. In all animals the activated partial thromboplastin time is prolonged three to four times control values. The prothrombin time, however, typically remains within normal limits. There is significant depletion of prothrombin; factors V, VIII, IX, and XI; and fibrinogen (to concentrations less than 100 mg/ml). In moribund animals, fibrin degradation products may be elevated extensively to levels exceeding 200 μg/ml, i.e., still detectable at serum dilutions greater than 1/256 in a latex agglutination test (Thrombo Wellco Kit, Wellcome Reagents Ltd.). The protamine sulfate test for fibrin monomers and early, clottable, fibrin degradation products is consistently positive. Schistocytes are noted occasionally on peripheral blood smears from animals with high concentrations of fibrin degradation products. In the terminal stage of disease, lactate dehydrogenase values frequently may exceed 30 000 IU/l with proportional elevation of all isoenzyme fractions, while creatine kinase values may be elevated to more than 10 000 IU/l. Alanine aminotransferase, aspartate aminotransferase, and γ-glutamyltransferase levels are moderately increased. Proteinuria ranging from 100 to

Table 6. Typical clinical pathology profile in simian hemorrhagic fever

Parameter	Normal range	End-stage infected macaque
Platelet count	230–650 × 10^3/μl	<100 × 10^3/μl
Platelet aggregation	100%	0%
Activated partial thromboplastin tine	20–30 s	60– > 100 s
Prothrombin time	10–13 s	12–15 s
Fibrinogen	180–400 mg/ml	<90 mg/ml
Fibrin degradation products	2 μg/ml (<1/8)	>200 μg/ml (1/256)
Protamine sulfate	Negative	Positive
Lactate dehydrogenase	42–426 IU/l[a]	>15000 IU/l
Creatine kinase	24–645 IU/l	>5000 IU/l
Alanine aminotransferase	5–61 IU/l	100–500 IU/l
Aspartate aminotransferase	12–63 IU/l	>1000 IU/l
γ-Glutamyltransferase	22–53 IU/l	>100 IU/l

[a] IU/l, International units per liter.

1000 mg/ml is present with a slight increase in blood urea nitrogen (e.g., 50 mg/dl). The serum creatinine concentration is typically within normal limits. Blood cultures are usually negative, but when septicemia is identified there are typically ulcers of the lower intestinal tract and other associated clinical and pathological changes, including prolonged prothrombin times.

Gross Appearance

Gross lesions associated with simian hemorrhagic fever virus infection generally are observed only in the final 24 h before death and, even then, their presence can be species-dependent. Southeast Asian macaques appear to die more quickly after infection and in many of these primates, gross lesions may be totally absent regardless of their clinicopathologic profile. In one epizootic, 50 imported cynomolgus monkeys (*Macaca fascicularis*) died or were euthanatized without significant gross lesions, even though rhesus monkeys infected while residing in the facility displayed more typical macroscopic changes (DM Renquist, personal communication). Although gross lesions of simian hemorrhagic fever are not pathognomonic, they are (and have been) critical in determining a differential diagnosis, establishing proper safety measures, and preventing continued transmission during the early stages of an epizootic. The most prominent and important gross lesions are congestion, hemorrhage, and necrosis of the proximal duodenal mucosa, starting abruptly at the pylorus and extending distally for 5–10 cm (Fig. 122) (Allen et al. 1968; Abildgaard

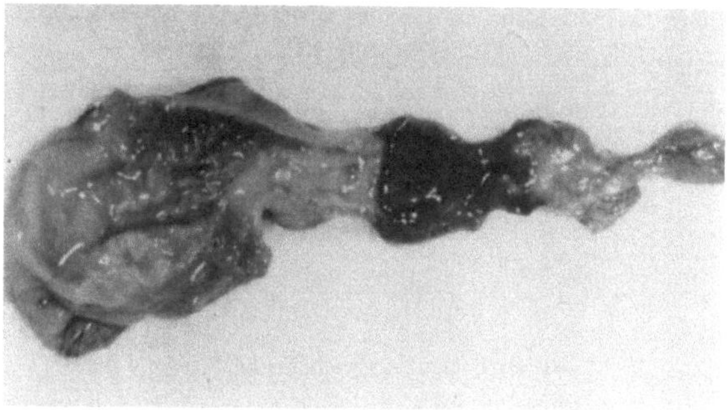

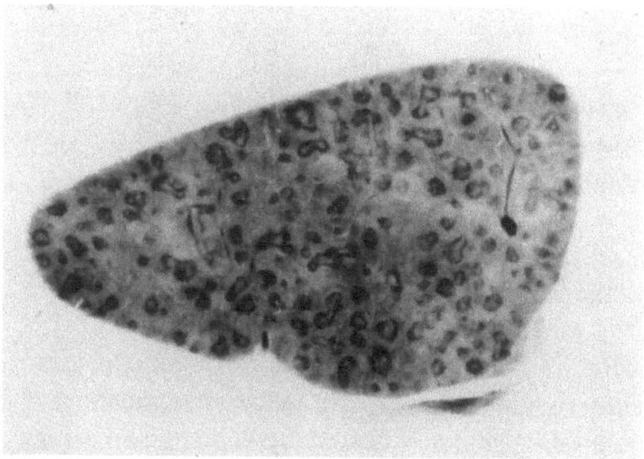

Fig. 122 (*above*). Stomach and duodenum, *Macaca mulatta*, simian hemorrhagic fever. Hemorrhage and necrosis of the proximal duodenal mucosa beginning abruptly at the pyloric valve. This lesion may be seen in other hemorrhagic fevers and probably indicates disseminated intravascular coagulation in macaques. (Photograph courtesy of Anton M. Allen)

Fig. 123 (*below*). Spleen, *Macaca mulatta*, simian hemorrhagic fever. After fixation, hemorrhage into the marginal sinuses around the white pulp (perifollicular hemorrhage) is more easily visualized. This lesion may be seen in other hemorrhagic fevers and probably indicates disseminated intravascular coagulation in macaques. (Photograph courtesy of Anton M. Allen) et al. 1975). Other large foci of mucosal hemorrhage and necrosis may be present randomly throughout the gastrointestinal tract. Discrete ulcers occasionally are observed. Splenic lesions are consistently present but are more difficult to interpret in fresh tissues. Spleens are enlarged and their weights are increased two to three times normal (15–30 g). The cut surface usually is a homogenous dark purple-red without visible white pulp, although, occasionally, faint red perifollicular haloes may be present, producing a mottled appearance. The red pulp is often dry and does not contain much free blood. After fixation dark red rims of hemorrhage within the marginal sinuses around the white pulp may be seen easily due to more efficient leaching of hemoglobin from the red pulp (Fig. 123). Hemorrhages throughout the body are a consistent finding. Typically, epistaxis and hematomas after venipuncture identify animals entering the final stage of disease. Petechiae usually are present in the skin and throughout the mucosa and serosa of thoracic and abdominal viscera. Periocular, nasal, perirenal, pulmonary, and retroperitoneal hemorrhages have been described. In many animals lymph nodes are enlarged two or four times normal and have a granular appearance along the surface of the capsule. The synovial membranes are often extremely congested without evidence of hemorrhage into the synovial fluid. In cases with meningoencephalitis, brain weights may be increased one and a half to twice normal, although grossly visible evidence of inflammation is absent.

Microscopic Features

There is no pathognomonic lesion in simian hemorrhagic fever virus-infected macaques and not all the changes described are seen in every case of the disease. Additionally, most of the following lesions may be observed in other hemorrhagic fevers and it is the lack of histologic changes in selected organs that allows differentiation. A definitive diagnosis can be made histologically but requires examination of tissues from more than one animal. Lesions are seen only in animals during the final stage of disease when selected serum enzyme levels and the activated partial thromboplastin time are elevated.

Microscopic changes are present most consistently in lymphoid tissues including the spleen, lymph nodes, thymus, and mucosa-associated lymphoid tissue. Typical histologic changes in the spleen from a moribund animal are illustrated in Figs. 124–127. There is severe necrosis of follicular center cells and marked depletion of lymphocytes in follicular mantles and in the periarteriolar lymphoid sheaths (Figs. 124–125). This change accounts for the inability to visualize white pulp on gross examination. However, if animals are not in extremis, most cells of the white pulp appear unaffected and prominent numbers of tingible body macrophages are present. Severe lymphocytic depletion is not observed in other lymphatic tissues and its presence in the spleen terminally may be due to ischemia. A characteristic feature in spleens is severe hemorrhage within the marginal sinuses around the white pulp, a change often described as perifollicular hemorrhage (Figs. 124–125) (Allen et al. 1968). This change, however, is not unique to simian hemorrhagic fever and may be observed in macaques infected with other hemorrhagic fever viruses or dying from endotoxic shock or disseminated intravascular coagulation. In the macrophage-lined cords of the red pulp, there is massive fibrin deposition and karyorrhectic debris associated with severe necrosis of the cord-lining cells. The venous sinusoids, however, lack fibrin thrombi, remain patent and congested, and are lined by hypertrophied endothelial cells (Fig.

Fig. 124 (*above*). Spleen, *Macaca mulatta*, simian hemorrhagic fever. Note the necrosis of follicular center cells and depletion of follicular mantle lymphocytes (*arrowheads*). The marginal sinus is filled with hemorrhage (*arrows*). The red pulp is pale due to fibrin deposition. This lesion is present uniformly during the final stage of infection. *Bar*, 100 μm; H and E, ×100 ▶

Fig. 125 (*lower left*). Spleen, *Macaca mulatta*, simian hemorrhagic fever. Higher magnification of Fig. 124. Karyorrhectic debris is extensive within the follicular center and extends into the mantle. Note that not all follicular center cells are affected and that a monomorphic population of cells with vesicular nuclei remain viable. *Bar*, 50 μm; H and E, ×200

Fig. 126 (*lower right top*). Splenic red pulp, *Macaca mulatta*, simian hemorrhagic fever. The red cords are filled with fibrin, karyorrhectic debris, and trapped erythrocyte and white blood cells. Red pulp cord lining cells are necrotic. The sinusoids (*astrisks*) remain patent and congested and are lined by hypertrophied endothelium. *Bar*, 10 μm; H and E, ×300

Fig. 127 (*lower right bottom*). Splenic red pulp, *Macaca mulatta*, simian hemorrhagic fever. Fibrin strands (dark fibrillar material) form concentric layers around the sinusoidal vessel walls (*S*) and surround and trap cells within the cord spaces. *Bar*, 10 μm; PTAH, ×400

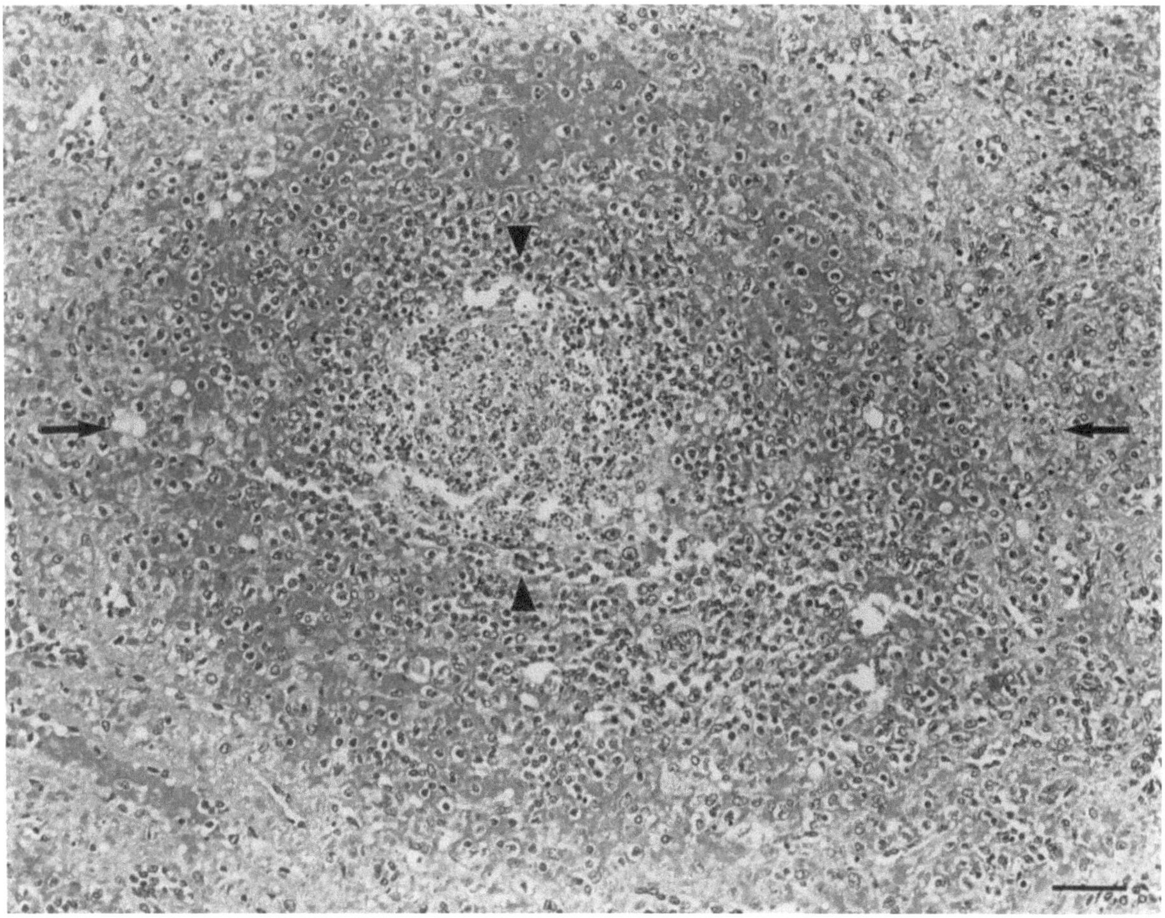

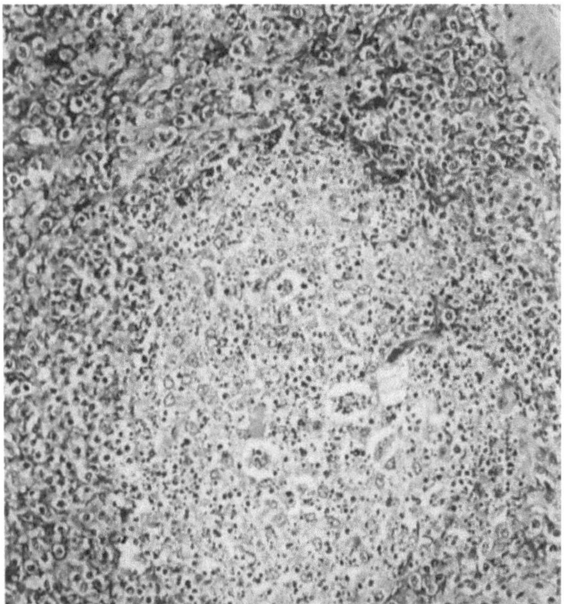

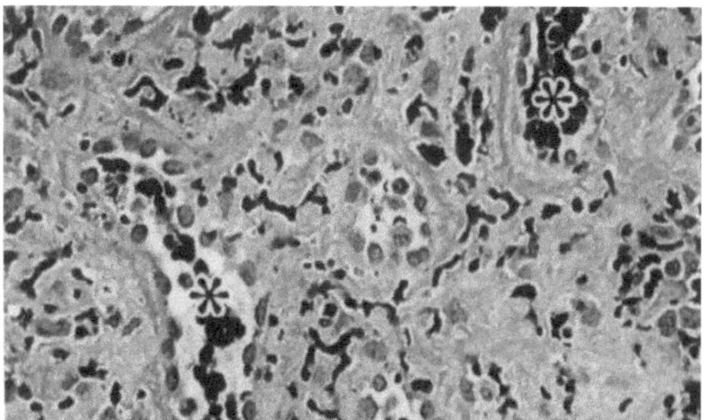

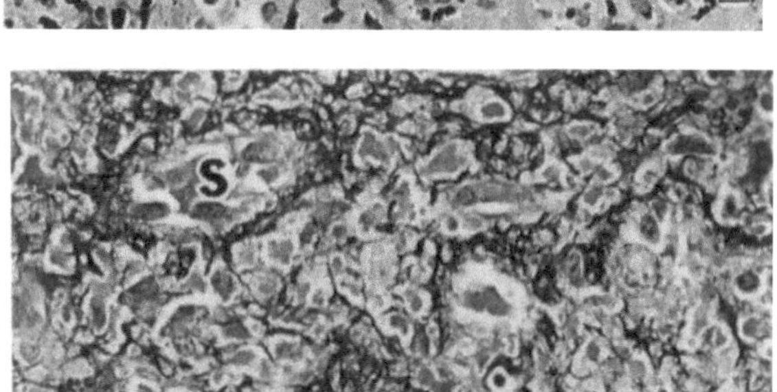
S
S

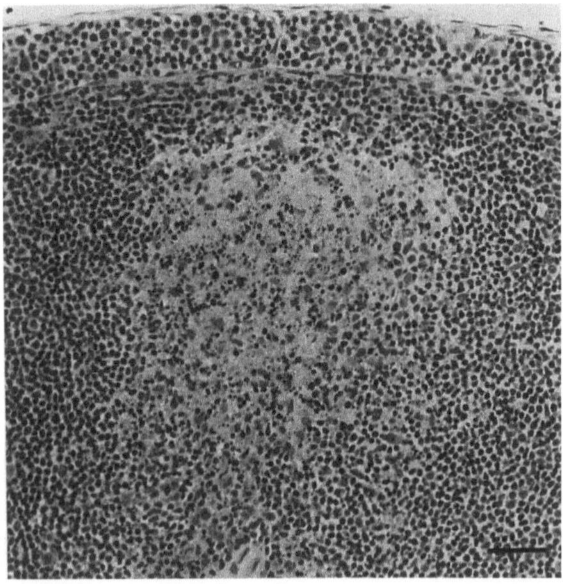

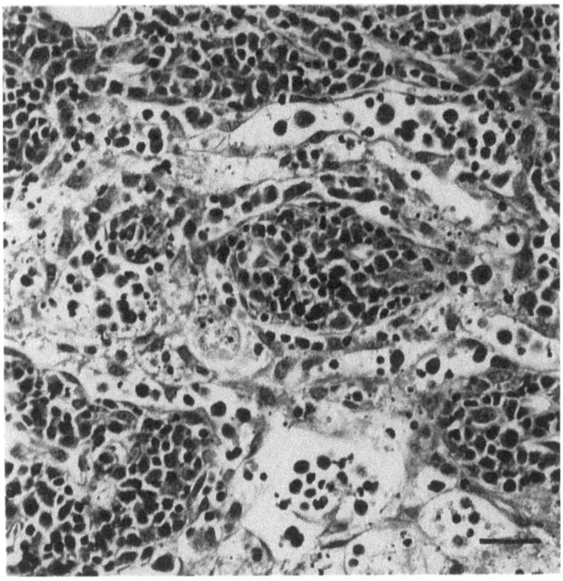

Fig. 128 (*above*). Lymph node, *Macaca mulatta*, simian hemorrhagic fever. Cortex; severe necrosis of follicular center cells. The mantle appears normal. *Bar*, 50 μm; H and E, ×150

Fig. 129 (*below*). Lymph node, *Macaca mulatta*, simian hemorrhagic fever. Medulla; karyorrhectic debris and degenerating cells are present within the sinusoidal spaces. Notice that cord cells and sinusoidal lining cells appear normal. *Bar*, 50 μm; H and E, ×150

126). This pattern is demonstrated readily with a phosphotungstic acid-hematoxylin (PTAH) stain for fibrin (Fig. 127).

Fibrin typically forms concentric layers around the sinusoidal walls and extends out into the cord spaces, trapping red and white cells and karyorrhectic debris. Some of the homogenous eosinophilic material deposited in the cords does not stain positively with PTAH and has been interpreted as trapped plasma. In other lymphatic tissues, including lymph nodes and mucosa-associated lymphoid tissue, there is severe necrosis of follicular center cells (Fig. 128). Tingible body macrophages may be scattered throughout the cortex and medulla, but necrosis of lymphoid cells, reticular cells, or sinusoidal lining cells is not observed. The sinuses, however, are filled with degenerating and necrotic cells and cellular debris (Fig. 129). The lesion in the thymus is unique. The thymic cortex undergoes complete necrosis while the medulla is spared and contains scattered tingible body macrophages (Fig. 130). This lesion has not been observed in other hemorrhagic fevers.

Microscopic changes in the proximal duodenum include severe congestion and hemorrhage in the villi, often accompanied by necrosis of villus epithelium (Fig. 131). There is edema in the submucosa and lamina propria. Vessels in the submucosa are congested; usually the large veins are distended with plasma and frequently contain strands of fibrin or even fibrin thrombi. Mild to moderate mucosal hemorrhages are observed sporadically elsewhere in the gastrointestinal tract and occasionally are accompanied by necrosis and ulceration. Other lesions include multifocal distention and necrosis of the villus tip lamina propria (Fig. 132) and deposition of karyorrhectic debris within intact, viable, mucosal epithelial cells. The latter change is noted most frequently along the villi and may represent necrosis of intraepithelial lymphocytes which are notably absent in affected regions (Fig. 133).

Fibrin thrombi may be observed within the capillaries of many organs, although fibrin deposition often is limited to the splenic red pulp. The most common location for fibrin thrombi is within the peritubular capillaries of the proximal renal medulla (Figs. 134–135).

Other sites of thrombosis include renal glomeruli, chorioid plexuses, hepatic sinusoids, adrenal medullae, venous plexuses of the nasal septum, episcleral vasculature, and placental vessels. Infarcts occasionally accompany thrombosed vessels and are often numerous in the kidney.

Meningoencephalitis is not present in all macaques infected with simian hemorrhagic fever but is an important lesion seen in animals surviving for longer periods and reaching the final stage of disease. Changes consist of lymphohistiocytic and neutrophilic infiltrates within the meninges, choroid plexus, and around vessels in

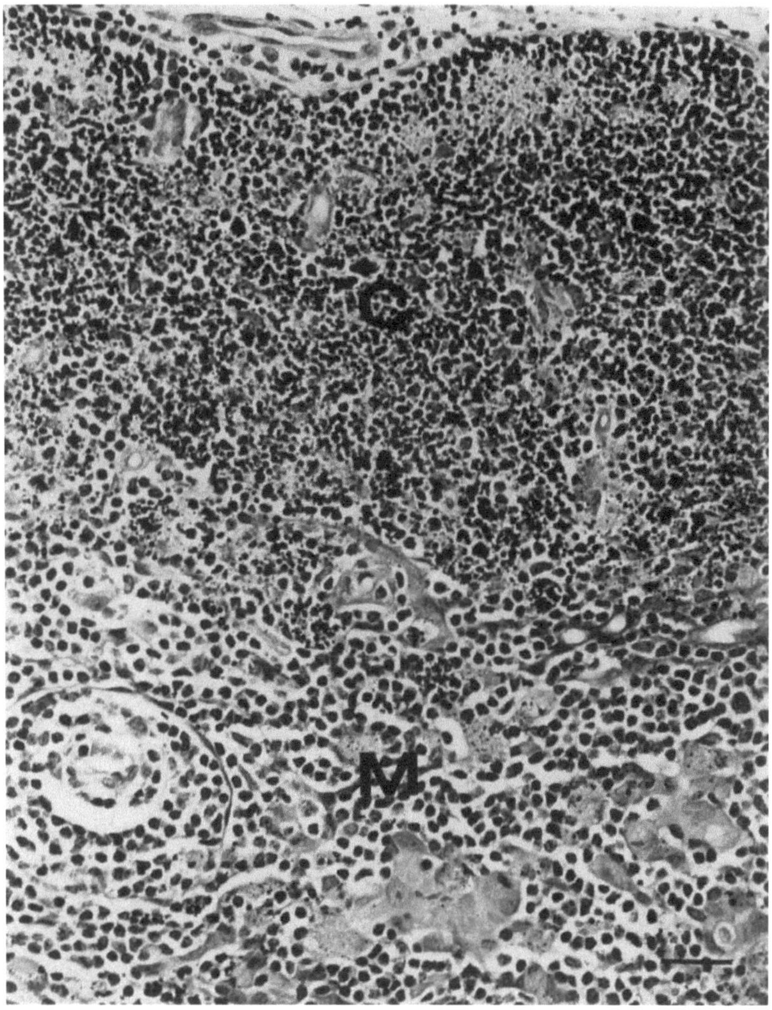

Fig. 130. Thymus, *Macaca mulatta*, simian hemorrhagic fever. The cortex (*C*) has undergone complete necrosis while the medulla (*M*) is spared. This lesion is not observed in other hemorrhagic disease viral infections. *Bar*, 50 μm; H and E, ×200

the neural parenchyma (Figs. 136–138). Focal glial nodules and perivascular foci of malacia may occasionally be observed. In some cases there is necrosis of selected cells, possibly macrophages, within a perivascular infiltrate, while lymphocytes appear to be spared.

Other lesions sporadically seen include lymphohistiocytic epididymitis, blepharitis, placental necrosis, and retinitis. Retinal lesions are located in the nerve fiber layer and appear identical to those seen in the central nervous system. They are present only in animals with meningoencephalitis. The absence of inflammation and necrosis (other than occasional infarcts) in the liver, adrenal glands, and lung is extremely important to rule out other hemorrhagic fever viral infections.

Ultrastructure

Even through moribund animals have high serum viral titers, identifying cells containing replicating virus in tissues is not rewarding and is not useful diagnostically. Ultrastructural examination of spleen from a macaque entering the terminal stage of infection revealed viral growth within rare macrophages of the red pulp cords (T.W. Geisbert, unpublished data). This replication pattern is duplicated in cell cultures in which virions are visualized primarily during a limited period shortly after inoculation, even though concentrations of intracellular viral antigens continue to increase until cell death. All strains of simian hemorrhagic fever virus grow extremely well in primary macaque macrophage cultures, while some strains may be

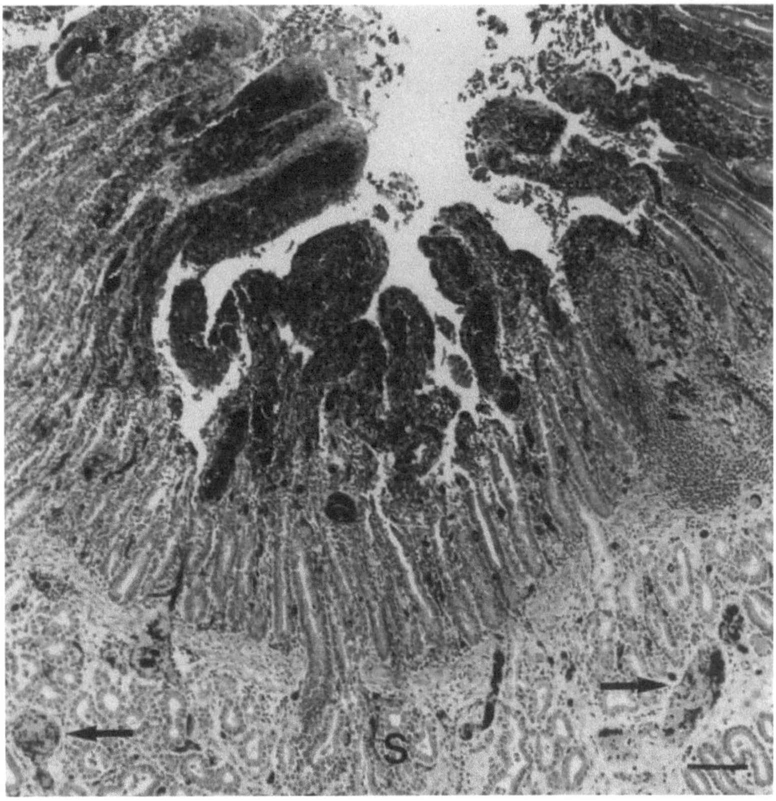

Fig. 131. Duodenum, *Macaca mulatta*, simian hemorrhagic fever. Note severe hemorrhage within the lamina propria of the villus and necrosis of villus epithelial cells. Other vessels in the mucosa and submucosa (*S*) are congested. *Bar*, 100 μm; H and E, ×40

propagated successfully in MA-104 embryonic, African green monkey kidney cell line, producing characteristic ultrastructural changes (Gravell et al. 1980a; Tauraso et al. 1971a,b; Wood et al. 1970). These are useful diagnostically.

Using MA-104 cells with the LVR strain of simian hemorrhagic fever virus, the earliest changes occur as early as 24 h after inoculation, at which time there are multifocal condensation of polyribosomes within the cytoplasm (Wood et al. 1970). During the next 24 h, double membrane lamellar structures form, ranging from 50 to 1000 nm in diameter, which are conical in three dimensions and contain clusters of polyribosomes centrally. The membranes eventually fuse to form long, circular, double-layered, filamentous ribbons 50–60 nm in diameter and several microns in circumference in cross-section (Fig. 139). These replicative structures are unique among viruses pathogenic for primates and are diagnostic in cell culture. By 72 h post-inoculation, mature virions are observed budding into dilated cisternae of the rough endoplasmic reticulum. Virions of simian hemorrhagic fever virus average 40–45 nm in diameter, are enveloped, and contain 22–25 nm cores with electron lucent centers (Fig. 140).

Differential Diagnosis

Since naturally occurring simian hemorrhagic fever infections are associated with epizootics, the differential diagnoses include all rapidly transmissible etiologic agents that produce disseminated intravascular coagulation in primates. Although macaques are susceptible to a number of viral hemorrhagic fevers experimentally, there are four agents that are associated with natural disease, and each represents a significant human health hazard. They include: Ebola, Marburg, yellow fever, and Kyasanur Forest disease viruses. Ebola and Marburg viruses produce both disseminated intravascular coagulation uniformly in macaques and many of the same lesions seen in simian hemorrhagic fever, includ-

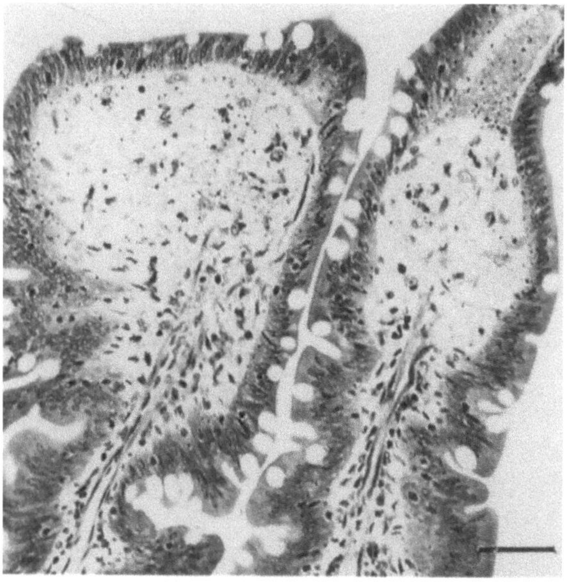

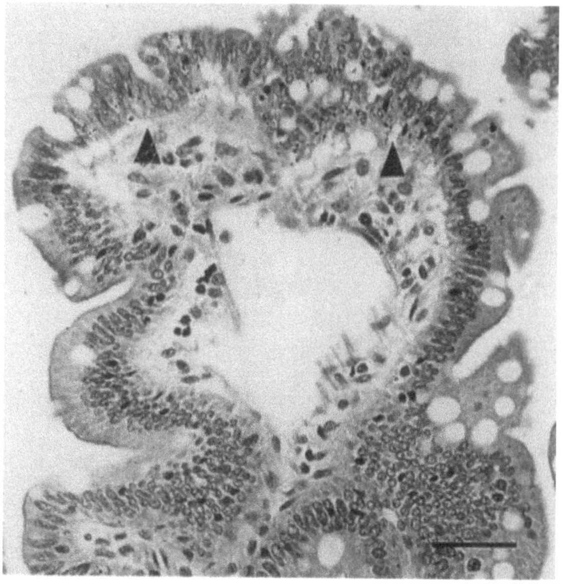

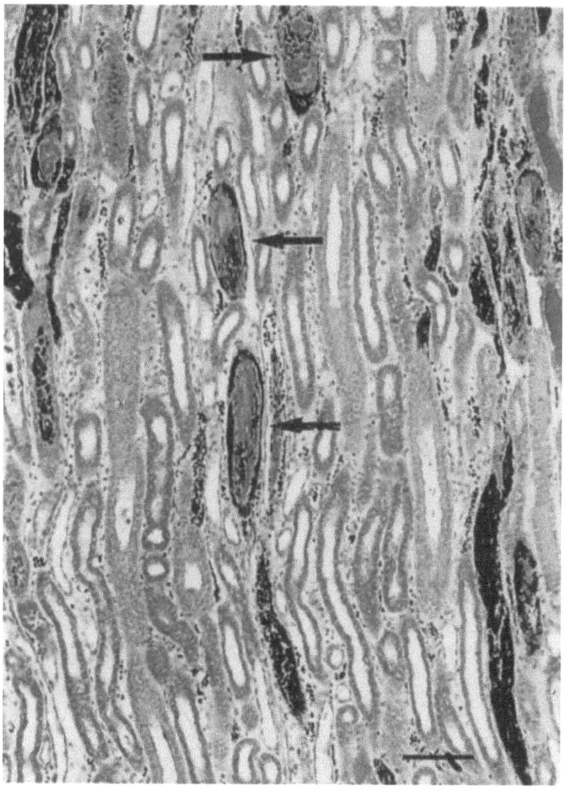

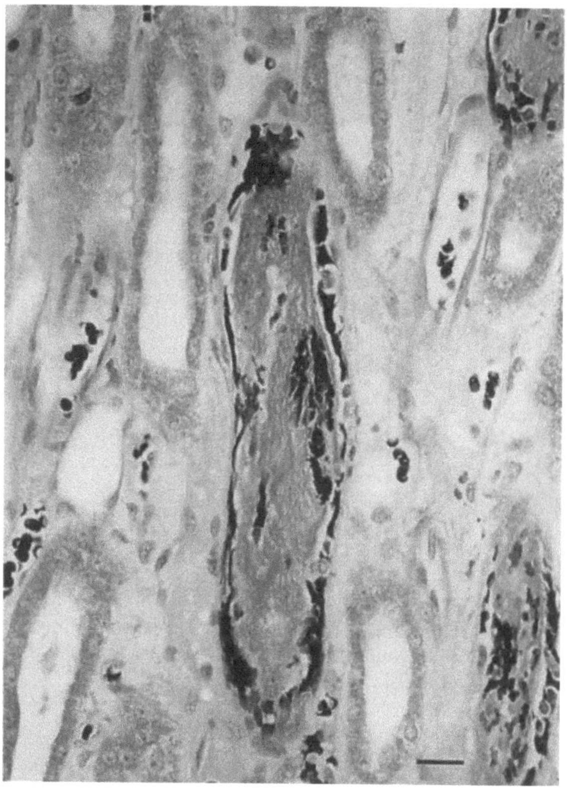

Fig. 132 (*upper left*). Small intestine, *Macaca mulatta*, simian hemorrhagic fever. The lamina propria at the villus tip is distended and filled with cellular debris and an eosinophilic coagulum that is not fibrin. This lesion may be seen in other hemorrhagic fevers. *Bar*, 50 μm; H and E, ×150

Fig. 133 (*upper right*). Small intestine, *Macaca mulatta*, simian hemorrhagic fever. Karyorrhectic debris is present within intact, normal-appearing villus epithelial cells (*arrowheads*). Its deposition may be due to necrosis of intraepithelial lymphocytes. *Bar*, 50 μm; H and E, ×150

Fig. 134 (*lower left*). Kidney, *Macaca mulatta*, simian hemorrhagic fever. Fibrin thrombi (*arrows*) occlude distended peritubular capillaries in the proximal renal medulla. This is a common location for thrombi in primates infected with simian hemorrhagic fever and other hemorrhagic disease viruses. *Bar*, 100 μm; H and E, ×40

Fig. 135 (*lower right*). Kidney, *Macaca mulatta*, simian hemorrhagic fever. High magnification of peritubular capillary thrombus. *Bar*, 100 μm; H and E, ×300

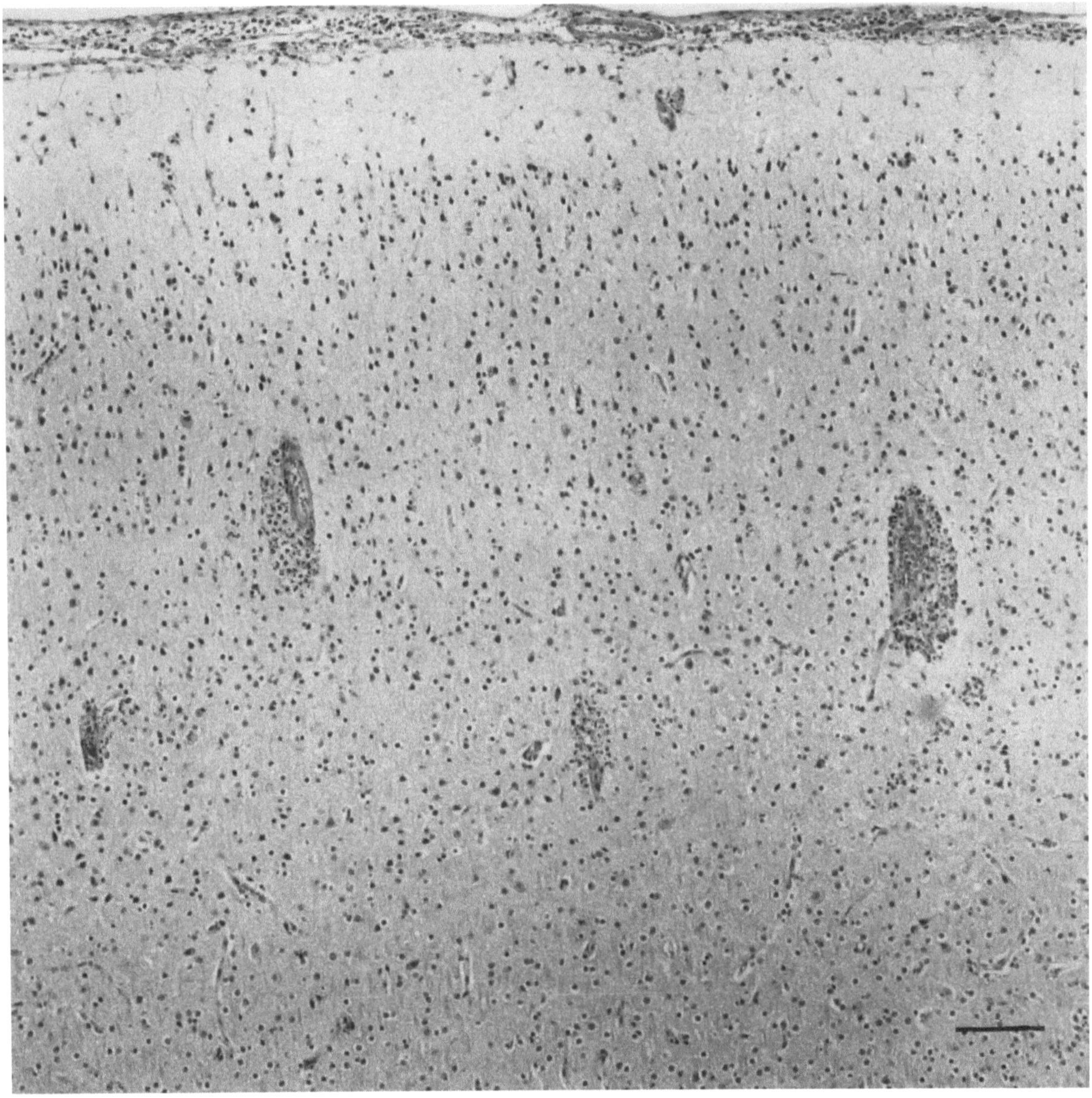

Fig. 136. Cerebrum, *Macaca mulatta*, simian hemorrhagic fever. Lymphohistiocytic infiltrates are present in the meninges and around vessels in the neural parenchyma. This lesion is not observed in other hemorrhagic viral infections. *Bar*, 100 μm; H and E, ×50

ing those involving the spleen and duodenum. Several major features allow differentiation between simian hemorrhagic fever and other viral hemorrhagic fevers. Prothrombin times re prolonged in Marburg, but not Ebola, viral infections. Hepatic necrosis with Councilman's bodies (apoptotic bodies) and higher levels of alanine aminotransferase are features of all the hemorrhagic fevers except simian hemorrhagic fever.

The alanine aminotransferase typically exceeds the aspartate aminotransferase in these infections while the reverse occurs in simian hemorrhagic fever. Necrosis of adrenal cortical cells and of pulmonary alveolar and bronchiolar epithelium occurs in cases of Marburg and Ebola viral hemorrhagic fevers. Ebola viral infections are pantropic and produce large, amphophilic, intracytoplasmic, inclusion bodies in many different tissues including the liver, adrenal gland, and spleen.

There are differences in the epidemiology of the various hemorrhagic fevers. Yellow fever and Kyasanur Forest disease are less likely to occur in epizootics in a laboratory setting because they

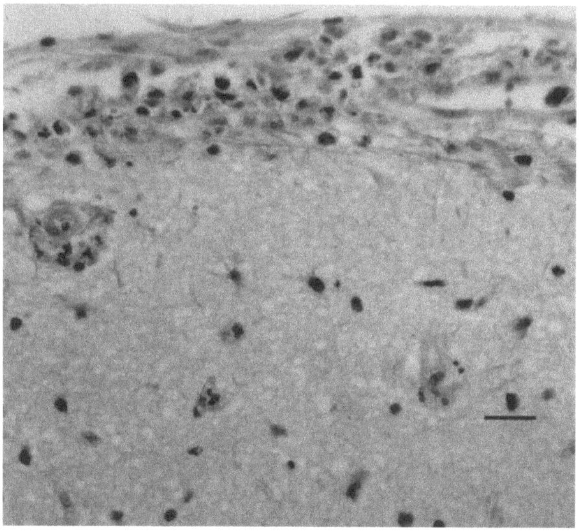

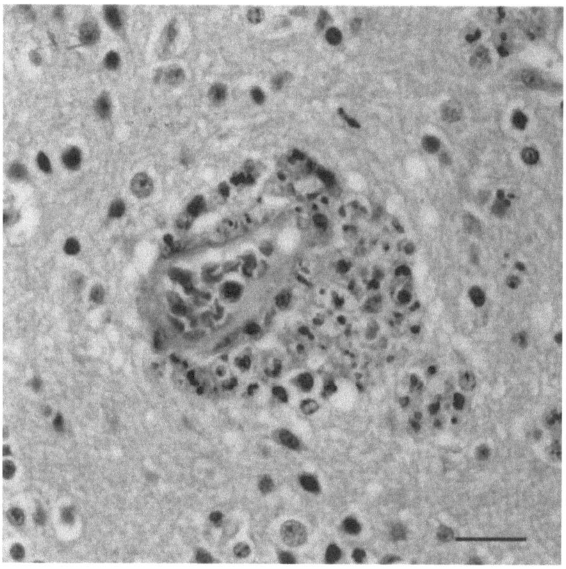

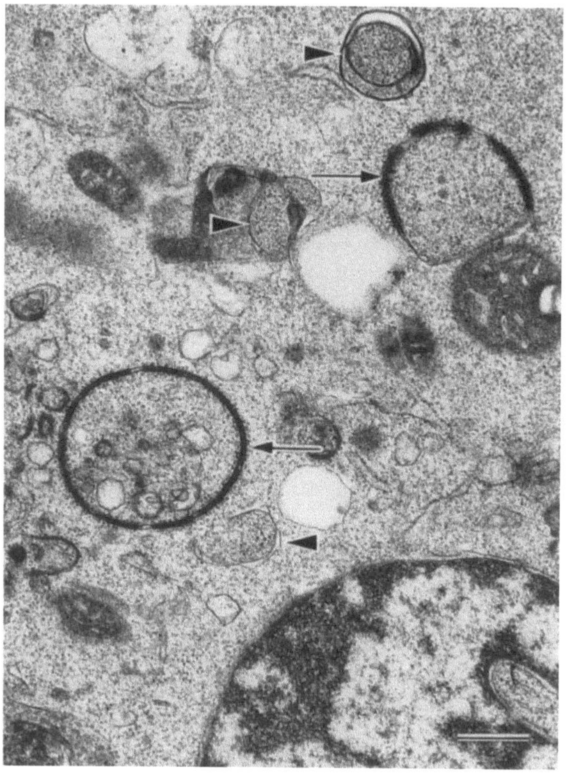

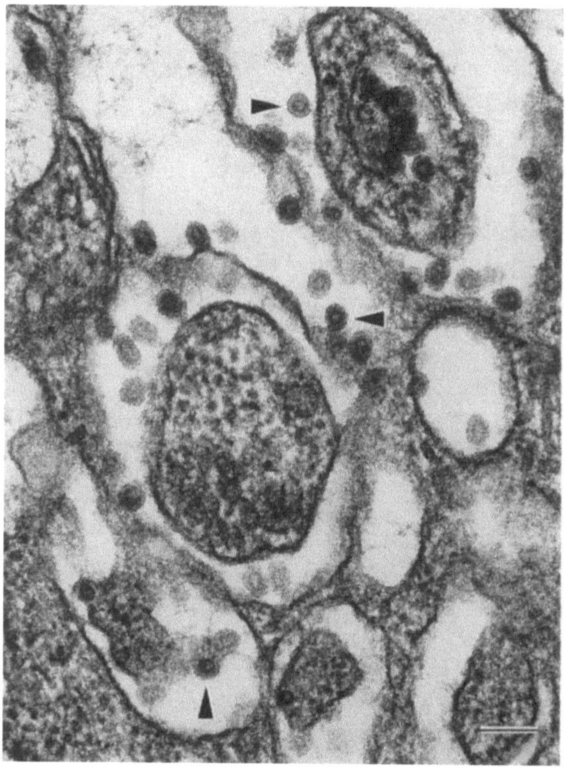

Fig. 137 (*upper left*). Cerebrum, *Macaca mulatta*, simian hemorrhagic fever. Meningeal infiltrate consisting of lymphocytes, histiocytes, and neutrophils. *Bar*, 20 μm; H and E, ×300

Fig. 138 (*lower left*). *Cerebrum, Macaca mulatta*, simian hemorrhagic fever. Perivascular cuff containing lymphocytes, histiocytes, neutrophils and karyorrhectic debris. *Bar*, 20 μm; H and E, ×400

Fig. 139 (*upper right*). MA-104 cells infected with simian hemorrhagic fever virus. Note double membrane lamellar structures containing central condensations of ribosomes (*arrowheads*) and long, circular, electron-dense, filamentous ribbons (*arrows*). Bar, 500 nm; TEM, ×19 175. (Electron micrograph courtesy of Thomas W. Geisbert)

Fig. 140 (*lower right*). MA-104 cells infected with simian hemorrhagic fever virus. Virions replicate by budding into the cisternae of the rough endoplasmic reticulum. Viral particles average 40–45 nm in diameter, are enveloped, and contain 22–25 nm cores with electron-lucent centers (*arrowheads*). Bar, 100 nm; TEM, ×75 900. (Electron micrograph courtesy of Thomas W. Geisbert) usually require biological or mechanical vectors for transmission. However, there has been at least one unreported epizootic of yellow fever attributed to aerosol transmission of virus between monkeys, some of which were originally infected from contaminated mouse cages. Additional differential histopathologic features include midzonal hepatic necrosis and nephrosis in yellow fever, and intestinal villus blunting with crypt necrosis and atrophy in Kyasanur Forest disease. Meningoencephalitis occasionally may be seen in either of these two viral infections. Electron microscopic examination of tissues or cell cultures easily differentiates the filoviruses from the agent of simian hemorrhagic fever. However, experience and/or immunoelectron microscopy are required to distinguish Ebola virus from Marburg virus. Finally, immunohistochemistry assays, which may be performed on paraffin-embedded tissues, exist for all the hemorrhagic fever viruses except simian hemorrhagic fever (Pathology Division, US Army Medical Institute of Infectious Disease, Ft. Dietrick, MD). Unfortunately, none of these assays is commercially available.

Biologic Features

Viral Characteristics

Simian hemorrhagic fever virus is currently classified as a *Flavivirus*. Recent studies of its genome have demonstrated that it has genetic structural similarities with some members of the coronavirus family (M. Brinton, personal communication). Interestingly, it shares homology with two other unclassified togaviruses, lactate dehydrogenase-elevating virus of mice and equine viral arteritis virus. The extensive elevation of several serum enzymes, including lactate dehydrogenase, observed in simian hemorrhagic fever, has also been described in lactate dehydrogenase-elevating virus (Brinton 1982). Simian hemorrhagic fever virus may be propagated in vitro in either macaque peritoneal macrophages or in the African green monkey kidney cell line MA-104 (Gravell et al. 1980a).

Epidemiology

Simian hemorrhagic fever virus is only lethal for macaques. Wild populations of patas monkeys are enzootically infected with the virus (an incidence of 48.6% in one study) and serve as the main reservoir host (Gravell et al. 1980b). Many of these animals carry the virus for life and experience persistent viremia without associated disease (London 1977). Asymptomatic infection in a small number of African green monkeys and baboons has also been reported. As some African primates are viremic without demonstrable circulating antibody, they cannot be screened or separated on this basis and represent a hidden threat as a virus reservoir (Gravell et al. 1980b). Other laboratory animals and human beings are not susceptible to this disease. Parenteral exposure to blood or tissue fluids appears to be required for the initial transfer of virus from patas monkeys to macaques. Common housing of patas monkeys and macaques, or employment of the same personnel to care without adequate decontamination for both species, has proved to predispose colonies to the disease. One epizootic was traced to a multidose vial used for animals of both species without changing the injection needle. Once infected, macaques shed the virus and transmission may occur via direct contact, fomites, droplets, or aerosol (London 1977). Epizootics of simian hemorrhagic fever in laboratories are characterized by explosive outbreaks of disease in colonies of macaques, with nearly 100% mortality. This disease was first reported in 1964 at NIH, Bethesda, Maryland, USA, and at the Sukhumi Institute, U.S.S.R. These two outbreaks occurred within weeks of each other after importation of animals from a common source. Subsequent to these reports, epizootics have occurred sporadically in the U.S., most recently at a colony in New Mexico in 1988 (Espana 1971; London 1977; Shelokov et al. 1971; Zack et al. 1989).

Pathogenesis

Limited information is available on the pathogenesis of simian hemorrhagic fever. Experimental studies have been performed and the clinicopathologic and pathologic lesions of the disease syndrome characterized (Abildgaard et al. 1975; Palmer et al. 1968). However, several key questions remain unanswered. For example, the incubation period ranges from 3–7 days, but most deaths occur 7–13 days after inoculation. This delay between inoculation and onset of clinical signs is unexplained. Even though macaques are viremic and have measurable levels of fibrin degradation products within 24 h after inoculation, several days elapse until serum enzymes become elevated, and there is evidence of disseminated intravascular clotting. Gross and histologic changes are not present until the final stage of disease, suggesting that lesions are largely sequelae of the coagulopathy. This interpretation is supported by the identification of essentially similar changes, including proximal duodenal hemorrhage and red pulp fibrin deposition, in macaques experiencing fatal disseminated intravascular coagulation due to various causes, including other hemorrhagic disease viruses. The target cells for simian hemorrhagic fever infection in vivo are unknown. Current data suggest that cells of the mononuclear phagocyte system are capable of supporting viral replication. In vitro studies indicate that primary rhesus peritoneal macrophages are highly sensitive to infection. Limited evidence also supports macrophage infection in vivo, including ultrastructural data on viral replication in a single macaque in splenic red pulp cord cells and in circulating monocytes within the liver (T.W. Geisbert, unpublished data). Antigen has been identified in impression smears from various organs by using immune sera from an African green monkey in an indirect fluorescent antibody assay (Shevtsova et al. 1975). In this study, virus-specific antigen was interpreted to be present within capillary endothelium, neurons, and glial cells of the brain and in the reticuloendothelial cells and macrophages of the spleen and liver.

The virus is lytic for macrophages in culture and there is histologic evidence for macrophage lysis within lymphatic tissues and even within lymphohistiocytic infiltrates in other organs, such as the gastrointestinal tract and brain. However, necrosis is not seen during the early stage of disease and is never observed in certain subpopulations of the mononuclear phagocytic system, such as Kupffer's cells or alveolar macrophages. The related lactate dehydrogenase elevating virus of mice may provide clues to the pathogenesis of lesions of simian hemorrhagic fever. Tropism for selected cells of the mononuclear phagocytic system is described for this murine virus (M. Brinton, personal communication). Increased concentrations of all lactate dehydrogenase isoenzymes, creatine kinase, and all liver enzymes occur in both viral infections. The elevation of these enzymes has been interpreted to result from reduced clearance by the mononuclear phagocytic systems (Brinton 1982). Marked necrosis of thymic lymphocytes is also evident in murine lactate dehydrogenase elevating virus infection and has been ascribed to an increase in the levels of circulating adrenal cortical hormones. Similarly, necrosis of cortical thymic lymphocytes in simian hemorrhagic fever could be the result of the extensive increase in serum glucocorticoids observed in all but moribund animals (Goncharov et al. 1971). This change, however, is not seen in any other viral infections or conditions associated with severe stress.

Limited published information is available characterizing the immune response to simian hemorrhagic fever. Serum antibodies are present in monkeys dying later in the course of disease and are high in the few animals that survive after intensive supportive treatment (Tauraso et al. 1968, 1971a). Antibodies have also been detected in a few asymptomatic macaques exposed to sick animals during an epizootic. Antibodies could be an important contributing factor in disease pathogenesis. The time required for antibody production could explain the delay between initial viremia and onset of the coagulopathy. Immune complex activation of the intrinsic coagulation pathway would prolong the activated partial thromboplastin time but not the prothrombin time. Additionally, the presence of meningoencephalitis in animals surviving longer suggests that antibody may be necessary for viral transit across the blood-brain barrier.

Treatment

Polyriboinosinic-polyribocytidylic acid-poly-L-lysine is effective in preventing death when administered before experimental inoculation of the LVR strain of simian hemorrhagic fever virus (Levy et al. 1976). Heparin treatment was not apparently efficacious but the dose and method of administration in the reported trials might not have been adequate (Tauraso et al. 1971b). Similarly, prophylactic ribavirin therapy was not effective in preventing death in experimental monkeys infected with the LVR strain (P.B. Jahrling, personal communication).

Recommended Diagnostic and Safety Procedures

Although epizootics of viral hemorrhagic fever occur infrequently, they represent a potential human health threat, so that pathologists and laboratory animal veterinarians need to be familiar with the most efficient methods for identification and containment of the agent. The maximum biohazard safety level possible should be instituted any time there is high mortality associated with signs of hemorrhagic diathesis in a primate colony. Such action is not only important for human safety but also helps to limit spread of the agent within the colony. Access should be restricted immediately, and personnel entering the facility should use respirators fit with HEPA filters. If possible, a room for changing clothes and decontaminating personnel leaving animal areas needs to be identified. The exterior of all containers, as well as bags, sample boxes, clothing, etc., must be decontaminated before removing them from the facility. Necropsies should be performed with safety in mind at all times. A dedicated postmortem area located within the contaminated area/building that can be decontaminated is ideal. if necessary to restrict transmission and spread, necropsies may need to be performed in the affected animal room. Constant emphasis must be placed upon avoiding puncture wounds and cuts. Scalpels and/or sharp instruments are not recommended. Two pairs of gloves should be used and the outer pair changed as necessary. A virucidal solution must be available to disinfect cuts or needle-sticks, and respirators are necessary. Gross lesions are particularly important and close attention should be given to the duodenum, spleen, and liver. Initially, sample collection may be limited to serum, spleen, liver, adrenal gland, kidney, and lung. Tissues should be sampled for both virus isolation (frozen) and pathology (formalin). Clinical pathology poses a biohazard problem and can be instituted safely when human pathogens have been excluded as etiologic agents. Carcasses must be incinerated and all medical waste autoclaved or sterilized by incineration.

A reliable diagnosis may be made by histopathology and this method is safe as complete fixation will inactivate the viruses of concern. For this reason tissue samples should be no more than 3–4 mm thick and fixed by immersion for several days before processing. Particular attention to the identification of any hepatic and adrenal lesions is critical. If present, the agent may represent a significant human health threat. Any primate epizootic of hemorrhagic disease with high mortality should be reported to the special pathogens branch at the Centers for Disease Control, Atlanta, GA. Investigators there will be able to provide guidance and assistance to arrive at a specific diagnosis. Current CDC requirements should be followed for animals under quarantine (Centers for Disease Control 1990).

References

Abildgaard C, Harrison J, Espana C, Spangler W, Gribble D (1975) Simian hemorrhagic fever: studies of coagulation and pathology. Am J Trop Med Hyg 24:537–544

Allen AM, Palmer AE, Tauraso NM, Shelokov A (1968) Simian hemorrhagic fever: II. Studies in pathology. Am J Trop Med Hyg 17:413–421

Brinton MA (1982) Lactate dehydrogenase-elevating virus. In: Foster HL, Small JD, Fox JG (eds) The mouse in biomedical research, vol II. Academic, New York, pp 193–208

Centers for Disease Control (1990) Interim guidelines for handling nonhuman primates during transit and quarantine. MMWR 39:22–24

Espana C (1971) Review of some outbreaks of viral disease in captive nonhuman primates. Lab Anim Sci 21:1023–1031

Goncharov NP, Verberger K, Shubert K, Shevtsova ZV (1971) The secretory function of the adrenal cortex in Macacus Rhesus with simian hemorrhagic fever (in Russian). Patol Fiziol Eksp Ter 15:31–37

Gravell M, Palmer AE, Rodriguez M, London WT, Hamilton RS (1980a) Method to detect asymptomatic carriers of simian hemorrhagic fever virus. Lab Anim Sci 30:988–991

Gravell M, London WT, Rodriguez M, Palmer AE, Hamilton RS, Curfman BL (1980b) Studies on simian hemorrhagic fever virus infection of patas monkeys I. Serology. In: Montali RJ, Migaki G (eds) The comparative pathology of zoo animals. Smithsonian Institution, Washington, pp 167–170

Levy HB, London W, Fuccillo DA, Baron S, Rice J (1976) Prophylactic control of simian hemorrhagic fever in monkeys by an interferon inducer, polyriboinosinic-polyribocytidylic acid-poly-L-lysine. J Infect Dis 133 [Suppl]:A256–259

London WT (1977) Epizootiology, transmission and approach to prevention of fatal simian hemorrhagic fever in rhesus monkeys. Nature 268:344–345

Palmer AE, Allen AM, Tauraso NM, Shelokov A (1968) Simian hemorrhagic fever: I. Clinical and epizootiologic aspects of an outbreak among quarantined monkeys. Am J Trop Med Hyg 17:404–412

Shelokov A, Tauraso NM, Allen AM, Espana C (1971) Epizootic, clinical and pathological aspects of simian hemorrhagic fever. In: Martini GA, Siegert R (eds) Marburg virus disease. Springer, Berlin Heidelberg New York, pp 203–207

Shevtsova ZV, Karmysheva VI, Chumakov MP (1975) Virological study of simian hemorrhagic fever (in Russian). Vopr Virusol 4:471–476
Tauraso NM, Shelokov A, Palmer AE, Allen AM (1968) Simian hemorrhagic fever: III. Isolation and characterization of a viral agent. Am J Trop Med Hyg 17:422–431
Tauraso NM, Myers MG, McCarthy K, Tribe GW (1970) Simian hemorrhagic fever. In: Balner H, Beveridge WIB (eds) Infections and immunosuppression in subhuman primates. Williams and Wilkins, Baltimore, pp 101–109
Tauraso NM, Aulisio CG, Espana CD, Wood OL, Liebhaber H (1971a) Simian hemorrhagic fever virus. In: Martini GA, Siegert R (eds) Marburg virus disease. Springer, Berlin Heidelberg New York, pp 208–215
Tauraso NM, Kalter SS, Ratner JJ, Heberling RL (1971b) Simian hemorrhagic fever. In: Goldsmith EI, Moor-Jankowski J (eds) Medical primatology 1970. Proc. 2nd conf exp med surg primates, New York 1969. Karger, Basel, pp 660–670
Wood O, Tauraso N, Liebhaber H (1970) Electron microscopic study of tissue cultures infected with simian hemorrhagic fever virus. J Gen Virol 7:129–136
Zack PM, Trotter RW, Vogel AP, Hall WC, Jahrling PB, Parrish D, Renquist DM, Frost P (1989) An epizootic of simian hemorrhagic fever virus in macaques. Proceedings of the American College Veterinary Pathologists annual meeting, Baltimore, p 226

Chromobacteriosis, Nonhuman Primates

Harold M. McClure, Sherry A. Klumpp, and Daniel C. Anderson

Synonyms. Chromobacterium violaceum infection.

Gross Appearance

Gross pathologic changes associated with *Chromobacterium violaceum* infection may be evident in multiple organ systems, including the lungs, liver, spleen, lymph nodes, and kidneys. Affected organs are often swollen and congested with scattered petechial hemorrhages, and contain variable sized areas of necrosis that range from pinpoint to 3–4cm in diameter. When sectioned, larger lesions are often cavitary and contain a yellowish-white semifluid exudate (Fig. 141).

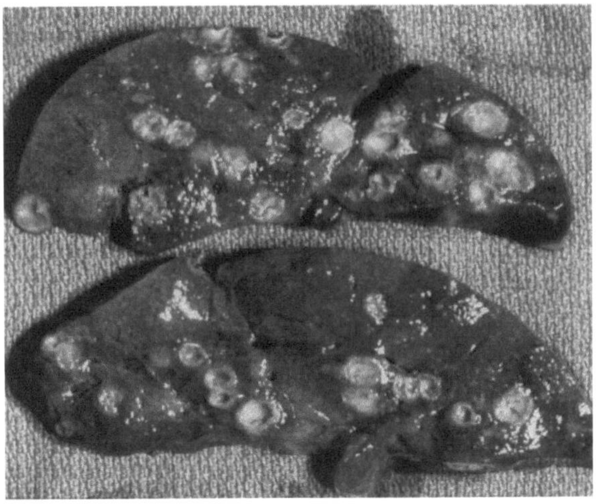

Fig. 141. Cut surfaces of liver of *Macaca assamensis* infected with *Chromobacterium violaceum*. Cavitation of the nodules is apparent on cut section. Lighter colored hepatic parenchyma interspersed between the nodules represents areas of hepatic necrosis. (From McClure and Chang 1976)

Microscopic Features

Histologically, extensive areas of necrosis are present throughout the parenchyma of affected organs, most frequently the liver, with larger lesions containing central areas of liquefaction necrosis surrounded by areas of coagulation necrosis (Fig. 142). Minimal inflammatory cell infiltrate is associated with the areas of necrosis. Large numbers of gram-negative bacilli can be demonstrated in areas of necrosis (Fig. 143). Smaller pinpoint lesions noted grossly are found histologically to represent foci of necrosis with varying degrees of inflammation and numerous gram-negative bacilli (Fig. 144).

Differential Diagnosis

Grossly, the lesions associated with *Chromobacterium violaceum* infection in nonhuman primates are similar to lesions seen with tuberculosis, yersiniosis (pseudotuberculosis), or mycotic infections. Cultures should be taken from the lesions to confirm or rule-out infection with specific organisms. Impression smears, made from representative lesions and stained with Wright-Giemsa, Gram's, PAS, and acid-fast methods, should be done to assist in the differential diagnosis. Numerous gram-negative bacilli may be evident in both chromobacteriosis and yersiniosis, and culture and isolation of the etiologic agent will be required to differentiate these organisms. The characteristic violet pigment produced in culture media by *Chromobacterium violaceum*, evident within 24h or less, immediately suggests a tentative diagnosis. The identification of acid-fast bacilli or fungal agents, either by special stains of impression smears, tissue sections or culture and isolation, will differentiate tuberculosis or mycotic disease from chromobacteriosis.

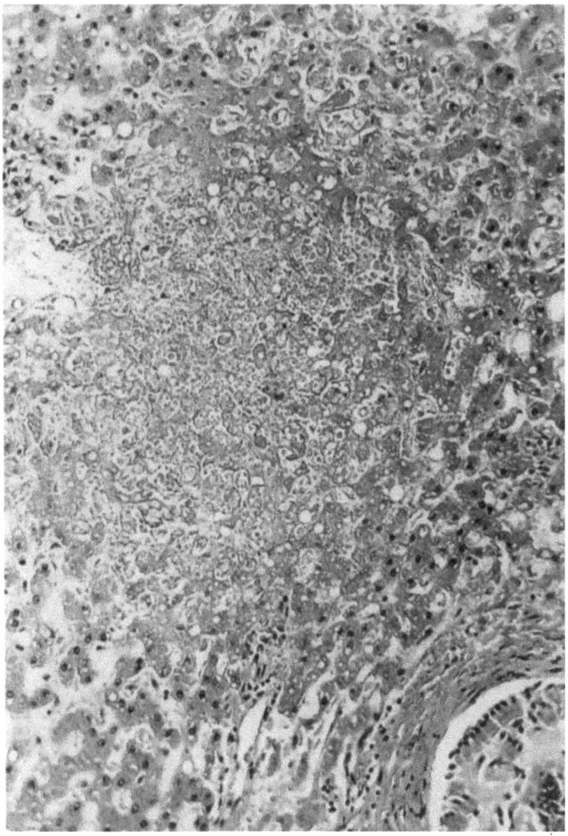

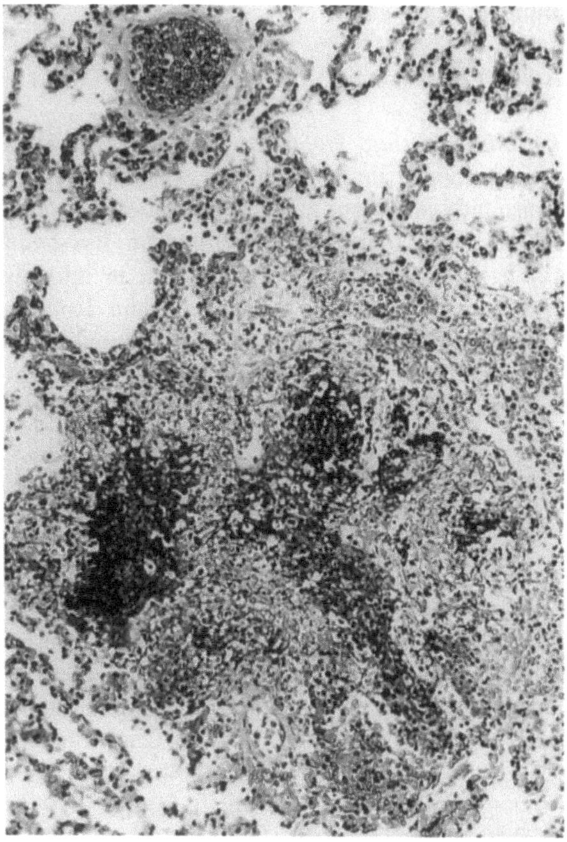

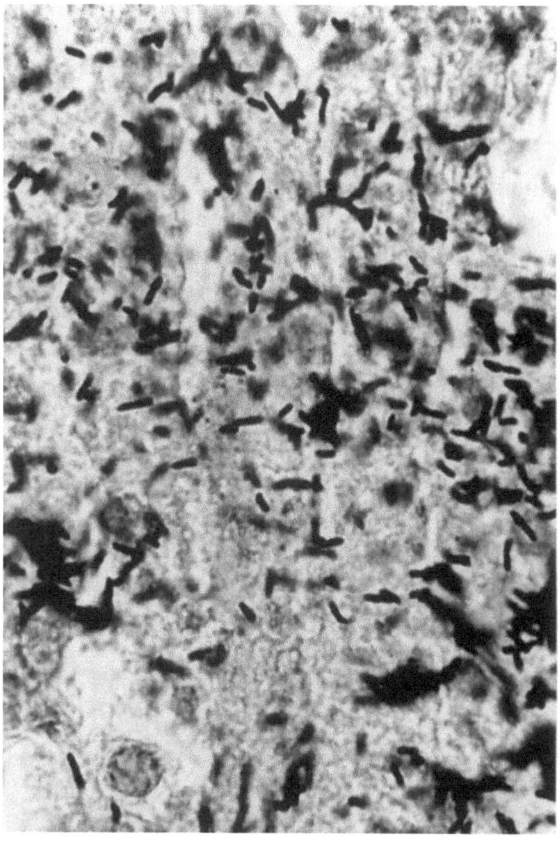

Fig. 142 (*upper left*). Liver, *Macaca assamensis*, containing a focus of hepatic necrosis without an associated inflammatory response. H and E, ×138

Fig. 143 (*upper right*). Lung of *Macaca assamensis* containing foci of necrosis (dark stained areas) with an inflammatory cell infiltrate. H and E, ×138

Fig. 144 (*lower right*). Higher magnification of section shown in Fig. 143. Note the rod-shaped bacilli typical of *Chromobacterium violaceum*. Modified Steiner silver stain, ×1376

Biologic Features

Natural History

Chromobacterium violaceum was first identified as a pathogen in 1904 when it was reported to cause fatal infections in water buffaloes (Woolley 1905). Since that time the organism has been reported as an infrequent pathogen in humans (Johnson et al. 1971; Dauphinais and Robben 1968; Ognibene and Thomas 1970; Nunnally and Dunlop 1968; Soule 1939; Black and Shahan 1938), domestic swine (Joseph et al. 1971; Sippel et al. 1954), cattle (Joseph et al. 1971; Sippel et al. 1954), a Malayan sun bear (Groves et al. 1969), and nonhuman primates (Groves et al. 1969; Johnsen et al. 1970; Joseph et al. 1971; Audebaud et al. 1954; McClure and Chang 1976). At least six cases of chromobacteriosis have been reported in humans in the United States; four in Florida and two in Louisiana (Johnson et al. 1971). One nonhuman primate case has been documented in the United States (McClure and Chang 1976), with other nonhuman primate cases occurring in Southeast Asia or Africa (Groves et al. 1969; Audebaud et al. 1954).

Pathogenesis

Chromobacterium violaceum is widely distributed in nature, especially in tropical and subtropical climates, and is generally considered a nonpathogenic saprophyte (Johnson et al. 1971; Dauphinais and Robben 1968). In susceptible hosts, the route of infection may be by way of the respiratory tract, the gastrointestinal tract, or the skin following some type of injury (Sneath 1960; Johnson et al. 1971; Groves et al. 1969; Joseph et al. 1971; Sippel et al. 1954). In some cases the route of infection cannot be determined (Nunnally and Dunlop 1968; McClure and Chang 1976). Infection usually results in a septicemia with prominent and extensive hepatic necrosis and abscess formation, with less extensive necrosis occurring in a variety of other organs including the lungs, spleen, lymph nodes, and kidneys (Kaufman et al. 1986; McClure and Chang 1976; Lee and Wright 1981).

Sivendra and Tan (1977) have experimentally infected mice with both pigmented and nonpigmented strains of *C. violaceum* and noted that the pathogenicity was similar for each strain. They also noted that the virulence for both strains varied widely, with the more virulent strains having a 50% lethal dose for mice of about 10^6 viable organisms. In experimentally infected mice, lesions were often seen in the liver but were seen less frequently in the lungs, spleen, kidneys, and testes (Sivendra and Tan 1977).

Etiology

Chromobacteriosis is caused by infection with *Chromobacterium violaceum*, a motile, gram-negative, aerobic bacillus that grows on most standard laboratory media with the production of a violet, alcohol-soluble, water and chloroform-insoluble pigment (Sneath 1960). Nonpigmented strains of *C. violaceum* have been shown to be comparable to pigmented strains with respect to their virulence and pathogenicity in mice (Sivendra and Tan 1977), although there have been no reports of disease caused by natural infection with nonpigmented strains. Nonpigmented variants are known to arise from subculture of pigmented strains (Sneath 1960), and both pigmented and nonpigmented strains can be isolated from soil and fresh water (Sivendra et al. 1975; Moss et al. 1978). Sivendra and Tan (1977) have noted that cases of natural infection with nonpigmented strains of *C. violaceum* may have occurred but were misdiagnosed as due to infection with the closely related organisms *Aeromonas hydrophila* or *Pseudomonas* sp.

Frequency

Chromobacteriosis has been reported very infrequently in nonhuman primates. Reported infections have been limited to gibbons in Southeast Asia (Groves et al. 1969; Johnsen et al. 1970), a guenon monkey in Africa (Audebaud et al. 1954), and a *Macaca assamensis* that was housed in the southeast United States (McClure and Chang 1976). The disease has also been infrequently reported in humans (Johnson et al. 1971; Petrillo et al. 1984; Kaufman et al. 1986), domestic swine, and cattle (Joseph et al. 1971; Sippel et al. 1954).

Comparison with Other Species

The gross and histologic features of chromobacteriosis in nonhuman primates are essentially the same as those reported in natural infections in humans, domestic swine and cattle and experimental infections in mice.

Acknowledgements. This work was supported in part by NIH grant RR-00165 from the National Center for Research Resources to the Yerkes Regional Primate Research Center. The Yerkes Center is fully accredited by the American Association for Accreditation of Laboratory Animal Care.

References

Audebaud F, Ganzin M, Ceccaldi J et al. (1954) Isolement d'un Chromobacterium violaceum à partir de lésions hépatiques observées chez un singe Cercopithecus cephus: étude et pouvoir pathogène. Ann Inst Pasteur 87:413–417

Black ME, Shahan J (1938) Bacillus violaceous infection in a human being. JAMA 110:1270–1271

Dauphinais RM, Robben GG (1968) Fatal infections due to Chromobacterium violaceum. Am J Clin Pathol 50:592–597

Groves MG, Strauss JM, Abbas J, Davis CE (1969) Natural infections in gibbons with a bacterium producing violet pigment (Chromobacterium violaceum). J Infect Dis 120:605–610

Johnsen DO, Pulliam JD, Tanticharoenyos P (1970) Chromobacterium septicemia in the gibbon. J Infect Dis 122:563

Johnson WM, DiSalvo AF, Steuer RR (1971) Fatal Chromobacterium violaceum septicemia. Am J Clin Pathol 56:400–406

Joseph PG, Sivendra R, Anwar M (1971) Chromobacterium violaceum infection in animals. Kajian Vet 3:55–56

Kaufman SC, Ceraso D, Schugurensky A (1986) First case report from Argentina of fatal septicemia caused by Chromobacterium violaceum. J Clin Microbiol 23:956–958

Lee TS, Wright BD (1981) Fulminating chromobacterial septicaemia presenting as respiratory distress syndrome. Thorax 36:557–559

McClure HM, Chang J (1976) Chromobacterium violaceum infection in a nonhuman primate (Macaca assamensis). Lab Anim Sci 26:807–810

Moss MO, Ryall C, Logan NA (1978) The classification and characterization of chromobacteria from a lowland river. J Gen Microbiol 105:11–21

Nunnally RM, Dunlop WH (1968) Fatal septicemia due to Chromobacterium janthinum. J La State Med Soc 120:278–280

Ognibene AJ, Thomas E (1970) Fatal infection due to Chromobacterium violaceum in Vietnam. Am J Clin Pathol 54:607–610

Petrillo VF, Severo V, Santos MM, Edelweiss EL (1984) Recurrent infection with Chromobacterium violaceum: first case report from South America. J Infect 9:167–169

Sippel WL, Medina G, Atwood MB (1954) Outbreaks of disease in animals associated with Chromobacterium violaceum: I. The disease in swine. J Am Vet Med Assoc 124:467–471

Sivendra R, Tan SH (1977) Pathogenicity of nonpigmented cultures of Chromobacterium violaceum. J Clin Microbiol 5:514–516

Sivendra R, Lo HS, Lim KT (1975) Identification of Chromobacterium violaceum: pigmented and nonpigmented strains. J Gen Microbiol 90:21–31

Sneath PHA (1960) A study of the bacterial genus Chromobacterium. Iowa State J Sci 34:243–500

Soule M (1939) A study of two strains of B violaceus isolated from human beings. Am J Pathol 15:592–595

Woolley PG (1905) Bacillus violaceus manilae (a pathogenic microorganism). Johns Hopkins Med J 16:89–93

Listeriosis

Daniel C. Anderson and Harold M. McClure

Synonyms. Listerellosis.

Gross Appearance

Listeria infections are most common in stillborn and neonatal infants. Gross lesions are usually not apparent in these animals due to death in utero and advanced autolysis at the time of delivery. When apparent, lesions usually consist of pinpoint to small gray-white foci which are visible on the surface and within the parenchyma of the liver (Fig. 145). Similar foci and occasional prominent macroabscesses may be seen in the placenta (Fig. 146). These lesions are less frequently observed in other tissues. Lesions in the brain are infrequently noted grossly in order primates but have not been reported in the stillborn (Fig. 147). Purulent meningeal exudate has been reported in an adult chimpanzee (*Pan*

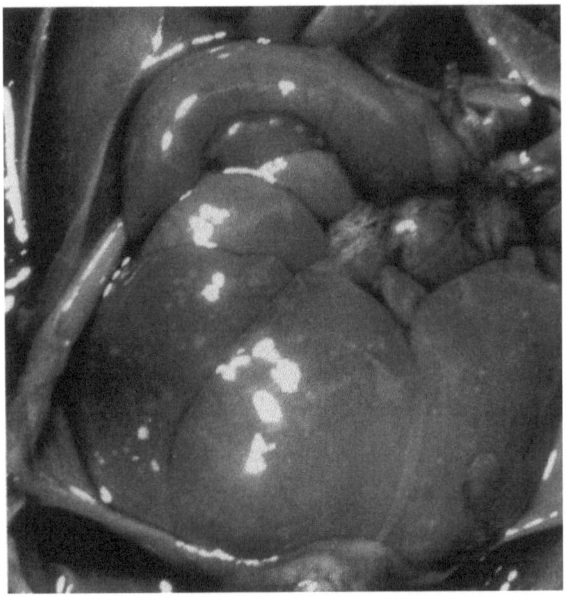

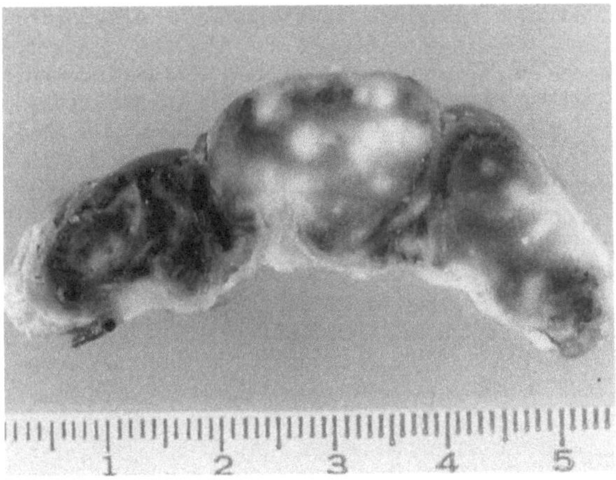

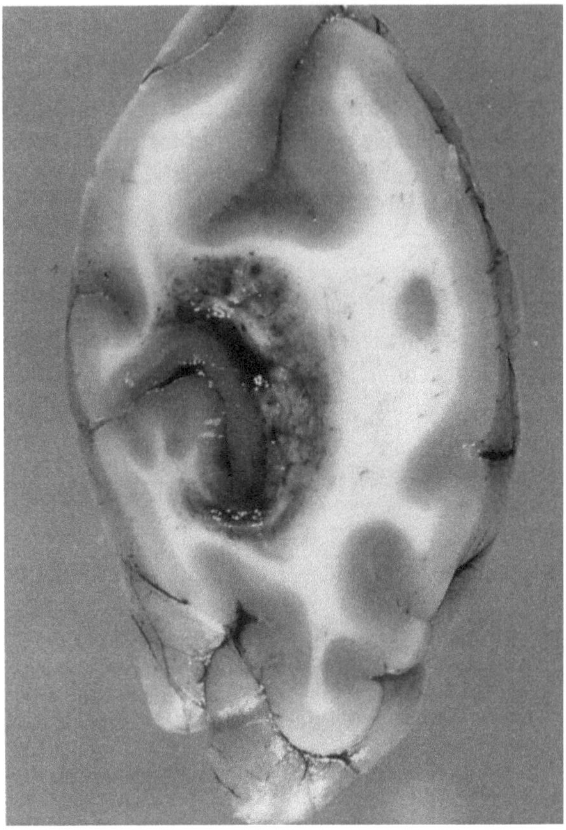

Fig. 145 (*upper left*). Listeriosis, liver of an aborted mangabey (*Cercocebus atys*). Note scattered gray foci of necrosis. *Listeria monocytogenes* type 4b was isolated from the liver

Fig. 146 (*lower left*). Listeriosis, placenta from pig-tailed macaque (*Macaca nemestrina*) from which *L. monocytogenes* type 1/2a was isolated. Note large white foci (abscesses) throughout the tissue

Fig. 147 (*upper right*). Listeriosis, brain from 2 year old rhesus monkey (*Macaca mulatta*) with listeric encephalitis. *Listeria monocytogenes* type 1/2a was isolated from this brain

troglodytes) and in a 3 day old cynomolgus monkey (*Macaca fascicularis*) (Chalifoux and Hajema 1981; Heldstab and Ruedi 1982).

Microscopic Features

Microscopic lesions consist of necrotic foci in parenchymous organs, in particular, in the liver (Fig. 148). Although these foci may contain neutrophils or mononuclear cells, areas of necrosis with little or no inflammatory response is the common finding in infants who have died

Fig. 148 (*upper left*). Listeriosis, liver from squirrel ▶ monkey with multifocal hepatic necrosis. H and E, ×160

Fig. 149 (*lower left*). Listeriosis, liver from aborted mangabey. Note necrotic focus in autolyzed tissue. Same animal as in Fig. 145. H and E, ×160

Fig. 150 (*upper right*). Listeriosis, placenta, pigtail macaque (*Macaca nemestrina*), autolysis and abscess. Same animal in Fig. 146. H and E, ×136

Fig. 151 (*lower right*). Listeriosis, liver from aborted rhesus monkey. Note numerous gram-positive organisms in a necrotic focus. Gram's stain, ×560

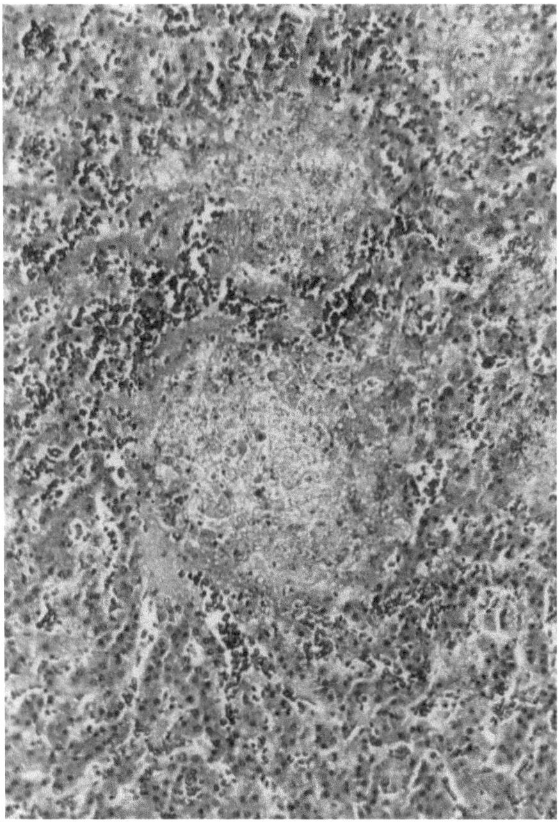

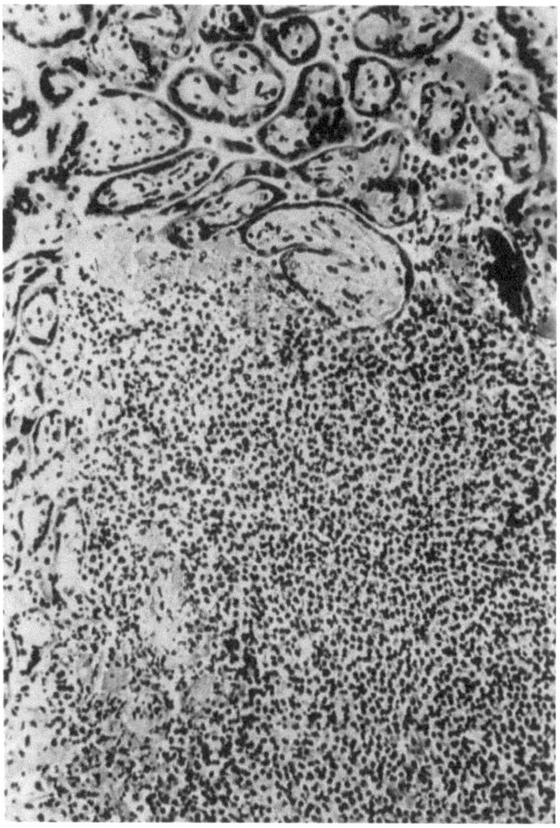

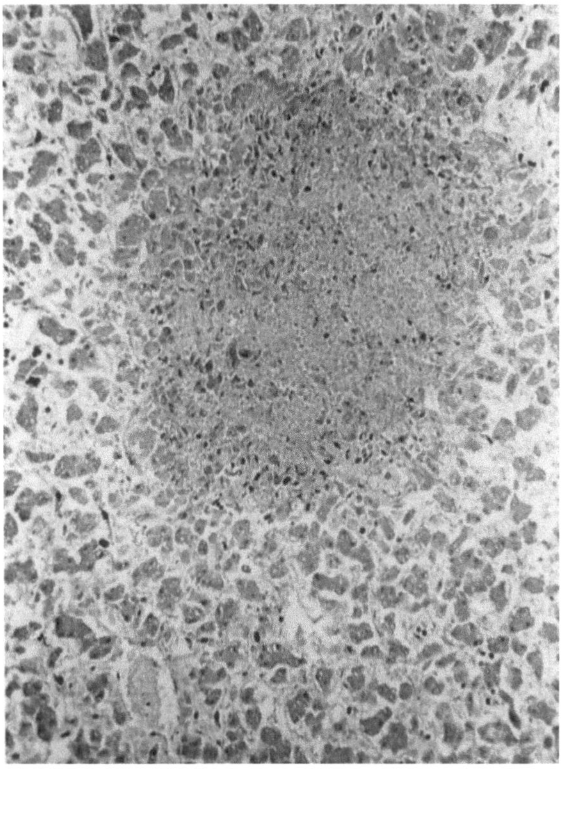

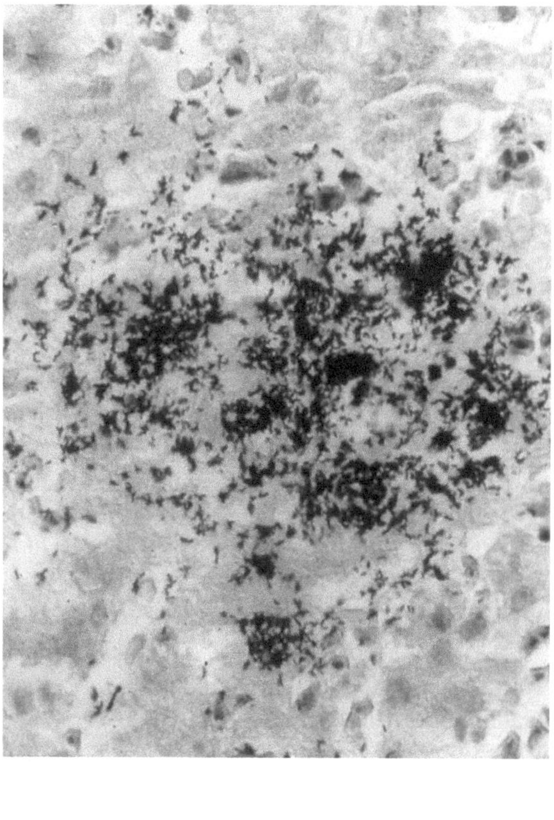

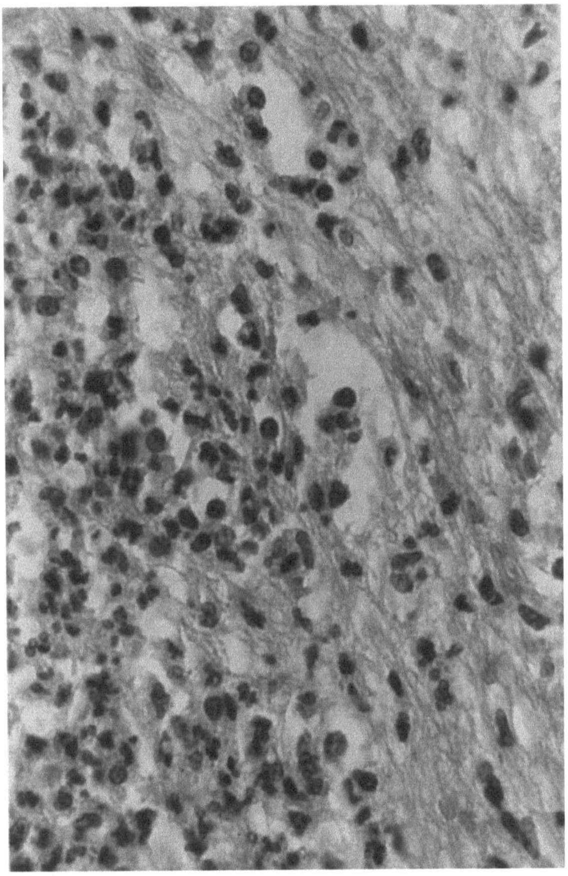

Fig. 152. Listeriosis, brain from rhesus monkey in Fig. 147 with severe, acute, suppurative periventriculitis and extension of inflammation into the surrounding parenchyma. H and E, ×544

in utero. Frequently, areas of necrosis are apparent only because they are less densely stained than are the surrounding tissues (Fig. 149).

In general, necrotic or inflammatory foci are well demarcated by histologically normal adjacent tissues. Placental specimens are not always available for evaluation; however, in those specimens examined, lesions are invariably present and consist of a diffuse, acute necrotizing, fibrinopurulent placentitis (Fig. 150). Inflammatory foci have been less frequently described in other tissues such as the lung, kidney, and adrenal. This may be due in part to limited histologic evaluation of stillborn infants in advanced stages of autolysis. Gram-positive organisms are easily found in necrotic foci and are often found in multiple tissues including the lung and the brain in the absence of an inflammatory response (Fig. 151).

We have found only two reports of lesions in the central nervous system associated with *Listeria* infections in nonhuman primates (Chalifoux and Hajema 1981; Heldstab and Ruedi 1982). One of these cases was a neonatal cynomolgus monkey (*Macaca fascicularis*) which also had evidence of septicemia; the other was an adult chimpanzee (*Pan troglodytes*) in which histologic lesions were limited to the brain. The findings consisted of a severe meningoencephalitis with fibrinopurulent exudate in the leptomeninges and ventricles with extension into periventricular areas. These microscopic lesions were similar to those seen in a rhesus monkey (Figs. 147, 152). Gram-positive rods were easily found in areas of inflammation. *Listeria monocytogenes* type 4 was isolated from the brain and heart blood of the infant cynomolgous monkey and *L. monocytogenes* type 1/2B was isolated from the brain, cerebrospinal fluid, and the tonsils of the chimpanzee.

Ultrastructure

Ultrastructural features of diagnostic significance have not been described. Experimental studies have revealed the organism within phagocytic vacuoles in Kupffer's cells, peritoneal macrophages, and polymorphonuclear leukocytes and in intestinal, corneal, conjunctival, and urinary bladder epithelium (North and Mackaness 1963a,b; Racz et al. 1972; Siddique et al. 1978).

Differential Diagnosis

Although infrequently reported, listeriosis should be considered a potential cause of fetal or neonatal loss in nonhuman primates. The diagnosis is not difficult, provided adequate bacteriological support is included in fetal and neonatal work-ups. The organism is readily isolated in most cases since most are associated with septicemia. In all likelihood, the diagnosis is most often not made due to the advanced stage of autolysis in the infant and because specimens are not submitted for bacterial culture. In the absence of cultures, Gram's stain of multiple tissues, including the liver, lung, and placenta, often reveals numerous, small, gram-positive rods. Typical lesions and the detection of characteristic bacilli are strong presumptive evidence of listeriosis.

Biologic Features

Natural History

Listeria species are ubiquitous in nature and have been frequently isolated from many environmental sources including soil, plant material, sewage, and river sludge (Weis and Seeliger 1975; Watkins and Sleath 1981). In addition, *Listeria* have been isolated from silage and from vegetable and dairy products that were incriminated in food-born outbreaks (Gellin and Broome 1989). *Listeria monocytogenes* can also be readily isolated from meat and poultry products (Skovgaard and Morgan 1988; Bailey et al. 1989) and from the feces of healthy persons and from patients with diarrheal disease (Muller 1990). These latter findings support the hypothesis that most cases of sporadic and epidemic listeriosis are due to the ingestion of contaminated food products. This is also supported by animal experimental studies in which the organism can be demonstrated in epithelial cells and in Peyer's patches in the small intestine (Racz et al. 1972). The disease was first reported in laboratory rabbits (designated *Bacterium monocytogenes*) in 1926 and was characterized by a large mononuclear leukocytosis, hence the designation "monocytogenes" (Murray et al. 1926; Gray and Killinger 1966). This was followed by a report the next year of an outbreak in rodents in South Africa (the organism was designated *Listerella hepatolytica* in honor of Joseph Lister (Seeliger 1988)). The organism subsequently was referred to by several other names until it was finally named *Listeria monocytogenes* in 1940 (Gray and Killinger 1966).

Etiology

Listeria monocytogenes is a small, gram-positive, catalase-positive, facultatively anaerobic rod that is β-hemolytic on blood agar and grows over a wide temperature range on ordinary media (Bortolussi et al. 1985). *Listeria monocytogenes* was the only recognized species in the genus *Listeria* until 1961 (Rocourt et al. 1982). The genus now includes six other species. Of these, *L. monocytogenes* is the only significant pathogen and includes at least 16 serovars (serotypes) (Gellin and Broome 1989) of which three of these (4b; 1/2b; 1/2a) account for more than 90% of the reported clinical infections in humans. *Listeria monocytogenes* has been isolated from at least 60 animal species, including ruminants, laboratory animals, amphibians and fowl (Seeliger 1988). The organism is readily isolated from clinical septicemias (e.g., brain and liver) and from apparently healthy animals and humans. Isolation from contaminated food, the environment, or feces requires the use of special media and cold enrichment techniques (Gellin and Broome 1989; Bortolussi et al. 1985). The organism grows at refrigeration temperatures ranging between 3° and 45° Fahrenheit and may survive high-temperature, short-time pasteurization conditions (Kvenberg 1988). This allows the organism to persist for long periods in the environment and also facilitates isolation of the organism from heavily contaminated specimens by incubation at low temperatures, which suppresses the growth of other organisms.

Pathogenesis

The primary route of listeric infection is currently considered to be oral, with the intestine the usual site of intitial colonization. Rácz et al. (1972), using a guinea pig model of enteritis, showed that *L. monocytogenes* was able to penetrate and multiply within intestinal epithelial cells shortly after oral challenge and before phagocytosis by macrophages. They had previously demonstrated an epithelial phase of infection in corneal, conjunctival, and urinary bladder epithelium. Additional evidence that the intestine probably serves as the primary site of infection includes recent epidemiologic findings and demonstration of in vitro intracellular multiplication of *L. monocytogenes* in enterocytic cell lines (Berche et al. 1988). Experimental evidence in sheep suggests that infection of the brain stem may occur by migration of the organism along nerve tracts following local oral trauma (Charlton and Garcia 1977).

Frequency

There are few reports in the literature of listeriosis in nonhuman primates (Vetesi et al. 1972; McClure and Strozier 1975; Chalifoux and Hajema 1981; Heldstab and Ruedi 1982; McClure et al. 1986). One report on the disease in nonhuman primates describes an endemic problem in group-housed colony-reared macaques in which *Listeria* infections are a significant cause of stillborn and neonatal deaths (McClure et al. 1986). This suggests that listeric infections may not be unusual in outdoor-housed nonhuman primates and that the small numbers of reported cases may reflect inadequate bacteriologic cultures of abortuses, stillborn infants, and neonatal deaths.

Listeria sp. have also been isolated from blood samples in two of 100 apparently healthy wild baboons and were found in normal marmosets in a zoo (Brack 1987; Zwart and Donker-Voet 1959). Attempts to reproduce the disease have been limited and have been generally unsuccessful with either no effect or with clinical signs limited to transient monocytosis and fever (Kautter et al. 1963; Osebold and Inouye 1954). The literature is replete with reports of natural infections in domestic animals and experimental studies in ruminants and in laboratory animal species (Gray 1962; Gray and Killinger 1966; Kautter et al. 1963). Numerous clinical reports of *Listeria* infection in humans may be found and extensive work has been done with isolation of the organism from environmental sources and from dairy, meat, and other food products.

Comparison with Other Species

Listeria monocytogenes is of public health concern and is of considerable economic importance in domestic food animal species. The incidence of listeriosis in humans is not known. Listeriosis is most frequently documented during pregnancy and neonatal morbidity and mortality are common; however, the infection is rarely life-threatening to the mother (Gellin and Broome 1989). After the neonatal period, infections are most commonly associated with advanced age and immunosuppression. However, 30%–54% of infections occur in persons without underlying disease or predisposing conditions (Gellin and Broome 1989).

The clinical manifestations are diverse in all species, including humans. In general, three forms are recognized: abortion or septicemia in the early neonatal period; meningitis or encephalitis; or septicemia. The disease in nonhuman primates occurs predominantly in late gestation or in the neonatal period with the affected mothers showing little or no clinical evidence of disease. Similarly, abortion in ruminant species is usually not associated with clinical disease in the adult female. Bacteremia occurs in the early neonatal period, usually within a week of age, and listeric encephalitis is almost solely restricted to adult ruminants. In ruminants *Listeria* have a remarkable affinity for the brain stem (Jubb et al. 1985). In pregnant rabbits, the organism has a predilection for the gravid uterus and encephalitis is not a feature. Chickens and guinea pigs occasionally have characteristic myocardial lesions. Chinchillas, in particular, are reported to be extremely susceptible (Gray and Killinger 1966).

Acknowledgements. This work was supported in part by NIH grant RR-00165 from the National Center for Research Resources to the Yerkes Regional Primate Research Center. The Yerkes Center is fully accredited by the American Association for Accreditation of Laboratory Animal Care.

References

Bailey JS, Fletcher DL, Cox NA (1989) Listeria monocytogenes colonization of broiler chickens. Poult Sci 69:457–461

Berche P, Gaillard JL, Richard S (1989) Invasiveness and intracellular growth of Listeria monocytogenes. Infection 16S [Suppl 2]:145–148

Bortolussi R, Schlech III WF, Albritton WL (1985) Listeria. In: Lennette EH, Balows A, Hausler WH Jr, Shadomy HJ (eds) Manual of clinical microbiology, 4th edn. American Society of Microbiology, Washington, pp 205–208

Brack M (1987) Agents transmissible from simians to man. Springer, Berlin Heidelberg New York, pp 198–202

Chalifoux LV, Hajema EM (1981) Septicemia and meningoencephalitis caused by Listeria monocytogenes in a neonatal Macaca fascicularis. J Med Primatol 10: 336–339

Charlton KM, Garcia MM (1977) Spontaneous listeric encephalitis and neuritis in sheep. Light microscopic studies. Vet Pathol 14:297–313

Gellin BG, Broome CV (1989) Listeriosis. JAMA 261/9: 1313–1320

Gray ML (1962) Listeria monocytogenes and listeric infection in the diagnostic laboratory. Ann NY Acad Sci 98:686–699

Gray ML, Killinger AH (1966) Listeria monocytogenes and listeric infections. Bacteriol Rev 30:309–382

Heldstab A, Ruedi D (1982) Listeriosis in an adult female chimpanzee (Pan troglodytes). J Comp Pathol 92: 609–612

Jubb KVF, Kennedy PC, Palmer N (1985) In: Jubb KVF et al (eds) Pathology of domestic animals, 3rd edn. Academic, London, pp 286–288

Kautter DA, Silverman SJ, Roessler WG (1963) Virulence of Listeria monocytogenes for experimental animals. J Infect Dis 112:167–180

Kvenberg JE (1988) Outbreaks of listeriosis/Listeria-contaminated foods. Microbiol Sci 5:355–358

McClure HM, Strozier LM (1975) Perinatal listeric septicemia in a Celebese black ape. J Am Vet Med Assoc 167:637–638
McClure HM, Brodie AR, Anderson DC, Swenson RB (1986) Bacterial infections of nonhuman primates. In: Benirschke K (ed) Primates: the road to self-sustaining populations. Springer, Berlin Heidelberg New York, pp 531–556
Muller HE (1990) Listeria isolations from feces of patients with diarrhea and from healthy food handlers. Infection 18:97–99
Murray EGD, Webb RA, Swann MBR (1926) A disease of rabbits characterized by a large mononuclear leukocytosis, caused by a hitherto undescribed bacillus, Bacterium monocytogenes (n sp). J Pathol Bacteriol 29:407–439
North RJ, Mackaness GB (1963a) Electronmicroscopical observations on the peritoneal macrophages of normal mice and mice immunised with Listeria monocytogenes. I. Structure of normal macrophages and the early cytoplasmic response to the presence of ingested bacteria. Br J Exp Pathol 44:601–607
North RJ, Mackaness GB (1963b) Electronmicroscopical observations on the peritoneal macrophages of normal mice and mice immunised with Listeria monocytogenes: 1. Structure of macrophages from immune mice and early cytoplasmic response to the presence of ingested bacteria. Br J Exp Pathol 44:608–611
Osebold JW, Inouye T (1954) Pathogenesis of Listeria monocytogenes infections in natural hosts. J Infect Dis 95:52–66
Racz P, Tenner K, Mero E (1972) Experimental Listeria enteritis: I. An electron microscopic study of the epithelial phase in experimental Listeria infection. Lab Invest 26:694–700
Rocourt J, Grimont F, Grimont PAD, Seeliger HPR (1982) DNA relatedness among serovars of Listeria monocytogenes sensu lato. Curr Microbiol 7:383–388
Seeliger HPR (1988) Listeriosis-history and actual developments. Infection 16 [Suppl]:80–84
Siddique IH, Mckenzie BE, Sapp WJ, Rich P (1978) Light and electron microscopic study of the livers of pregnant mice infected with Listeria monocytogenes. Am J Vet Res 39:887–892
Skovgaard N, Morgen C-A (1988) Detection of Listeria spp. in faeces from animals, in feeds, and in raw foods of animal origin. Int J Food Microbiol 6:229–242
Vetesi F, Balsai A, Kemenes F (1972) Abortion in Gray's monkey (Cercopithecus mona) associated with Listeria monocytogenes. Acta Microbiol Acad Sci Hung 19: 441–443
Watkins J, Sleath KP (1981) Isolation and enumeration of Listeria monocytogenes from Sewage, Sewage Sludge and River Water. J Appl Bacteriol 50:1–9
Weis J, Seeliger HPR (1975) Incidence of Listeria monocytogenes in nature. Appl Microbiol 30:29–32
Zwart P, Donker-Voet J (1959) Listeriosis bij in gevangenschap gehouden dieren. Tijdschr Diergeneesk 84:712–716

Tuberculosis

Norval W. King, Jr.

Synonyms. *Mycobacterium tuberculosis* infection, *M. bovis* infection, *M. africanum* infection, *M. kansasii* infection, *M. simiae* infection, *M. scrofulaceum* infection, *M. chelonei* (*abscessus*) infection, mycobacteriosis, TB, simian tuberculosis, Potts' disease.

Gross Appearance

The gross findings in nonhuman primates with tuberculosis can vary from no detectable lesions to widely disseminated, 1.0–10.0 mm, yellow-white, focal to confluent granulomas affecting all major organs (Fig. 153). On cut-section, the granulomas contain a central core of yellow-white caseous material. Organs most often affected include the lungs, hilar lymph nodes, spleen, liver, kidney, intestine, and mesenteric nodes, but any organ may be involved. Large granulomas in the lung may be cavitary as the result of expulsion of the caseous exudate into adjacent airways. Adhesions may also occur between the affected portions of the lung and the adjacent parietal pleura. Tuberculous lesions involving the vertebra and adjacent spinal cord, termed "Potts' disease" in humans, have also been described in nonhuman primates (Innes 1963; Lapin and Yakovleva 1963; Martin et al. 1968; Fox et al. 1974; Tryphonas and Wong 1982).

Microscopic Features

The microscopic findings in animals with tuberculosis can also vary depending upon the duration and extent of the disease. Early in the disease, one may find only a few, widely scattered, microscopic granulomas consisting of circumscribed collections of epithelioid cells and an

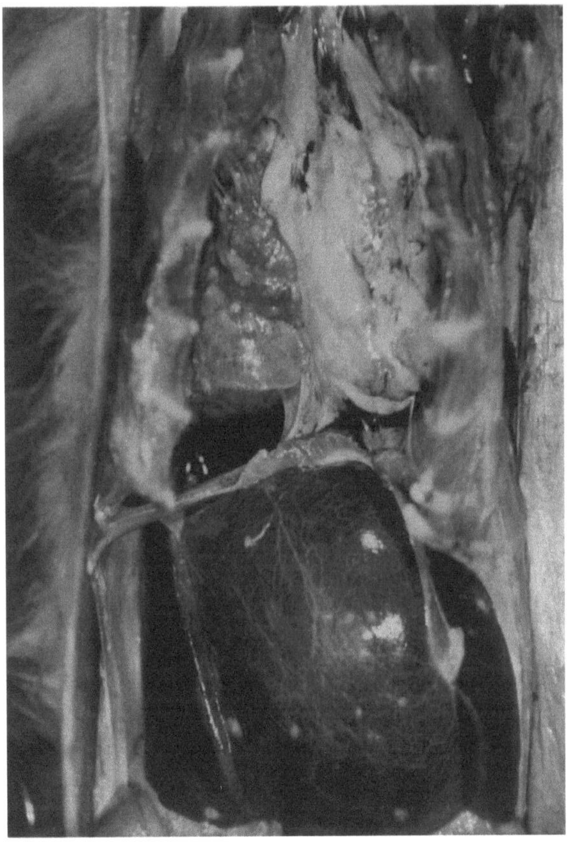

Fig. 153. Opened abdomen and thorax of a rhesus monkey with disseminated *M. tuberculosis* infection. Note multiple foci in the lungs and liver

occasional Langhans' giant cell. There may even be focal collections of neutrophils in the centers of some early microgranulomas. Initially these lesions may be confined to the lungs or intestinal tract, the two major portals of entry of the tubercle bacilli. At this stage, acid-fast stains may reveal only a rare acid-fast bacillus within an epithelioid cell.

The classic lesion of advanced tuberculosis is the tubercle. It consists of a central core of acellular necrotic debris surrounded by a zone of epithelioid cells and widely scattered Langhans-type multinucleate giant cells (Fig. 154). Rarely, the necrotic centers of tubercles may be partially calcified. The periphery of a tubercle is generally surrounded by variable amounts of fibrous connective tissue and infiltrates of lymphocytes. Individual tubercles often coalesce to form larger granulomas. In the lung, larger granulomas may break into major airways and become cavitary. Acid-fast stains, such as Ziehl-Nielsen, Fite-Faraco Kinyoun's stains, will usually reveal acid-fast bacilli in the epithelioid and multinucleate giant cells, although their numbers can vary considerably (Fig. 155). In some instances, numerous tissue sections may have to be examined before tubercle bacilli are found. Microscopic examination of the tuberculous lesions alone is not sufficient to identify which of the several species of *Mycobacterium* is responsible for the disease, as the lesions caused by each of them are similar.

Differential Diagnosis

Clinically, the diagnosis of tuberculosis relies heavily on the tuberculin skin test, which uses a component of the tubercle bacillus to elicit a delayed type hypersensitivity reaction. In nonhuman primates, the preferred antigen is mammalian old tuberculin (OT) prepared from *M. tuberculosis*, whereas in human patients a purified protein derivative (PPD) of the same organism is the antigen of choice (McLaughlin and Marrs 1978). This test involves the injection of 0.1 ml of OT (25 mg protein) into the upper palpebrum of the eye followed by examination of the injection site 24, 48, and 72 h postinjection. A positive reaction is indicated by erythema and swelling of the upper eyelid that may lead to closure of the lids of the test eye (Fig. 156). Verification of an equivocal positive reaction may be attempted by repeating the test in the lid of the opposite eye after a one week interval. Additionally, weak positive reactions to OT may occur in animals with *M. avium-intracellulare* infection, in which case retest using avian tuberculin will produce a more intense reaction. Radiographs of the thorax may confirm the presence of pulmonary lesions in advanced cases, but these may not be evident early in the course of the disease. Animals suspected of having tuberculosis based upon a positive tuberculin test or suspicious chest film should be removed from the colony and submitted for postmortem examination. False positive reactions do occur occasionally, as do false negative reactions. A satisfactorily explanation for false positive reactors is not always attainable, but prior exposure to a nonpathogenic saprophytic mycobacterium has been suggested. False negative reactions are commonly associated with anergy resulting from overwhelming widely disseminated infection.

At necropsy, tuberculosis must be differentiated from other granulomatous diseases, including

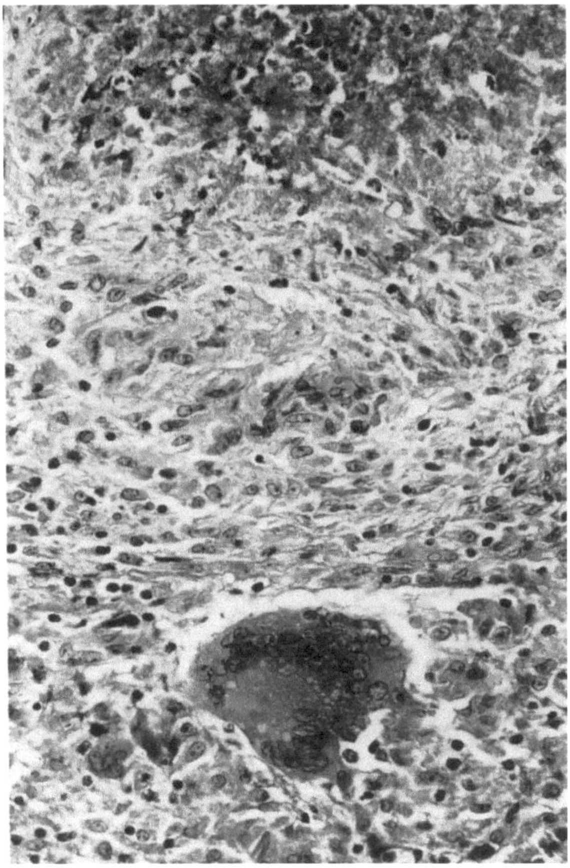

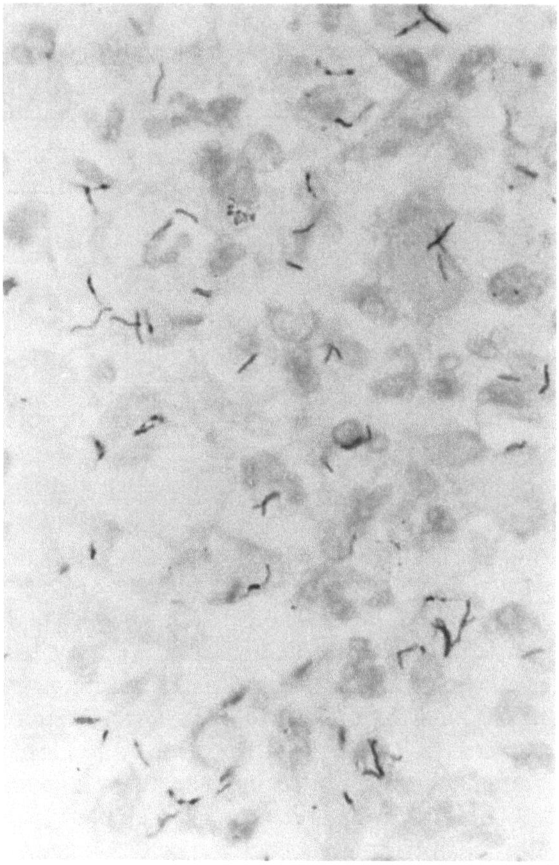

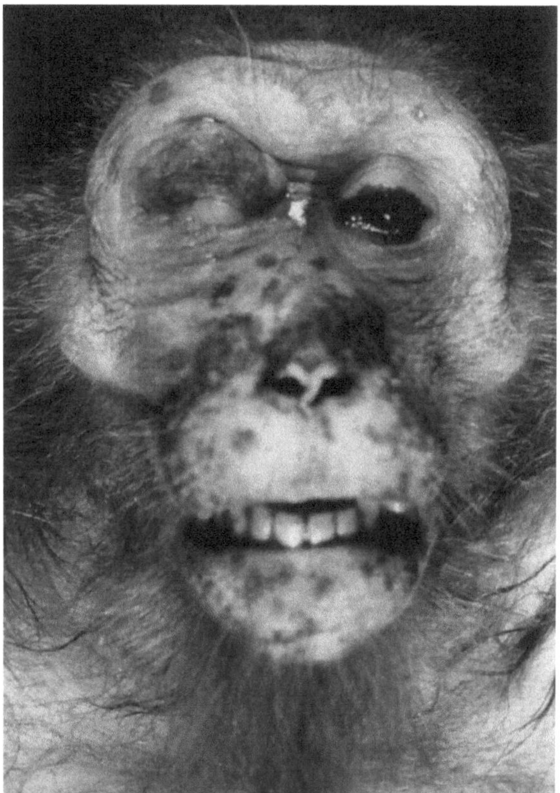

Fig. 154 (*upper left*). Tuberculosis, lung, *Macaca mulatta*. Portion of a tubercle with central area of caseous necrosis (*top*) surrounded by a zone of epithelioid cells containing a multinucleate giant cell. H and E, ×450

Fig. 155 (*upper right*). Tuberculosis, lung, *Macaca mulatta*. Epithelioid cell component of a tubercle containing many, scattered, acid-fast bacilli. Fite-Faraco stain, ×1500

Fig. 156 (*lower right*). A positive tuberculin reaction in the right upper eyelid of a stump-tail macaque, *Macaca arctoides*. The animal had an intrapalpebral injection of 0.1 ml of old tuberculin 72 h earlier those caused by foreign bodies, such as kaolin granulomas (once seen as a complication of administration of anti-diarrheal medications) and certain mycotic, protozoan, and parasitic organisms. The latter include nocardiosis, *Pneumonyssus simicola*, *Hepatocystis kochi*, and oesophagostomiasis. The definitive means of diagnosing tuberculosis is by isolation and identification of a specific *Mycobacterium sp.* from lesions in which it is suspected. In the absence of fresh material for culture, special stains can be used to demonstrate acid-fast bacilli in granulomas or classic tubercles. Although this method is sufficient for the diagnosis of tuberculosis, it does not definitively identify the causative organism. The lesions caused by *Pneumonyssus simicola*, *Hepatocystis*, and *Oesophagostomum* are easily distinguishable microscopically from tuberculosis. Although *Nocardia* can be confused with mycobacteria because of their partially acid-fast nature, they readily stain gram-positive with Gram's stain. Furthermore, the lesions produced by *Nocardia* invariably have a prominent neutrophilic component and thus are purulent or pyogranulomatous rather than typical tubercles.

Biologic Features

Natural History

Simian tuberculosis has been recognized most often in laboratory and zoological collections of Asian monkeys, particularly those of the genus *Macaca*, i.e., *M. mulatta* (Schroeder 1938; Carpenter 1940; Habel 1947; Benson et al. 1955; Ruch 1959; Keeling et al. 1969), *M. arctoides* (Wolf et al. 1967), *M. nemestrina* (Lau et al. 1972), and *M. fascicularis* (Habermann and Williams 1957), and to a lesser extent in African cercopithecoids, i.e., *Papio*, *Cercopithecus*, *Erythrocebus*, and *Cercocebus* (Hammer 1965; Thoen et al. 1977; Silberman 1978; Quimby et al. 1979; Renquist and Potkay 1979; Zumpe et al. 1980; Fourie and Odendaal 1983) and pongids, i.e., *Pongo*, *Gorilla*, and *Pan* (Griffith 1928; Kennard and Willner 1941; Benson et al. 1954; Savoy 1965; McClure and Guilloud 1971; Fineg et al. 1966). There are relatively few reports of tuberculosis in New World monkeys and these have been mainly in *Saimiri*, *Aotus*, and *Ateles* (Fiennes 1965; Chrisp et al. 1968; Hessler and Moreland 1968; Bone and Soave 1970; Duffy et al. 1970; Karlson et al. 1970; Moreland 1970; Snyder et al. 1970; Leathers and Hamm 1976; Griner 1978; Mayhall et al. 1981). Reports of tuberculosis in prosimians are likewise rare (Knezevic and McNulty 1967).

Etiology

The two most common causes of tuberculosis in nonhuman primates are *Mycobacterium tuberculosis* and *Mycobacterium bovis* (Renner and Bartholomew 1974; Kaufman and Anderson 1978; McLaughlin 1978; Thoen et al. 1977). There are, however, isolated reports of tuberculosis in monkeys from which atypical mycobacterial agents have been isolated. These agents include: *M. africanum* (McClure and Silcox 1970; Thorel 1980), *M. kansasii* (Valerio et al. 1979), *M. simiae* (Karasseva et al. 1965; Weiszfeiler et al. 1970; Krasnow and Gross 1975; Donovan et al. 1976; Weiszfeiler et al. 1981), *M. scrofulaceum* (Renquist and Potkay 1979), or *M. chelonei* (*abscessus*) (Weiszfeiler et al. 1969; Karlson et al. 1970; Quimby et al. 1979). In addition, agents belonging to the *Mycobacterium avium-intracellulare* complex have been frequently isolated from macaque species with mycobacterial infection in which tubercle formation is not a feature. These have occurred most often as opportunistic infections in animals with immuno-deficiency associated with infection by simian immunodeficiency virus (SIV) or the type D retrovirus (SRV-1). Brack (1987) has reviewed the literature on all of the above described simian mycobacterial infections. One confirmed report of *M. paratuberculosis* infection in stump-tailed macaques, *Macaca arcotides*, has also appeared (McClure et al. 1987). Since the lesions caused by *M. avium-intracellulare* and *M. paratuberculosis* in nonhuman primates are so distinctive and decidedly different from those seen in classic tuberculosis, these diseases are discussed separately in this monograph.

Epidemiology

Tuberculosis is extremely rare in natural populations of nonhuman primates living in areas remote from human beings (Schroeder 1938; Carpenter 1940; Fremming et al. 1955). The source of infection in recently imported monkeys is usually tuberculous human beings, cattle, or other ruminants in the country of origin where these diseases are endemic (Kleeberg 1975).

In macaque monkeys, and especially rhesus monkeys, tuberculosis is a highly contagious rapidly progressive disease. A single infected animal, when included in a shipment of animals or introduced into an established colony, may result in rapid spread of the disease throughout the entire group. When tuberculosis occurs spontaneously in an established closed colony, the source of infection is thought to be inapparently infected personnel with close contact with the susceptible species (Cappucci et al. 1972). Tuberculosis was once the most frequently reported infectious disease of captive nonhuman primates and accounted for up to 30% of deaths at some institutions. The adoption of strict quarantine procedures and routine tuberculin testing of nonhuman primates and animal care personnel have dramatically reduced the prevalence of tuberculosis in zoos and biomedical research facilities. The imposition of glass barriers between primates on display at zoos and the visiting public has also markedly reduced the exposure of these collections to outside sources of infection. Although great strides have been made in the control of simian tuberculosis over the last several decades, this disease remains a constant threat to captive colonies of nonhuman primates for which a rigorous surveillance program is essential (Renquist and Whitney 1978). Transmission of tuberculosis occurs most often via the respiratory tract through inhalation of infected aerosols and, to a lesser extent, via the alimentary tract through ingestion of infected materials. Less common modes of transmission include contact with infected fomites or percutaneously through use of contaminated hypodermic needles. From a public health point of view, documented reports of transmission of tuberculosis from monkeys to humans have been exceedingly rare (Duffy et al. 1970).

Etiology

Mycobacteria are aerobic, slightly curved or straight, occasionally beaded, rod-shaped bacilli that stain poorly by Gram's stain. They have a high lipid content that includes high molecular weight, esterified, mycolic acid residues that impart an affinity of the organism for carbolfuchsin after treatment with acid-alcohol. This characteristic is the basis for the acid-fast nature of mycobacteria as demonstrated by the classic Ziehl-Nielsen procedure. Similarly, the affinity of mycobacteria for the fluorochrome dyes auramine-rhodamine after treatment with acid-alcohol, as demonstrated by the fluorescence technique originally described by Truant et al. (1962), is also another method to demonstrate acid-fastness which does not depend upon color recognition. Culture of mycobacteria requires the use of agents that destroy contaminating nonacid-fast bacteria and special nutrient medium such as Lowenstein-Jensen or Middlebrook and Cohn 7H10 or 7H11 media. Once isolated, speciation of mycobacterial isolates is based upon specific growth characteristics and biochemical and metabolic differences between the species (Wayne and Kupica 1986). The two most common causes of tuberculosis in nonhuman primates are *Mycobacterium tuberculosis* and *Mycobacterium bovis* (Renner and Bartholomew 1974). There are, however, isolated reports of tuberculosis in monkeys from which atypical mycobacterial agents have been isolated. These agents include: *M. africanum* (McClure and Silcox 1970; Thorel 1980), *M. kansasii* (Valeric et al. 1979), *M. simiae* (Karasseva et al. 1965; Weiszfeiler et al. 1970, 1981; Krasnow and Gross 1975; Donovan et al. 1976), *M. scrofulaceum* (Renquist and Potkay 1979), or *M. chelonei* (*abscessus*) (Weiszfeiler et al. 1969; Karlson et al. 1970; Quimby et al. 1979). In addition, agents belonging to the *Mycobacterium avium-intracellulare* complex have been frequently isolated from macaque species with mycobacterial infections in which tubercle formation is not a feature. These have occurred most often as opportunistic infections in animals with immunodeficiency associated with infection by simian immunodeficiency virus (SIV) or the type D retrovirus SRV-1. Brack (1987) has reviewed the literature on all of the above described simian mycobacterial infections. Also one case report of *M. paratuberculosis* infection in stump-tailed macaques, *Macaca arcotides* has been confirmed (McClure et al. 1987). Since the lesions caused by *M. avium-intracellulare* and *M. paratuberculosis* in nonhuman primates are so distinctive and decidedly different from those seen in classic tuberculosis, these diseases are discussed separately in this monograph.

Pathogenesis

The pathogenesis of tuberculosis in nonhuman primates and other species has been reviewed in detail by Dannenberg (1978). In summary, after tubercle bacilli enter the lung or intestine they are phagocytized by resident macrophages which vary considerably in their microbicidal ability. The waxy nature of the cell wall of mycobacteria confers a relative resistance to lysosomal enzymes and bacteriocidal oxidative processes and thus the organisms are able to survive and replicate within phagosomes of some macrophages which they eventually kill. Other macrophages are able to kill the tubercle bacilli, process their antigens, and present them to T lymphocytes which they have recruited to the site of infection through secretion of various cytokines. Monocytes from the peripheral blood are also recruited to the site of infection where they become activated by products of the tubercle bacilli and by cytokines released by reactive lymphocytes. They too phagocytize bacilli released from dying macrophages and become transformed into immobile epithelioid cells with ruffled membranes that interdigitate with other cells of this type to form an epithelioid granuloma. Those epithelioid cells at the center of the granuloma die and others move in to replace them and surround their necrotic remains. Other macrophages remain mobile and transport the tubercle bacilli to regional lymph nodes and to other tissues where the process begins again. As the host develops delayed hypersensitivity to the tubercle bacilli through the generation of antigen-specific cytotoxic T lymphocytes, these cells play a major role in enhancing the recruitment, activation, and eventual destruction of tubercle bacilli-laden epithelioid cells leading to formation of central areas of caseous necrosis that characterize a tubercle. Some of the activated macrophages fuse to form multinucleated Langhans type giant cells. Release of mediators of inflammation from the dead and dying tissues also promotes the proliferation of fibroblasts which partially encapsulate the lesion.

Comparison with Other Species

In contrast to humans, the lesions of tuberculosis in nonhuman primates rarely become arrested as static fibrocalcific nodules. Instead, in rhesus monkeys in particular, tuberculosis tends to be a rapidly progressive disease characterized by early and widespread dissemination of lesions. Tubercles in monkeys undergo extensive caseation necrosis, but calcification is rare. Of the various macaque species, it has been suggested that cynomolgus monkeys (*Macaca fascicularis*) are more resistant to tuberculosis than rhesus monkeys (Ruch 1959). This would appear to be supported by the relatively small number of reports of tuberculosis in cynomolgus monkeys, despite their widespread use in biomedical research. The literature also suggests that New World monkeys and great apes are more resistant to tuberculosis than Old World monkeys, but this appears to be based more upon epidemiologic data than transmission experiments.

Acknowledgement. This work was supported by Public Health Service Award RR00168 from the National Institutes of Health.

References

Benson RE, Remming BG, Young RJ (1954) Care and management of chimpanzees at the radiobiological laboratory of the University of Texas and the United States Air Force. Proc Anim Care Panel 5:27–36

Benson RE, Fremming BG, Young RJ (1955) A tuberculosis outbreak in a Macaca mulatta colony. Am Rev Tuberc 72:204–209

Bone JF, Soave OA (1970) Experimental tuberculosis in owl monkeys (Aotus trivirgatus). Lab Anim Care 20: 946–948

Brack M (1987) Mycobacteriaceae – tuberculosis type. In: Brack M (ed) Agents transmissible from simians to man. Springer, Berlin Heidelberg New York, pp 214–243

Cappucci DT Jr, O'Shea JL, Smith GD (1972) An epidemiologic account of tuberculosis transmitted from man to monkey. Am Rev Resp Dis 106:819–823

Carpenter CR (1940) Rhesus monkeys (Macaca mulatta) for American laboratories. Science 92:284–286

Chrisp CE, Cohen BJ, Ringler DH, Adams GD (1968) Tuberculosis in a squirrel monkey (Saimiri sciureus) (1968) J Am Vet Med Assoc 153:918–922

Dannenberg AM (1978) Pathogenesis of pulmonary tuberculosis in man and animals: protection of personnel against tuberculosis. In: Montali RJ (ed) Mycobacterial infections of zoo animals. Smithsonian Institution, Washington, pp 65–75

Donovan WN, Krasnow I, Donowho EM Jr, Johanson WG Jr (1976) Mycobacterium simiae. Am J Resp Dis 113, part 2:55 (abstr)

Duffy M, Frederickson L, Linnabary RD (1970) Tuberculosis in a squirrel monkeys with an associated tuberculin conversion in a handler. Primate Zoonosis Surveil Rep 1:9–10

Fiennes RNT-W (1965) Incidence of T.B. in squirrel monkey and marmosets. Lab Primate Newsl 4:10

Fiennes RNT-W (1972) Tuberculosis. In: Fiennes RNT-W (ed) Pathology of simian primates, part II. Karger, Basel, pp 314–334

Fineg J, Hanly WC, Prine JR, Van Riper DC, Day PW (1966) Isoniazid therapy in the chimpanzee. Lab Anim Sci 16:436–446

Fourie PB, Odendaal MW (1983) Mycobacterium tuberculosis in a closed colony of baboons. Lab Anim 17: 125–128

Fox JG, Campbell LH, Snyder SB, Reed C, Soave OA (1974) Tuberculous spondylitis and Potts' paraplegia in a rhesus monkey, Macaca mulatta. Lab Anim Sci 24: 335–339

Fremming BD, Benson RE, Young RJ (1955) Procurement of monkeys for radiobiological research. Sci Mon NY 80:260–262

Griffith AS (1928) Tuberculosis in captive wild animals. J Hyg 28:198–218

Griner LA (1978) Occurrence of tuberculosis in the zoo collection of the Zoological Society of San Diego, 1964–1975

Habel K (1947) Tuberculosis in a laboratory monkey colony; its spread and its control. Am Rev Tuberc 55:77–92

Habermann RT, Williams FP Jr (1957) Diseases seen in necropsy of 708 Macaca mulatta (rhesus monkey) and Macaca philippinensis (cynomolgus monkey). Am J Vet Res 18:419–426

Hammer JE III (1965) Pathological findings in a tuberculous baboon. Lab Anim Care 15:140–150

Hessler JR, Moreland AF (1968) Pulmonary tuberculosis in a squirrel monkey (Saimiri sciureus). J Am Vet Med Assoc 153:923–927

Innes JRM (1963) Tuberculosis of the central nervous system in Macaca mulatta. Three cases with vertebral and dural involvement. Br Vet J 119:393–398

Karasseva V, Weiszfeiler JG, Krasnay E (1965) Occurrence of atypical mycobacteria in Macacus rhesus. Acta Microbiol Acad Sci Hung 12:275–282

Karlson AG, Seibold HR, Wolf RH (1970) Mycobacterium abscessus infection in an owl monkey (Aotus trivirgatus). Pathol Vet 7:448–454

Kaufman AF, Anderson DC (1978) Trends in the incidence of simian tuberculosis. In: Montali RJ (ed) Mycobacterial infections of zoo animals. Smithsonian Institution, Washington, pp 51–53

Keeling ME, Froehlich RE, Ediger RD (1969) An epizootic of tuberculosis in a rhesus monkey conditioning colony. Lab Anim Care 19:629–634

Keeling ME, Bommer WB, McClure HM (1971) Preliminary note on an epizootic of tuberculosis in a great ape population. Primate Zoonoses Surveil Rep 4:9–10

Kennard MA, Willner MD (1941) Findings at autopsies of seventy anthropoid apes. Endocrinology 28:967–976

Kleeberg HH (1975) Tuberculosis and other mycobacterioses. In: Hubbert WT, McCulloch WF, Schnurrenber PR (eds) Diseases transmitted from animals to man, 6th edn, Thomas, Springfield, pp 301–360

Knezevic AL, McNulty WP (1967) Tuberculosis in Lemur mongoz. Folia Primatol (Basel) 6:153–159

Krasnow I, Gross W (1975) Mycobacterium simiae infection in the United States. A case report and discussion of the organism. Am Rev Respir Dis 111:357–360

Lapin BA, Yakovleva LA (1963) Tuberculosis. In: Lapin BA (ed) Comparative pathology in monkeys. Thomas, Springfield, pp 65–94

Lau DT, Fuller JM, Sumner PE (1972) Tuberculosis in a pig-tailed macaque. J Am Vet Med Assoc 161:696–699

Leathers CW, Hamm TE Jr (1976) Naturally occurring tuberculosis in a squirrel monkey and a cebus monkey. J Am Vet Med Assoc 169:909–911

Martin JE, Cole WC, Whitney RA (1968) Tuberculosis of the spine (Potts' disease) in a rhesus monkey (M. mulatta.). J Am Vet Med Assoc 153:914–917

Mayhall CG, Lamb VA, Coleman PH (1981) Infection in rhesus (Macaca mulatta) and squirrel (Saimiri sciureus) monkeys due to Mycobacterium tuberculosis phage type B. J Med Primatol 10:302–311

McClure HM, Guilloud NB (1971) Comparative pathology of the chimpanzee. In: Bourne GH (ed) The chimpanzee: 4. Behavior, growth, and pathology of chimpanzees. University Park Press, Baltimore

McClure HM, Silcox VA (1970) Possible Mycobacterium africanum infection in a rhesus monkey. Primate Zoonoses Surveill Rep 1:7–8

McClure HM, Chiodini RJ, Anderson DC, Swenson RB, Thayer WR, Coutu JA (1987) Mycobacterium paratuberculosis infection in a colony of stumptail macaques (Macaca arctoides). J Infect Dis 155:1011–1019

McLaughlin RM (1978) Mycobacterium bovis in nonhuman primates. In: Montali RJ (ed) Mycobacterial infections of zoo animals. Smithsonian Institution, Washington, pp 151–155

McLaughlin RM, Marrs GE (1978) Tuberculin testing in nonhuman primates: OT vs PPD. In: Montali RJ (ed) Mycobacterial infections of zoo animals. Smithsonian Institution, Washington, pp 123–128

Moreland AF (1970) Tuberculosis in New World monkeys. Lab Anim Care 20:262–264

Quimby FW, Miller K, Valeri CR (1979) Atypical mycobacteriosis in a baboon. J Am Vet Med Assoc 174:954

Renner M, Bartholomew WR (1974) Mycobacteriologic data from two outbreaks of bovine tuberculosis in nonhuman primates. Am Rev Respir Dis 109:11–16

Renquist DM, Potkay S (1979) Mycobacterium scrofulaceum infection in Erythrocebus patas monkeys. Lab Anim Sci 29:97–101

Renquist DM, Whitney RA (1978) Tuberculosis in nonhuman primates – an overview. In: Montali RJ (ed) Mycobacterial infections of zoo animals. Smithsonian Institution, Washington, pp 9–16

Ruch TC (1959) Tuberculosis. In: Ruch TC (ed) Diseases of laboratory primates. Saunders, Philadelphia, pp 199–253

Savoy JC (1965) Tuberculosis in a zoo collection of great apes. Zoo Yearbook 5:189–192

Schroeder CR (1938) Acquired tuberculosis in the primate in laboratories and zoological collections. Am J Public Health 28:469–475

Silberman MS (1978) Tuberculosis in a band of imported hamadryas baboons. In: Montali RJ (ed) Mycobacterial infections of zoo animals. Smithsonian Institution, Washington, pp 161–162

Snyder RL (1978) Historical aspects of tuberculosis in the Philadelphia zoo. In: Montali RJ (ed) Mycobacterial infections of zoo animals. Smithsonian Institution, Washington, pp 33–44

Snyder S, Peace T, Soave O, Lund J (1970) Tuberculosis in an owl monkey (*Aotus trivirgatus*). J Am Vet Med Assoc 157:712–713

Thoen CO, Beluhan FZ, Himes EM, Capek V, Bennett T (1977) Mycobacterium bovis infection in baboons (*Papio papio*). Arch Pathol Lab Med 101:291–293

Thorel MF (1980) Isolation of Mycobacterium africanum from monkeys. Tubercle 61:101–104

Truant JP, Brett WA, Thomas W Jr (1962) Fluorescence microscopy of tubercle bacilli stained with auramine and rhodamine. Henry Ford Hosp Med Bull 10:287–296

Tryphonas L, Wong J (1982) Potts' paraplegia in a tuberculous rhesus monkey, Macaca mulatta. J Med Primatol 11:211–220

Valerio DA, Dalgard DW, Voelker RW, McCarrol NE, Good RC (1979) Mycobacterium kansasii infection in rhesus monkeys. In: Montali RJ (ed) Mycobacterial

infections in zoo animals. Smithsonian Institution, Washington, pp 145–150
Wayne LG, Kubica GP (1986) The mycobacteria. In: Sneath PHA (ed) Bergey's manual of systematic bacteriology, vol 2. Williams and Wilkins, Baltimore, pp 1436–1457
Weiszfeiler JG, Karasseva V, Karczag E (1969) Comparative studies of the taxonomic relationship between Mycobacterium abscessus and *Mycobacterium borstelense*. Incidence of Mycobacterium abscessus in monkeys. Acta Microbiol Acad Sci Hung 16:371–379
Weiszfeiler JG, Karasseva V, Karczag E (1970) Studies on Mycobacterium simiae and some other atypical mycobacteria isolated from monkeys. Proc Microbiol Res Group Hung Acad Sci III:31–39
Weiszfeiler JG, Karasseva V, Karczag E (1981) Mycobacteria simiae and related mycobacteria. Rev Infect Dis 3:1040–1045
Wolf RH, Bullock BC, Clarkson TB (1967) Tuberculosis in the stumptailed macaque (Macaca speciosa). J Am Vet Med Assoc 151:914–917
Zumpe D, Silbermann MS, Michael RP (1980) Unusual outbreak of tuberculosis due to Mycobacterium bovis in a closed colony of rhesus monkeys (Macaca mulatta). Lab Anim Sci 30:237–240

Paratuberculosis, Nonhuman Primates

Daniel C. Anderson and Harold M. McClure

Synonyms. Johne's disease.

Gross Appearance

Gross lesions are often minimal. Animals usually have a history of chronic wasting and are generally culture negative for other potential enteric pathogens although concomitant bacterial infections may occur. Visible lesions are usually limited to enlarged mesenteric lymph nodes although distended lymphatics may be seen in advanced cases. The intestinal mucosa may be slightly thickened, however, it often appears normal.

Microscopic Features

In general, lesions are limited to the intestinal tract, mesenteric lymph nodes, and the liver. Other tissues are rarely or minimally involved. The characteristic lesion consists of granulomatous inflammation without caseous necrosis and involving the intestinal tract. Lesions are usually prominent in animals dying of chronic infection and may extend from the pyloric-duodenal junction to and including the anal-rectal junction (McClure et al. 1987). Granulomatous inflammation is usually most severe in the jejunum and ileum (Fig. 157); the inflammatory infiltrate is predominantly histiocytic with focal clusters of epithelioid cells in the outer portion of the intestinal mucosa in less severe cases or dense sheets of similar inflammation involving large areas or virtually effacing the lamina propria in more advanced cases (Fig. 158). Involvement of the cecum, colon, and rectum is usually less evident, however, lesions may range from very mild to focally severe. Syncytial and Langhans type giant cells are not a prominent feature in the intestinal tract, however, they may be rarely observed (McClure et al. 1987). Numerous acid-fast bacilli in the cytoplasm of histiocytes, although found at all levels of the small and large intestine, are much more prominent in the small intestine (Fig. 159). Occasionally, small numbers of organisms may be found in the tonsils.

The mesenteric, colonic, and ileocolic lymph nodes are often involved although histologic changes are variable in these tissues, even within the same animal. Changes range from mild histiocytosis to focal clusters of epithelioid cells in the cortex to total effacement of the node by dense sheets of histiocyes. Histiocytic infiltrates are most frequently seen within subcapsular sinuses with extension into the medullary sinuses (Fig. 160). Although Langhans' giant

Fig. 157 (*above*). Paratuberculosis, small intestine, stump-tailed macaque. Diffuse inflammatory cell infiltrate distending the lamina propria. H and E, ×74 ▶

Fig. 158 (*lower left*). Paratuberculosis, stump-tailed macaque. Higher magnification of Fig. 157. The infiltrate in the lamina propria consists of histiocytes. H and E, ×160

Fig. 159 (*lower right*). Paratuberculosis, *Macaca arctoides*, small intestine. Note numerous organisms within histiocytes in the basilar lamina propria. Kinyoun's acid-fast stain, ×544

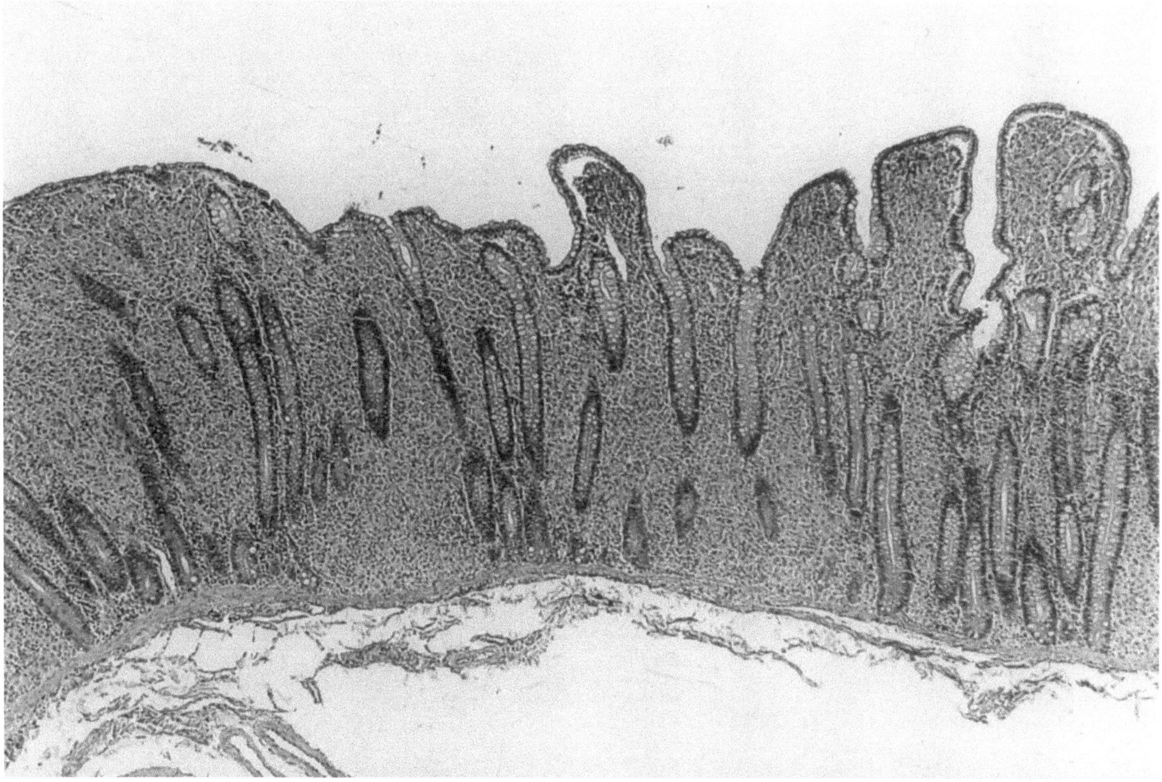

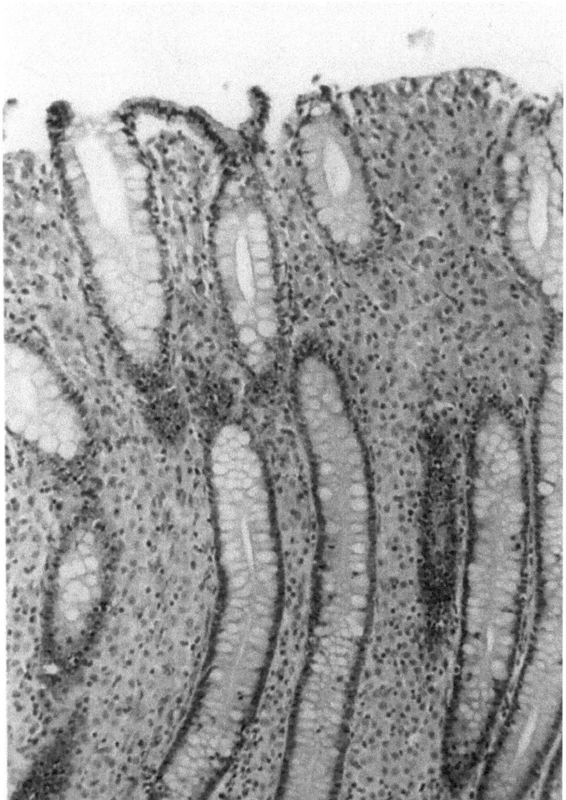

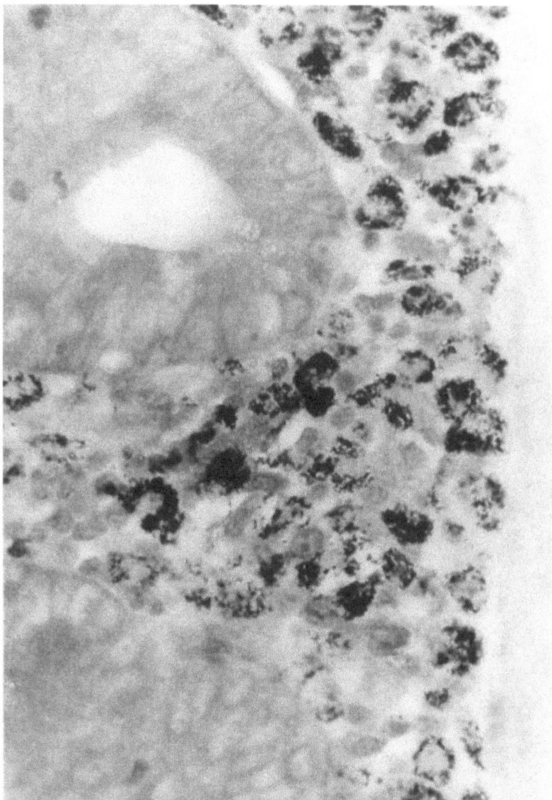

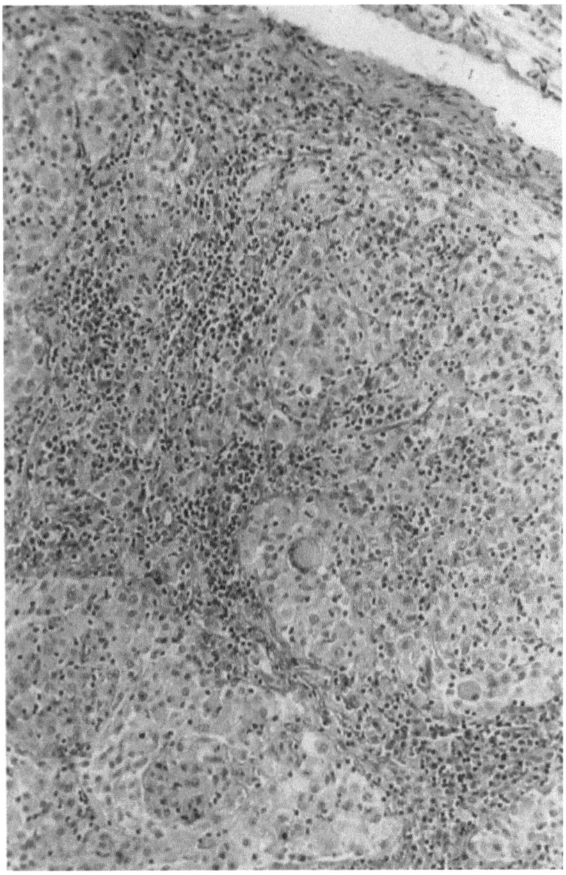

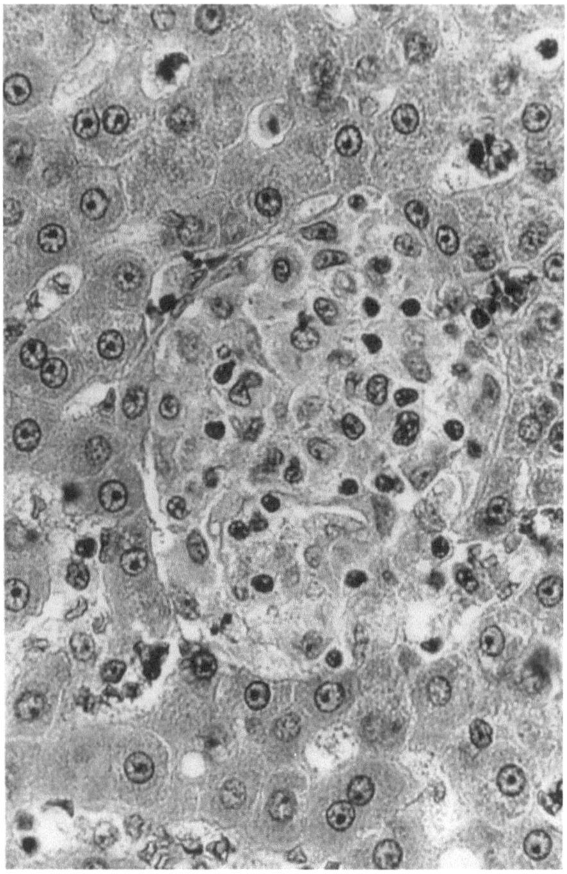

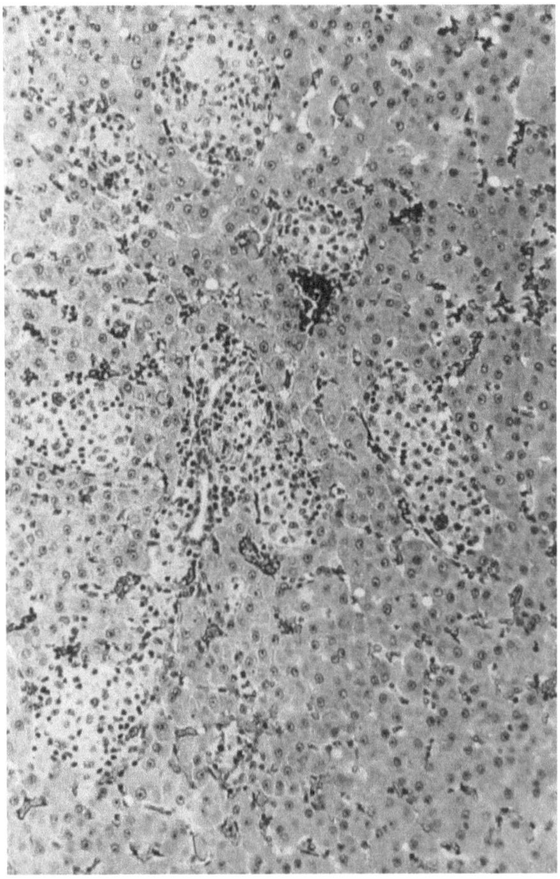

Fig. 160 (*upper left*). Paratuberculosis, mesenteric lymph node, stump-tailed macaque. Dense clusters of histiocytes and occasional Langhans' giant cell extend from the subcapsular sinus into the medulla. H and E, ×160

Fig. 161 (*lower left*). Paratuberculosis, liver, stump-tailed macaque. Note numerous granulomas throughout the hepatic parenchyma. H and E, ×160

Fig. 162 (*upper right*). Paratuberculosis, liver, stump-tailed macaque. Higher magnification of hepatic granuloma in Fig. 161. Note dense histiocytic infiltrate with fewer lymphocytes. H and E, ×544 cells are commonly noted, acid-fast bacilli are much less apparent in lymph nodes than within the intestine.

Hepatic granulomas are often seen, however, they may be infrequent or numerous if present (Fig. 161). Granulomas consist predominantly of histiocytes with fewer lymphocytes (Fig. 162). When numerous, scattered granulomas may contain Langhans' giant cells. An acid-fast stain may be negative or reveal only small numbers of acid-fast becilli within hepatic granulomas.

Ultrastructure

Ultrastructural features are not of diagnostic significance.

Differential Diagnosis

Chronic weight loss and diarrhea are frequent diagnostic problems in nonhuman primate colonies. Although intestinal mycobacteriosis may occur sporadically in outdoor-housed animals, it is infrequently reported. Progressive weight loss and diarrhea associated with intestinal amyloidosis or chronic nonspecific colitis are much more commonly observed. Nonetheless, *Mycobacterium avium* and *M. paratuberculosis* infections have resulted in significant morbidity and mortality associated with chronic diarrheal illness in some colonies (Holmberg et al. 1982, 1985; McClure et al. 1987). Histologic evaluation will rule out other enteric infections, however, a definitive diagnosis is dependent on culture of the organism (Chiodini et al. 1984a) in conjunction with other criteria used to separate *M. paratuberculosis* from other slow-growing mycobacteria (Whipple et al. 1987). The diagnosis of paratuberculosis is hindered by a lack of accurate and sensitive tests (Chiodini et al. 1984a). Numerous serologic and other diagnostic methods have been used and are of the most value on a herd basis; however, they are of limited value in evaluating individual animals due to the large numbers of false negative and false positive tests (Chiodini et al. 1984a). The recent introduction of DNA hybridization probes for *Mycobacterium paratuberculosis* may prove to be useful adjuncts for the rapid detection of infection and for differentiating between closely related species (Ambrosio et al. 1991; Hurley et al. 1989; McFadden et al. 1987a,b; Whipple et al. 1987).

Biologic Features

Natural History

In only one reported outbreak of this disease in nonhuman primates has the organism, *Mycobacterium paratuberculosis*, been precisely identified (McClure et al. 1987). In that study, the organism was implicated as the cause of a severe granulomatous enteritis in a colony of stump-tailed macaques (*Macaca arctoides*). The paucity of other studies of this disease in nonhuman primates makes it necessary to consider studies in other species.

Bovine paratuberculosis (Johne's disease) is a commonly diagnosed chronic enteric disease of ruminants throughout the world. Although descriptions of the disease were reported as early as 1826, acid-fast organisms were first described in association with the disease in 1895 (Johne and Frothingham 1895; Chiodini et al. 1984a), with isolation of the organisms in 1910 (Twort and Ingram 1912). Since that time, it has been recognized as economically important in cattle, but less important in other domestic species (Merkal et al. 1975; Riemann and Abbas 1983; Chiodini et al. 1984a). All ruminants are susceptible to infection and paratuberculosis has been recognized in wild ruminants and in zoo animals (Williams et al. 1983; Chiodini et al. 1984a).

Etiology

Mycobacterium paratuberculosis is a member of the family *Mycobacteriaceae*. It is a short, slender, gram-positive, acid-fast rod which is about 0.5 µm wide by 1–2 µm in length (Riemann and Abbas 1983). The bacillus can survive in the environment for extended periods in water and soil (Chiodini et al. 1984a) and can survive freezing for more than 1 year (Chiodini et al. 1984a; McClure et al. 1987). *Mycobacterium paratuberculosis* is closely related to *M. avium* and to mycobactin-dependent strains of *M. avium* and mycobacterial organisms isolated from wood pigeons and from some patients with Crohn's disease (Whipple et al. 1987; Hurley et al. 1988; McFadden et al. 1987b). *M. paratuberculosis* is usually distinguished from other slow-growing *Mycobacteria* by its dependence on exogenous mycobactin, an iron-chelating product of *Mycobacteria*, for in vitro growth. The addition of iron in high concentrations in media may circumvent the mycobactin dependency (Merkal and McCullough 1982; Chiodini et al. 1984a). Although mycobactins obtained from any species of *Mycobacteria* can be used for primary cultivation, the amount needed for cultivation varies with the species from which it was isolated. The standard mycobactin used for *M. paratuberculosis* was obtained from *M. phlei* (mycobactin P) (Merkal and McCullough 1982). Strains which require at least 8–12 weeks of growth on initial culture and which grow only in the presence of mycobactin are identified as *M. paratuberculosis* (Merkal and McCullough 1982; Chiodini et al. 1984a). The incorporation of mycobactin J, derived from *M. paratuberculosis*, in medium reduces the incubation period by about 3 weeks (Merkal and McCullough 1982) with primary isolation at 6–8 weeks (Chiodini et al. 1984a).

Pathogenesis

Transmission of the infection usually requires exposure at an early age (Chiodini et al. 1984a; Jubb et al. 1985). Clinical disease has been documented in cattle between 4 months to 15 years of age, however, it is most common in animals at 3–5 years of age. A significant number of infected animals either do not become chronically infected or become asymptomatic carriers (Chiodini et al. 1984a). Adult animals may become infected, however, they are less likely to develop the disease. Animals less than 30 days old are most susceptible to infection and the critical susceptible age is probably less than 6 months (Chiodini et al. 1984a; Jubb et al. 1985). Although bovine fetuses may be infected as early as 2 months gestation and the organism can be isolated from a high percentage of fetuses from terminally infected dams, transplacental exposure does not appear to be a significant cause of natural infection (Merkal et al. 1975; Chiodini et al. 1984a; Jubb et al. 1985). The route of infection is primarily by the fecal-oral route and occurs most commonly by the ingestion of contaminated feces during nursing (Chiodini et al. 1984a). Following oral infection, the organism enters the lymphatic system through the tonsils and the intestinal tract and localizes mainly in the small intestine (Jubb et al. 1985). Peyer's patches in the lower small intestine are considered to be the most common portal of entry. One report suggests that *M. paratuberculosis* crosses the intestinal barrier in the ileum by M cell uptake in dome epithelium of the Peyer's patches rather than by active invasion and may explain why early lesions are seen at this site (Momotani et al. 1988).

The disease is characterized by chronic or intermittent diarrhea that eventually results in severe weight loss and emaciation. At death, the severity of the clinical signs may not correlate with the severity of the gross and histologic changes. Lesions may be slight and easily missed unless they are specifically sought.

Granulomatous lesions in natural and experimental disease in sheep, goats, and wild ruminants have been reported in extraintestinal sites in multiple organs and lymph nodes and in the tonsils and the liver (Williams et al. 1983; Hines et al. 1987). Extraintestinal lesions probably reflect seeding of the organism during bacteremia.

Frequency

M. paratuberculosis infection in nonhuman primates has only been confirmed by culture on one occasion (McClure et al. 1987). An earlier report of paratuberculosis in rhesus monkeys (*Macaca mulatta*) was not substantiated by culture (Pitcock and Gisler 1961). The single report of documented *M. paratuberculosis* infection in nonhuman primates may be due to several factors. First, intestinal mycobacteriosis is not a frequent infection in colony-housed animals and is usually diagnosed at postmortem in nonhuman primates. Second, mycobacterial culture is often not done because tissues are either not available in isolated cases or because the disease is assumed to be due to *M. avium* infection.

Finally, other conditions, such as intestinal amyloidosis or nonspecific chronic diarrhea, are much more frequently associated with severe weight loss and are more likely to be considered in the diagnosis. Intestinal mycobacteriosis is probably considered in the differential only in endemic or outbreak situations.

Comparison with Other Species

The clinical and pathological features of *M. paratuberculosis* infection in stump-tailed macaques is virtually identical to Johne's disease in ruminants and to *M. avium* infection in nonhuman primates. The disease has been well characterized in natural and experimental infections in domestic and wild ruminants (Buergelt et al. 1978; Riemann and Abbas 1983; Williams et al. 1983; Chiodini et al. 1984a). Although many laboratory animals can be infected with *M. paratuberculosis*, detailed studies of the disease have been hampered by the lack of a reliable laboratory model (Riemann and Abbas 1983; Chiodini et al. 1984a; Hamilton et al. 1989). The organism will replicate after prolonged periods in some hosts and may produce inflammatory lesions in the intestine and lymph nodes, however, the changes seen are not consistent with those of paratuberculosis (Chiodini et al. 1984a). It has been demonstrated that persistent *M. paratuberculosis* infection can be established in athymic nude gnotobiotic mice (Hamilton et al. 1989). Reported lesions were similar to those observed in Johne's disease and in some cases of Crohn's disease. Although *M. paratuberculosis* has been isolated from several patients with Crohn's disease (Chiodini et al. 1984b; McFadden et al. 1987b), its role is controversial. Other investigators have failed to implicate any specific *Mycobacterium* in the pathogenesis of this human disease (Graham et al. 1987).

In the bovine, the classical lesions consist of a severely thickened corrugated ileum, mesenteric lymphadenopathy, and lymphangitis. However, lymphangitis may be the only recognizable gross lesion and changes may be minimal in goats and sheep (Williams et al. 1983; Jubb et al. 1985). In addition, while focal necrosis and mineralization are rare in cattle, they are more common in sheep and goats (Williams et al. 1983). Microscopic lesions are characterized by granulomatous inflammation of the intestinal tract and the regional lymph nodes (Williams et al. 1983; Chiodini et al. 1984a; Jubb et al. 1985). The lesions occur primarily in the distal small intestine and the mesenteric lymph nodes, however, granulomatous inflammation and acid-fast organisms may be found throughout the intestinal tract from the duodenum to the rectum. Despite the isolation of *M. paratuberculosis* from multiple organs in cattle, microscopic lesions, until recently, had not been reported in that species in sites other than in the intestinal tract, liver, and mesenteric and hepatic lymph nodes (Hines et al. 1987).

Acknowledgements. This work was supported in part by NIH grant RR-00165 from the National Center for Research Resources to the Yerkes Regional Primate Research Center. The Yerkes Center is fully accredited by the American Association for Accreditation of Laboratory Animal Care.

References

Ambrosio RE, Harris Y, Huchzermeyer HF (1991) A DNA probe for the detection of Mycobacterium paratuberculosis. Vet Microbiol 26:87–93

Buergelt CD, Hall C, McEntee K, Duncan JR (1978) Pathological evaluation of paratuberculosis in naturally infected cattle. Vet Pathol 15:196–207

Chiodini RJ, Van Kruiningen HJ, Merkal RS (1984a) Ruminant paratuberculosis (Johne's disease): the current status and future prospects. Cornell Vet 74: 218–262

Chiodini RJ, Van Kruiningen HJ, Thayer WR, Merkal RS, Coutu JA (1984b) Possible role of mycobacteria in inflammatory bowel disease: I. An unclassified Mycobacterium species isolated from patients with Crohn's disease. Dig Dis Sci 29:1073–1079

Graham DY, Markesich DC, Yoshimura HH (1987) Mycobacteria and inflammatory bowel disease. Results of culture. Gastroenterology 92:436–442

Hamilton HL, Follett DM, Siegfried LM, Czuprynski CJ (1989) Intestinal multiplication of Mycobacterium paratuberculosis in athymic nude gnotobiotic mice. Infect Immun 57:225–230

Hines SA, Buergelt CD, Wilson JH, Bliss EL (1987) Disseminated Mycobacterium paratuberculosis infection in a cow. J Am Vet Med Assoc 190:681–683

Holmberg CA, Henrickson RV, Malaga C, Schneider R, Gribble D (1982) Nontuberculous mycobacterial disease in rhesus monkeys. Vet Pathol [Suppl] 7:9–16

Holmberg CA, Henrickson R, Lenninger R, Anderson J, Hayashi L, Ellingsworth L (1985) Immunologic abnormality in a group of Macaca arctoides with high mortality due to atypical mycobacterial and other disease processes. Am J Vet Res 46:1192–1196

Hurley SS, Splitter GA, Welch RA (1988) Deoxyribonucleic acid relatedness of Mycobacterium paratuberculosis to other members of the family Mycobacteriaceae. Int J System Bacteriol 38:143–146

Hurley SS, Splitter GA, Welch RA (1989) Development of a diagnostic test for Johne's disease using a DNA hybridization probe. J Clin Microbiol 27:1582–1587

Johne HA, Frothingham L (1895) Ein eigenthümliche Fall von Tuberculose beim Rind. Dsch Z Thiermed Verg Pathol 21:438–454

Jubb KVF, Kennedy PC, Palmer N (1985) Pathology of domestic animals, vol 2. Academic, New York, pp 155–159

McClure HM, Chiodini RJ, Anderson DC, Swenson RB, Thayer WR, Coutu JA (1987) Mycobacterium paratuberculosis infection in a colony of stumptail macaques (Macaca arctoides). J Infect Dis 155:1011–1019

McFadden JJ, Butcher PD, Thompson J, Chiodini RJ, Hermon-Taylor J (1987a) The use of DNA probes identifying restriction-fragment-length polymorphisms to examine the Mycobacterium avium complex. Mol Microbiol 1:283–291

McFadden JJ, Butcher PD, Chiodini RJ, Hermon-Taylor J (1987b) Crohn's disease-isolated mycobacteria are identical to Mycobacterium paratuberculosis, as determined by DNA probes that distinguish between mycobacterial species. J Clin Microbiol 25:796–801

Merkal RS, McCullough WG (1982) A new mycobactin, mycobactin J, from Mycobacterium paratuberculosis. Curr Microbiol 7:333–335

Merkal RS, Larsen AB, Booth GD (1975) Analysis of the effects of inapparent bovine paratuberculosis. Am J Vet Res 36:837–838

Momotani E, Whipple DL, Theirmann AB, Cheville NF (1988) Role of M cells and macrophages in the entrance of Mycobacterium paratuberculosis into domes of ileal Peyer's patches of calves. Vet Pathol 25:131–137

Pitcock JA, Gisler DB (1961) Paratuberculosis (Johne's Disease) in the monkey (Macaca mulatta). Report 61-86. Brooks Air Force Base, Texas; School of Aerospace Medicine, USAF Aerospace Medical Center

Riemann HP, Abbas B (1983) Diagnosis and control of bovine paratuberculosis (Johne's disease). Adv Vet Sci Comp Med 27:481–506

Twort FW, Ingram GLY (1912) A method for isolating and cultivating the Mycobacterium enteriditis chronicae pseudotuberculosae bovis Jöhne, and some experiments on the preparation of a diagnostic vaccine for pseudo-tuberculosis enteritis of bovines. Proc R Soc Lond 84:517–545

Whipple DL, Le Febvre RB, Andrews RE Jr, Theirmann AB (1987) Isolation and analysis of restriction endonuclease digestive patterns of chromosomal DNA from Mycobacterium paratuberculosis and other Mycobacterium species. J Clin Microbiol 25:1511–1515

Williams ES, Snyder SP, Martin KL (1983) Pathology of spontaneous and experimental infection of North American wild ruminants with Mycobacterium paratuberculosis. Vet Pathol 20:274–291

Epidermal and Gastric Mucosal Metaplasias Caused by Polychlorinated Biphenyls, Dibenzo-*p*-Dioxins and Dibenzofurans, Rhesus Monkeys

Wilbur P. McNulty

Synonyms. Chloracne, poisoning due to polyhalogenated polycyclic aromatic hydrocarbons (PHPAHs).

Gross Appearance

The skin of the face is coarsely granular and dry, with small dark cysts and excoriations, especially about the lips (Fig. 163). Most facial hair is absent. The eyelids are thick and red, and on the inner surfaces of the everted lids the meibomian glands appear thick, white, and tortuous (Fig. 164). The nails are elevated by subungual accumulations of friable, pale material and may even be lost, except for the roots (Fig. 165). The gastric body wall is thick, and the normal rugal pattern of the mucosa is irregularly replaced by a flat granular surface (Fig. 166). The submucosa is thick and may contain small cysts, 1–3 mm in diameter.

Microscopic Features

Sebaceous glands in facial skin are absent or persist only as squamous nubbins along the walls of the pilary canals. Hair follicles are almost all in telogen phase, and some are devoid of shafts. Some sebaceous glands are converted to squamous cysts, especially the large glands associated with vibrissae about the lips (Fig. 167). This keratinizing metaplasia is most striking in the large racemose sebaceous glands (meibomian glands) radially disposed in the tarsal plates of the eyelids. Each gland has become a simple tubular squamous cyst (Fig. 168). An occasional cyst is ruptured, with attendant chronic inflammatory and giant cell reaction. During recovery after exposure to polyhalogenated polycyclic aromatic hydrocarbons ceases, the keratinizing squamous epithelium lining the former sebaceous glands is first replaced by conjunctival epithelium (Fig. 169); ultimately, histologically normal sebaceous glands are fully reconstituted.

In the nail, the layer of columnar cells capping the stratified nonkeratinizing epithelium beneath the plate and providing a sliding bond to the distally moving plate, which is growing from the root, is replaced by squamous cells, the upper layers of which now contain keratohyaline granules (Fig. 170). Squamous debris and keratin accumulate beneath the nail plate, raising and detaching it.

Chief and parietal cells are absent from the mucosa of the gastric body; tall columnar cells with large mucous vacuoles line foveolae and glands (Fig. 171). Epithelial nuclei are enlarged but not dysplastic. Superficial mucosal ulcers are frequent and sometimes extensive. Cysts lined with mucous epithelium are scattered through an edematous submucosa (Fig. 171); in advanced cases these cysts may be found in the muscularis and even in the serosa. The pyloric mucosa is not affected.

Ultrastructure

Electron microscopic studies of serial gastric mucosal biopsies in rhesus monkeys fed the polychlorinated biphenyl, Aroclor 1242 did not show diagnostic changes. Signs of injury-dilated endoplasmic reticulum, distorted mitochondria, and autophagic vesicles were present in all mucosal cell types. Vacuoles believed to be mucous appeared in zymogenic cells, but the

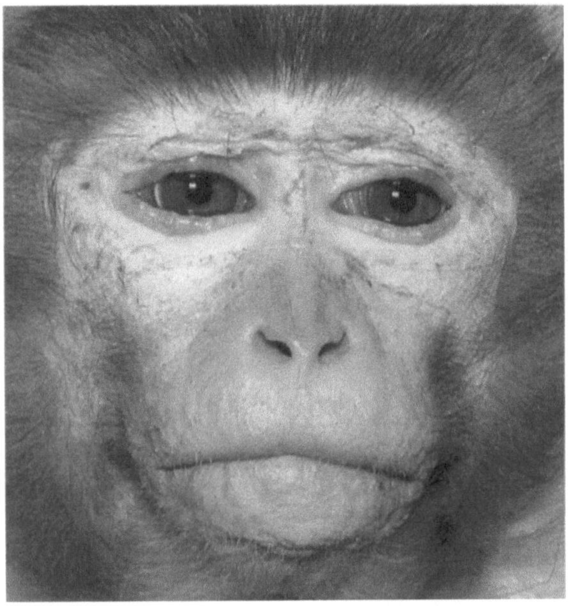

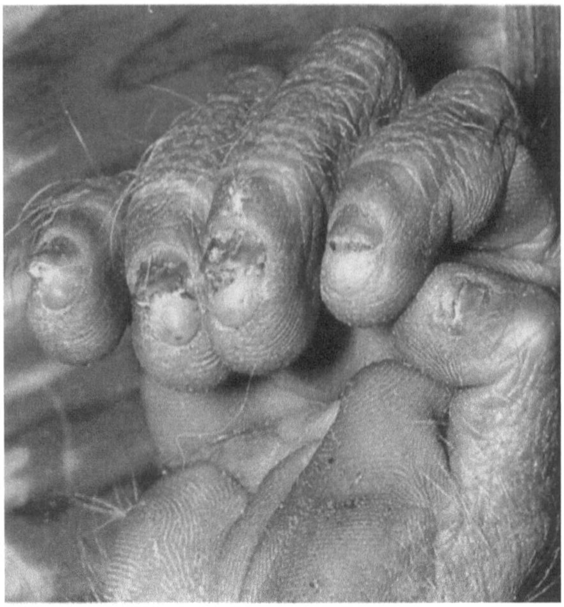

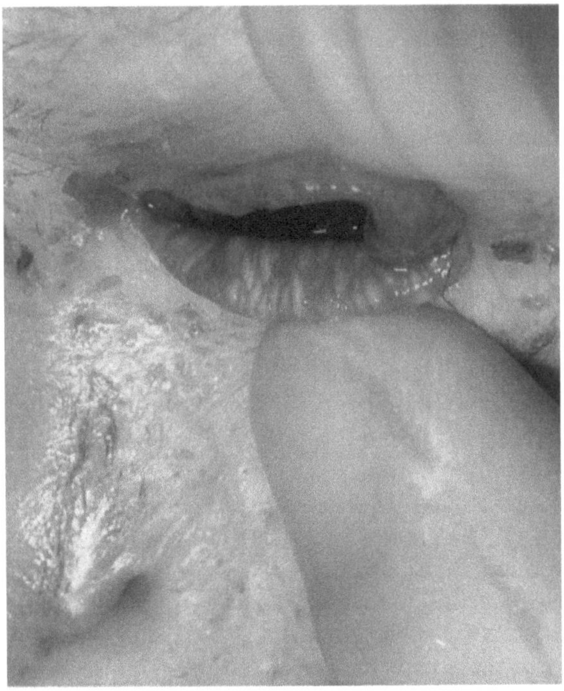

Fig. 163 (*upper left*). Young rhesus male consuming diet containing 2,3,7,8-tetrachlorodibenzofuran. Facial skin is rough, facial hair has been lost, and eyelids are red, thick, and beaded

Fig. 164 (*lower left*). Close-up of everted eyelids in monkey in Fig. 163, showing swollen meibomian glands. Small dark cysts are present in the skin of the nose and cheek

Fig. 165 (*upper right*). Elevated and lost fingernails in rhesus monkey fed 2,3,7,8-tetrachlorodibenzofuran (TCDF)

Fig. 167. *Above*: Lower lip of young rhesus male after ▶ consuming 3,4,3′,4′-tetrachlorobiphenyl. The normally abundant sebaceous glands attached to ordinary hair follicles and to vibrissae have disappeared, been reduced to squamous pegs, or formed squamous cysts. ×30. *Below*: Lower lip of young rhesus male control, H and E, ×30

issue of whether conversion to or replacement by (or both) mucous cells led to the almost exclusively mucous cellular population of the lesion has not been resolved (Becker et al. 1979). Synthesis of DNA occurs throughout the abnormal mucosa and in submucosal cysts, whereas normally it is restricted to the mucous necks of the glands (Becker and McNulty 1984).
There have been no ultrastructural studies of the epidermal lesions in primates.

Differential Diagnosis

These metaplasias in sebaceous glands and nails are pathognomonic for toxicity from some congeners of polychlorinated biphenyls (PCBs), 2,3,7,8-tetrachlorodibenzo-p-dioxin (TCDD), 2,-3,7,8-tetrachlorodibenzofuran (TCDF), and commercial mixtures of PCBS, such as Aroclors, which may contain contaminating polychlorinated dibenzofurans. Polybrominated analogues cause similar changes. Collectively these compounds

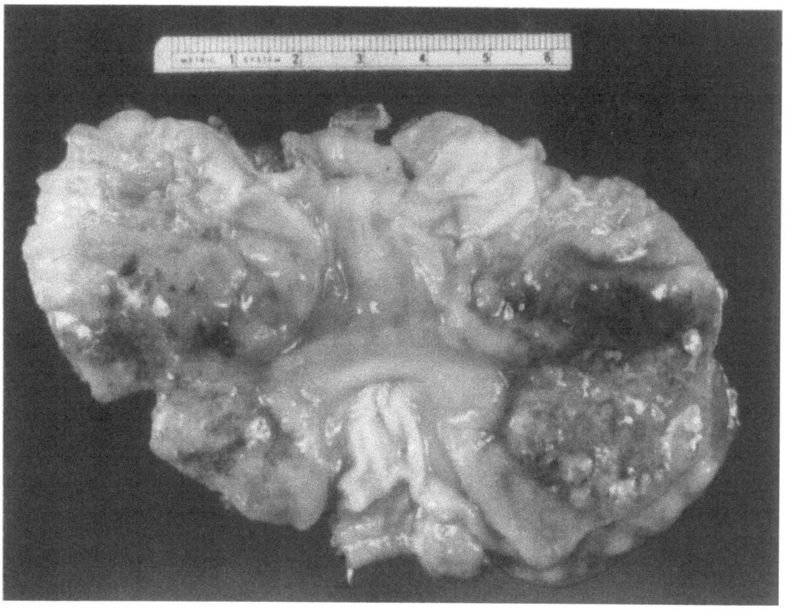

Fig. 166. Stomach of young rhesus male after consuming 3,4,3',4'-tetrachlorobiphenyl; the opening along the greater curvature passes through a large irregular, firm, hemorrhagic lesion with rolled mucosal margins

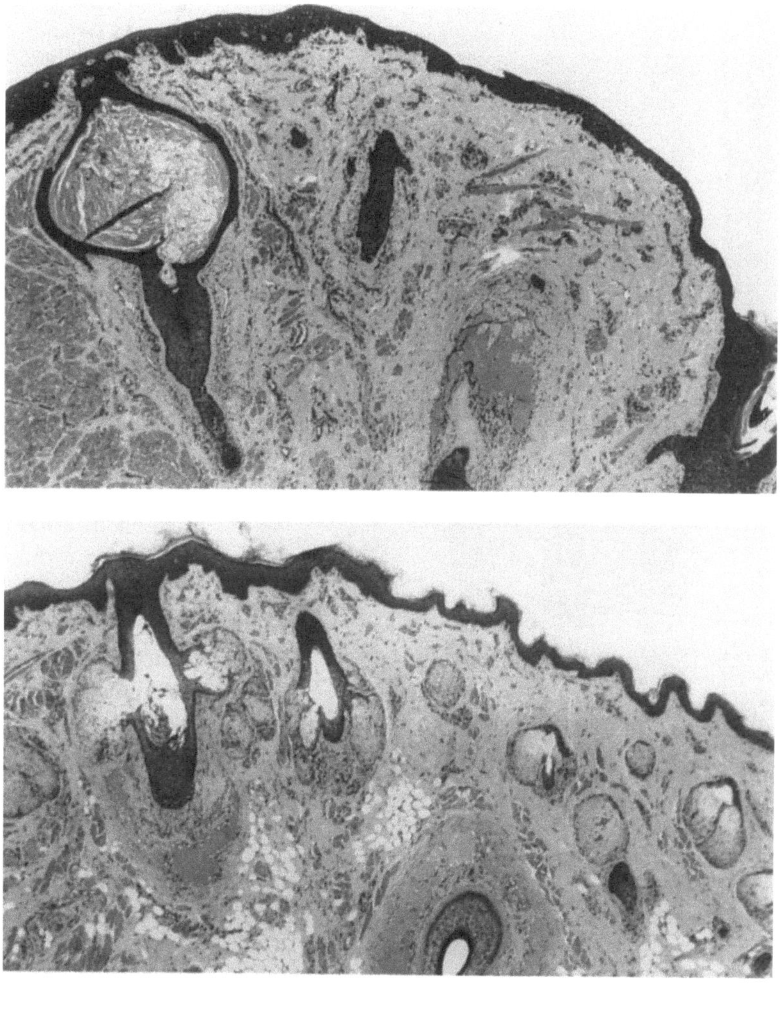

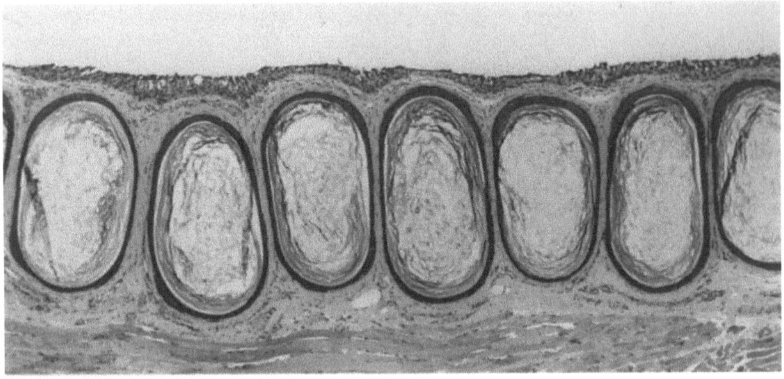

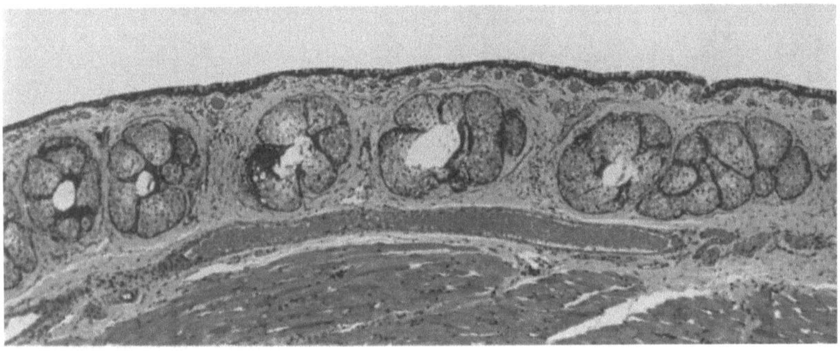

▲
Fig. 168. *Above*: Meibomian glands in eyelids of young rhesus male after consuming 3,4,5,3′,4′,5′-hexachlorobiphenyl. The normally racemose sebaceous glands have become simple tubular squamous cysts. H and E, ×30. *Below*: Eyelid of control rhesus male. H and E, ×30

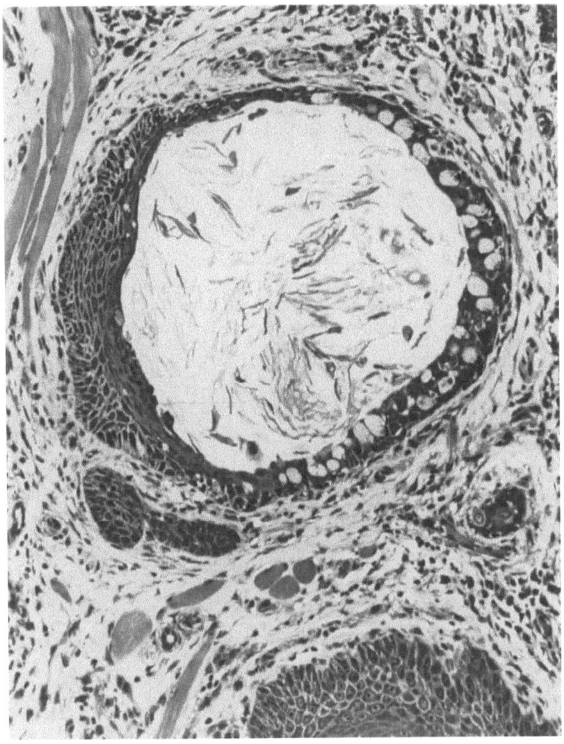

Fig. 169. Meibomian gland of young rhesus male recovering from dietary intake of 3,4,5,3′,4′,5′-hexachlorobiphenyl. The metaplastic, keratinizing, squamous epidermal lining has been partly replaced by a multilayered epidermis containing scattered goblet cells (conjunctival pattern). H and E, ×120

are here referred to as polyhalogenated polycyclic aromatic hydrocarbons (PHPAHs). Localized atrophy or squamous transformation of skin appendages and nails may of course be associated with infection or trauma, but such lesions are accompanied by inflammatory infiltrates, which are inconspicuous or absent in the toxic changes. The gastric mucosal lesion can be distinguished from papillary and sharply localized adenomatous hyperplasia in *Macaca fascicularis* due to *Nochtia nochti* (Bonne and Sandground 1939) by the nematode eggs seen deep in clefts of the growth. However, the toxic lesion does strongly resemble those described as adenomatous lesions (Lushbaugh 1949) and as Meniere's disease (Uemura et al. 1979). Since the monkeys in the earlier study were experimentally fed diesel motor lubricating oil and those in the latter lived in an environment where contamination with PCBs had been a problem, the lesions may have been due to the PHPAHs. In neither case were observations on the skin reported, and PHPAH-induced gastric metaplasia is always

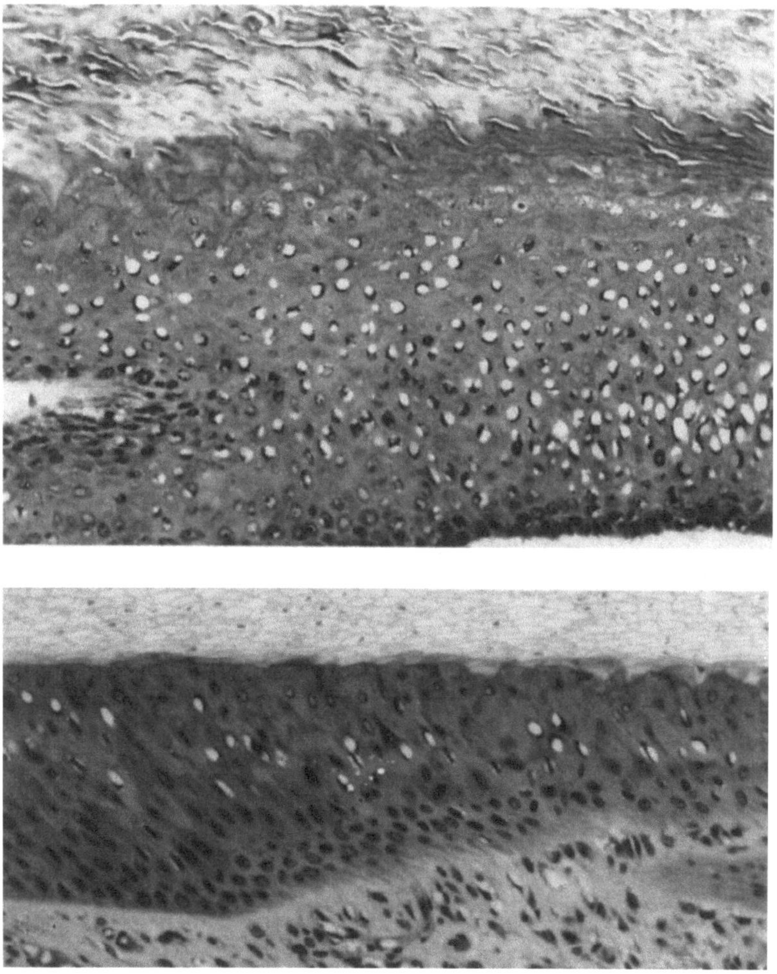

Fig. 170. *Above*: Fingernail bed from young rhesus male after consuming 3,4,3′,4′-tetrachlorobiphenyl. The polyhedral cells of the bed and keratinized in the upper layers acanthotic and keratinized cells desquamated. H and E, ×210. *Below*: Fingernail bed from control rhesus monkey. H and E, ×210

accompanied by metaplasias in sebaceous glands and nails.

Biologic Features

Nearly all spontaneous and experimental exposures to PHPAHs have involved rhesus monkeys. Experimental lesions in cynomolgus monkeys (*Macaca fascicularis*) (Tryphonas et al. 1986) and in common marmosets (*Callithrix jacchus*) (van den Berg et al. 1988) are qualitatively similar, although these species are less sensitive than rhesus monkeys.

Poisoning of rhesus monkeys by the toxic congeners of PHPAHS can be fatal (Allen 1975; Allen et al. 1977; McNulty 1977; McConnell et al. 1978; McNulty et al. 1980, 1981), though death can hardly be attributed to the metaplastic lesions here described. Indeed, the mechanism of death is obscure. Wasting is severe, and depression of the bone marrow often presages death, but no specific, diagnostic, morphological or biochemical abnormalities have been consistently found in vital organs. The epithelial metaplasias are, however, so characteristic, if not life-threatening, that a diagnosis of poisoning by PHPAHs can be made with confidence.

Although monkeys have usually been orally exposed to PHPAHS, accidentally or experimentally, gastric metaplasia is probably not caused by direct contact with ingested toxicants, since the lesion occurred in a rhesus monkey following repeated intraperitoneal injections of 3,4,3′,4′-tetrachlorobiphenyl (unpublished).

Not all congeners and isomers of PHPAHS are highly toxic (in fact most are not), but the lesions in rhesus monkeys caused by those congeners of

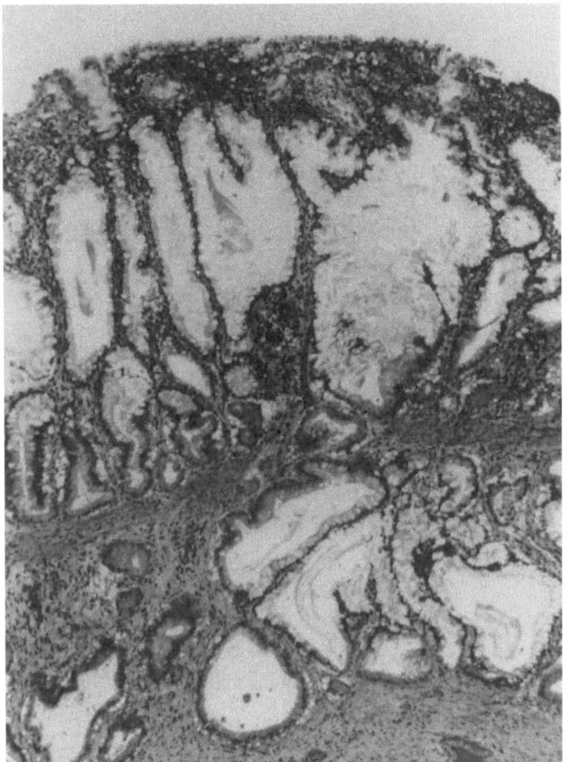

Fig. 171. *Above*: Mucosa of body of stomach, young rhesus males, after consuming 3,4,3′,4′-tetrachlorobiphenyl. Parietal and zymogenic cells are gone, and dilated distorted foveolae and glands are lined with mucous goblet cells. Irregular cystic glands have invaded the submucosa. H and E, ×40. *Below*: Mucosa of body of stomach, young male rhesus control. H and E, ×40

PCBs, TCDDs, dibenzo-p- and TCDFs that are toxic and by crude commercial mixtures of PCBs (Aroclors) are morphologically identical (Allen 1975; McConnell et al. 1978; Moore et al. 1979; McNulty 1985). The molecular requirements for toxic activity are absence of chlorination in the *ortho* positions of the linked benzene rings and chlorination in the *meta* and *para* positions for biphenyls, and in at least three of the 2,3,7, and 8 lateral positions of the rings for dioxins and furans. Molecules so constituted are in (dioxins and furans) or may assume (biphenyls) a planar configuration with electron dense atoms at both ends. Such compounds compete for stereospecific binding to a cytoplasmic receptor called Ah, and binding to this receptor is an obligatory condition for toxicity in mice (Poland and Knutson 1982; Whitlock 1987). This receptor is also present in the liver of *Macaca fascicularis* (Okey et al. 1984). Continuous ingestion of biphenyls heavily chlorinated in the *ortho* positions, such as 2,5,2′5′-tetrachlorobiphenyl and 2,4,5,2′,4′,5′-hexachlorobiphenyl (McNulty et al. 1980; McNulty 1985) have produced no toxic signs in rhesus monkeys, and these congeners do not bind to the Ah receptor. Commercial mixtures of polychlorinated biphenyls, Aroclors 1242, 1248, and 1254, do induce the characteristic toxic signs (Allen 1975; Geistfeld et al. 1982), but the offending congeners or contaminants have not been identified.

Other chlorinated aromatics with approximately the same molecular conformation, such as 3,4,3′4′-azo- and azoxybenzene and 4,5,6,7-naphthalene, also bind to the Ah receptor and cause toxic signs in mice similar to those from furans, dioxins, and biphenyls. Presumably they would do so also in nonhuman primates, but no tests of them have been reported.

Assessments of the relative intrinsive toxicities for rhesus monkeys of chlorinated dioxins, dibenzofurans, and biphenyls are confounded by marked differences in metabolism and excretion. TCDD has a biologic half-life of at least a year (McNulty 1984), whereas the very similar TCDF is rapidly metabolized and excreted (Birnbaum et al. 1981; King et al. 1983). When exposure stops, recovery from intoxication from the furan is likely (McNulty et al. 1981), but not from the dioxin. Indeed, a single oral dose of TCDD may be lethal, even though the first toxic signs are noticeable only many days later, with progression to death after weeks or even months (McConnell et al. 1978; McNulty

1984). The same comparison can be made between two toxic non-*ortho*-substituted PCBs: 3,4,5,3′,4′,5′-hexachlorobiphenyl (little metabolism, progressive severe toxicity and death) and 3,4,3′4′-tetrachlorobiphenyl (rapid metabolism, severe but reversible toxicity) (McNulty et al. 1980; McNulty 1985).

Low metabolism and cumulative storage in body tissues are neither sufficient nor required for toxicity. 2,5,2′,5′-tetrachlorobiphenyl is rapidly metabolized and excreted and does not accumulate, whereas 2,4,5,2′,4′,5′-hexachlorobiphenyl is hardly metabolized at all and accumulates to high levels in adipose tissue (McNulty et al. 1980; McNulty 1985). Neither, however, is toxic for rhesus monkeys.

The toxicity of PHPAHs for nonhuman primates is not only a laboratory phenomenon. Spontaneous poisonings by PCBs have been reported from the Oregon Regional Primate Research Center (McNulty and Griffin 1976), Bowman Gray School of Medicine, Winston-Salem, NC (Geistfeld et al. 1982), and the EPA laboratory in Perrine, Florida (Altman et al. 1979), and anecdotal reports of poisonings in other institutions have circulated by word of mouth. After the Oregon epidemic, the presence of PCBs was confirmed by analysis of stored tissues. The source of the toxicant was not established, but a concrete sealer was suspected. PCBs have not been manufactured or incorporated into commercial products in the United States for over a decade. Only PCBs in older electrical equipment (capacitors and transformers) and in waste disposal sites constitute an environmental hazard, and it seems unlikely that future spontaneous poisoning of monkeys will occur.

PHPAHs have acquired a reputation as potent carcinogens in rats and mice (Kimbrough and Linder 1974; Kimbrough et al. 1975; Safe 1989). No neoplasms have been seen in the spontaneous or experimental exposures of nonhuman primates, but in these long-lived animals the latency for chemical carcinogenicity is protracted, and PHPAH-exposed monkeys have not been held for observation for the years requisite to test carcinogenicity. Still, it is obvious that PHPAHs strongly perturb proliferation and differentiation in a variety of epithelia. In particular, the gastric lesion is ominous, because proliferating atypical glands invade the submucosa and muscularis. However, this growth disturbance is reversible; no trace of it is found in the monkeys that recovered from poisoning by 3,4,3′,4′-tetrachlorobiphenyl or TCDF (unpublished).

Other epithelial structures are also affected but are mentioned only briefly here (McNulty 1985). Mucous hyperplasia occurs in intra- and extrahepatic bile ducts, in Brunner's glands in the duodenum, in esophageal and bronchial glands, and in the ducts of salivary glands. An occasional mucous goblet cell may even be found in cystic Hassall's corpuscles in the thymus. The ameloblast of unerupted teeth in young rhesus monkeys converts to squamous stratified of oral type, with resulting periodontal cysts, and myriads of squamous nests appear in the gingiva, presumably from remnant cells from the dental ledge or the track of embryologic invagination of the oral epithelial contribution to tooth buds.

Comparison with Other Species

The toxicities of PHPAHs differ markedly among mammalian species, both quantitatively and qualitatively. The LD_{50} doses of TCDD for rats and mice are in the tens and hundreds of micrograms per kilogram body weight but only 1 µg/kg for guinea pigs and rhesus monkeys (Moore et al. 1979; McNulty 1985). The relative toxicities of TCDD, TCDF, and the non-*ortho*-substituted PCBs appear to be about the same in all species. Quantitative data are not available for humans.

The epidermal metaplasias so prominent in rhesus monkeys and humans (chloracne, Fig. 10) are minor or undetectable in the nonprimate species that have been tested. Dermal follicular hyperplasia follows topical application of PCB mixtures to the ears of rabbits (Vos and Beems 1971), and dermal squamous cysts develop in "hairless" mice (Knutson and Poland 1982). (Actually, the animals are not hairless (Montagna et al. 1952); abortive misdirected pilary structures are present in the dermis, and these are the source of the acneiform cysts).

The gastric mucosal metaplasia is almost unique for nonhuman primates; it has not been described among the findings in the few autopsies of people dying form incidental causes following exposure of PHPAHs sufficient to cause chloracne. A mucous gastric metaplasia progressing to adenocarcinoma in F344 rats has been attributed to Aroclor 1254 fed for 2 years (Morgan et al. 1981), but gastric mucosal lesions have not been noted in numerous other studies

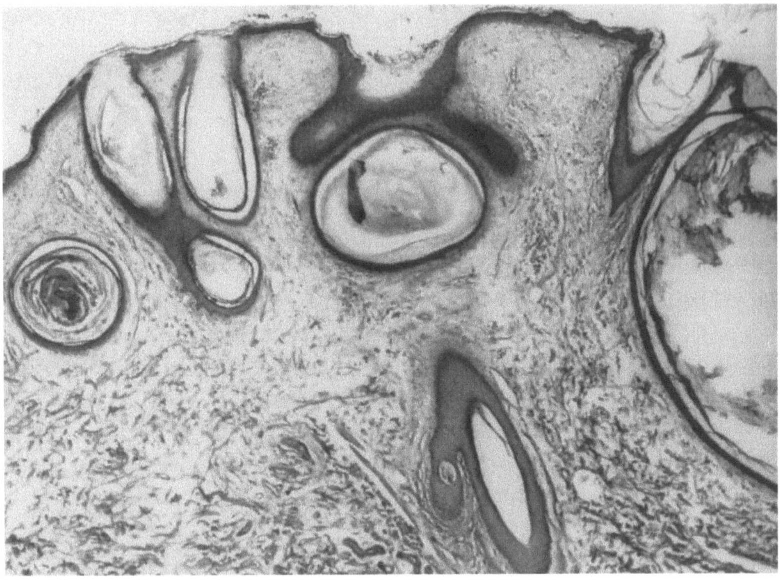

Fig. 172. Human chloracne: sebaceous glands are absent and pilary structures have mostly become squamous cysts. (Compare with skin of rhesus monkey lip, Fig. 147. H and E, ×30. (Courtesy of Dr. J. D. Crow)

of the effects of PHPAHs in rats and mice. A mucinous hyperplasia of the gastric mucosa was found in swine ingesting Aroclor 1254 (Hansen et al. 1976). Necrosis and giant cell formation are prominent in the hepatic parenchyma of murine species (Kimbrough et al. 1972; Jones and Butler 1974; Jones and Grieg 1975), whereas the livers of rhesus monkeys show no changes except for an increase in hepatocytic smooth endoplasmic reticulum and mucous hyperplasia in cholangial epithelium. The thymus undergoes profound atrophy in all species examined.

Although information about the pathological effects of PHPAHs on humans exposed in industrial operations or accidentally is meager, the known human responses are much like those in rhesus monkeys, though the susceptibility is probably less. The lesions in human skin (Fig. 172) – "chloracne" – histologically closely resemble those in rhesus monkeys, and evidence for liver damage has been minimal.

References

Allen JR (1975) Response of the nonhuman primate to polychlorinated biphenyl exposure. Fed Proc 34: 1675–1679

Allen JR, Barsotti DA, Van Miller JP, Abrahamson LJ, Lalich JJ (1977) Morphological changes in monkeys consuming a diet containing low levels of 2,3,7,8-tetrachlorodibenzo-p-dioxin. Food Cosmet Toxicol 15:401–410

Altman NH, New AE, McConnell EE, Ferrell TL (1979) A spontaneous outbreak of polychlorinated biphenyl (PCB) toxicity in rhesus monkeys (Macaca mulatta): clinical observations. Lab Anim Sci 29:661–665

Becker GM, McNulty WP (1984) Gastric epithelial cell proliferation in monkeys fed 3,4,3′,4′-tetrachlorobiphenyl. J Pathol 143:267–274

Becker GM, McNulty WP, Bell M (1979) Polychlorinated biphenyl-induced morphologic changes in the gastric mucosa of the rhesus monkey. Lab Invest 40:373–383

Birnbaum LS, Decad GM, Matthews HB, McConnell EE (1981) Fate of 2,3,7,8-tetrachlorodibenzofuran in the monkey. Toxicol Appl Pharmacol 57:189–196

Bonne C, Sandground JH (1939) On the production of gastric tumors, bordering on malignancy, in Javanese monkeys through the agency of Nochtia nochti, a parasitic nematode. Am J Cancer 37:173–185

Geistfeld JG, Bond MG, Bullock BC, Varian MC (1982) Mucinous gastri hyperplasia in a colony of rhesus monkeys (Macaca mulatta) induced by polychlorinated biphenyl (Aroclor 1254). Lab Anim Sci 32:83–86

Hansen LG, Wilson DW, Byerly CS (1976) Effects on growing swine and sheep of two polychlorinated biphenyls. Am J Vet Res 37:1021–1024

Jones G, Butler WH (1974) A morphological study of the liver lesion induced by 2,3,7,8-tetrachlorodibenzo-p-dioxin in rats. J Pathol 112:93–97

Jones G, Greig JB (1975) Pathological changes in the liver of mice given 2,3,7,8-tetrachlorodibenzo-p-dioxin. Experientia 31:1315–1317

Kimbrough RD, Linder RE (1974) Induction of adenofibrosis and hepatomas of the liver in BALB/cJ mice by polychlorinated biphenyls (Aroclor 1254). J Natl Cancer Inst 53:547–552

Kimbrough RD, Linder RE, Gaines TB (1972) Morphological changes in livers of rats fed polychlorinated biphenyls. Light microscopy and ultrastructure. Arch Environ Health 25:354–364

Kimbrough RD, Squire RA, Linder RE, Strandberg JD, Montali RJ, Burse VW (1975) Induction of liver tumors

in Sherman strain female rats by polychlorinated biphenyl (Aroclor 1260). J Natl Cancer Inst 55: 1453–1459
King FG, Dedrick RL, Collins JM, Matthews HB, Birnbaum LS (1983) Physiological model for the pharmacokinetics of 2,3,7,8-tetrachlorodibenzofuran in several species. Toxicol Appl Pharmacol 67:390–400
Knutson JC, Poland A (1982) Response of murine epidermis to 2,3,7,8-tetra-chlorodibenzo-p-dioxin: interaction of the Ah and hr loci. Cell 30:225–234
Lushbaugh CL (1949) Infiltrating adenomatous lesions of the stomach, cecum, and rectum of monkeys similar to early human carcinoma and carcinoma in situ. Cancer Res 9:385–394
McConnell EE, Moore JA, Dalgard DW (1978) Toxicity of 2,3,7,8-tetrachlorodi-benxo-p-dioxin in rhesus monkeys (Macaca mulatta) following a single oral dose. Toxicol Appl Pharmacol 43:175–187
McNulty WP (1977) Toxicity of 2,3,7,8-tetrachlorodibenzo-p-dioxin for rhesus monkeys: brief report. Bull Environ Contam Toxicol 18:108–109
McNulty WP (1984) Fetotoxicity of 2,3,7,8-tetrachlorodibenzo-p-dioxin (TCDD) for rhesus macaques (Macaca mulatta). Am J Primatol 6:41–47
NcNulty WP (1985) Toxicity and fetotoxicity of TCDD, TCDF, and PCB isomers in rhesus macaques (Macaca mulatta). Environ Health Perspect 60:77–88
McNulty WP, Griffin DA (1976) Possible polychlorinated biphenyl poisoning in rhesus monkeys (Macaca mulatta). J Med Primatol 5:237–246
McNulty WP, Becker GM, Cory HT (1980) Chronic toxicity of 3,4,3′,4′- and 2,5,2′,5′-tetrachlorobiphenyls in rhesus macaques. Toxicol Appl Pharmacol 56: 182–190
McNulty WP, Pomerantz I, Farrell T (1981) Chronic toxicity of 2,3,7,8-tetrachlorodibenzofuran for rhesus macaques. Food Cosmet Toxicol 19:57–65
McNulty WP, Nielsen-Smith KA, Lay JO Jr, Lippstreu DL, Kangas NL, Lyon PA, Gross ML (1982) Persistence of TCDD in monkey adipose tissue. Food Chem Toxicol 20:985–987
Montagna W, Chase HB, Melaragno HP (1952) The skin of hairless mice: I. The formation of cysts and the distribution of lipids. J Invest Dermatol 19:83–94
Moore JA, McConnell EE, Dalgard DW, Harris MW (1979) Comparative toxicity of three halogenated dibenzofurans in guinea pig, mice, and rhesus monkeys. Ann NY Acad Sci 320:151–163
Morgan RW, Ward JM, Hartman PE (1981) Aroclor 1254-induced intestinal metaplasia and adenocarcinoma in the glandular stomach of F344 rats. Cancer Res 41:5052–5059
Okey AB, Vella LM, Iverson F (1984) *Ah* receptor of primate liver: binding of 2,3,7,8-tetrachlorodibezo-p-dioxin and carcinogenic aromatic hydrocarbons. Can J Physiol Pharmacol 6:1292–1295
Poland A, Knutson JC (1982) 2,3,7,8-Tetrachlorodibenzo-p-dioxin and related halogenated aromatic hydrocarbons: examination of the mechanism of toxicity. Annu Rev Pharmacol Toxicol 22:517–554
Safe S (1989) Polychlorinated biphenyls (PCBs): mutagenicity and carcinogenicity. Mutat Res 220: 31–47
Tryphonas L, Charbonneau S, Tryphonas H, Zawidska Z, Mes J, Wong J, Arnold DL (1986) Comparative aspects of Aroclor 1254 toxicity in adult cynomolgus and rhesus monkeys: a pilot study. Arch Environ Contam Toxicol 15:159–169
Uemura E, Houser WD, Cupp CJ (1979) Menetrier's disease in a rhesus monkey (Macaca mulatta). J Med Primatol 8:252–256
van den Berg KJ, Zurcher C, Brouwer A, van Bekkum DW (1988) Chronic toxicity of 3,4,3′,4′-tetrachlorobiphenyl in the marmoset monkey (Callithrix jacchus). Toxicology 48:209–224
Vos JG, Beems RB (1971) Dermal toxicity studies of technical polychlorinated biphenyls and fractions thereof in rabbits. Toxicol Appl Pharmacol 19:617–633
Whitlock JP Jr (1987) The regulation of gene expression by 2,3,7,8-tetrachlorodibenzo-p-dioxin. Pharmacol Rev 39:147–161

Lead Poisoning in Nonhuman Primates

Bernard C. Zook

Synonyms. Plumbism, saturnism.

Gross Appearance

Nonhuman primates chronically poisoned by inorganic lead have pale mucous membranes as a result of microcytic hypochromic anemia. The anemia, a sequela of lead interference with hemoglobin synthesis, is usually associated with stippled immature erythrocytes (Fig. 173). The gums near the teeth may display a thin black line (Burtonian or lead line) in adults with poor dental hygiene (Fig. 174). The long bones of juvenile primates accumulate lead and may develop radiopaque, dense, gray-white bands in the metaphyses (Fig. 175 and Fig. 176). These lesions are not seen in adults poisoned after long bones cease growing. Where lead-containing paint is a source of poisoning, radiopaque paint chips may be seen in the digestive tract on abdominal films or in the ingesta at necropsy. The brains of juveniles poisoned by lead are often edematous with flattened gyri and narrowed sulci (Fig. 177).

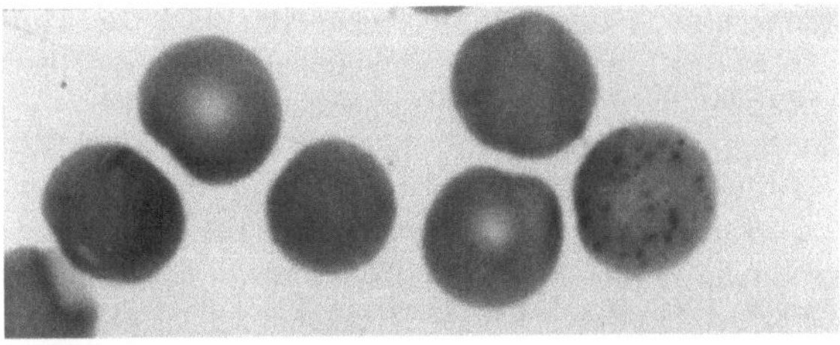

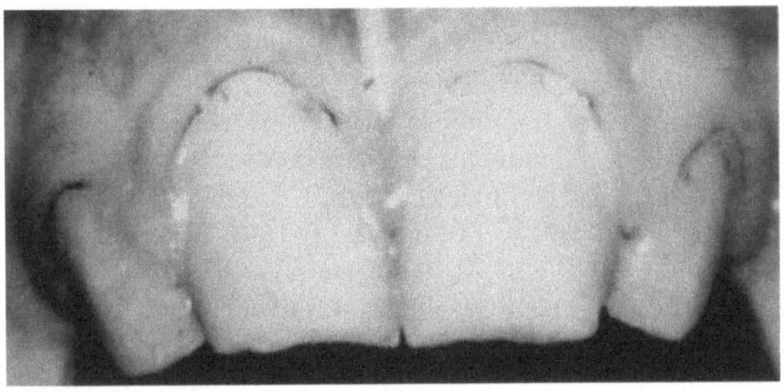

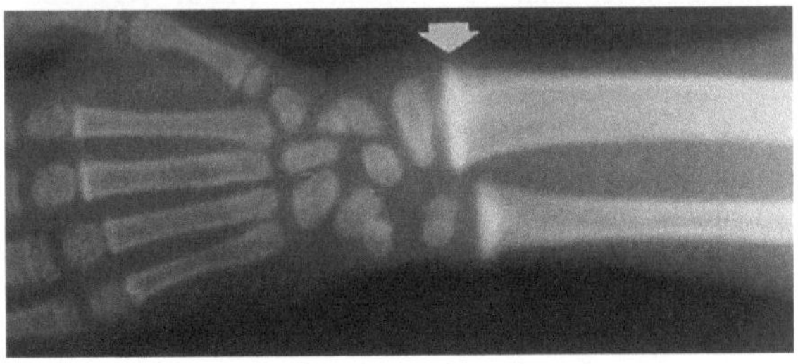

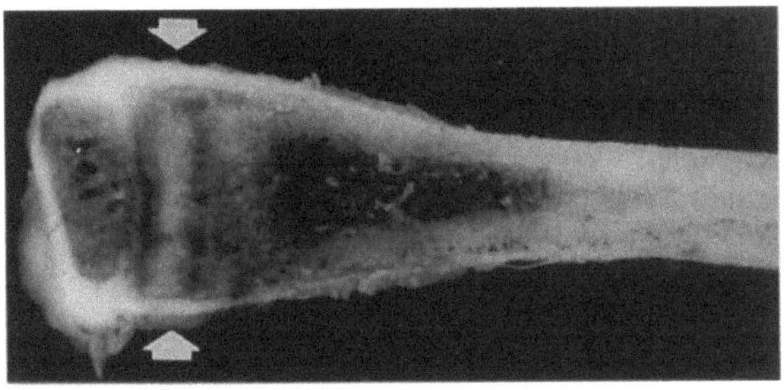

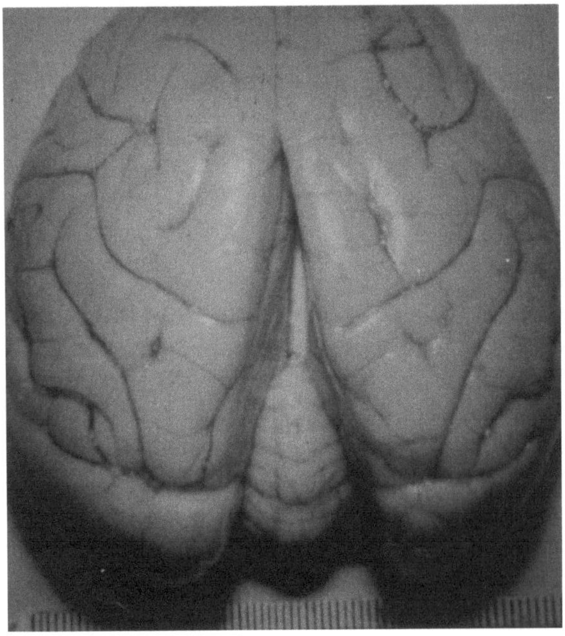

Microscopic Features

Lesions induced by lead in long bones of growing primates are characterized by a dense band in the metaphyses composed of many thick trabeculae of mineralized cartilage bearing a superficial layer of bone. Between the abnormal thick trabeculae are numerous large chondroclasts and osteoclasts (Fig. 178). The multiple nuclei of the altered osteoclasts may contain small eosinophilic and acid-fast inclusion bodies. The kidneys also contain a high concentration of lead, which primarily affects the proximal tubules often causing proteinuria and sometimes glycosuria. Renal tubular cells become swollen and many nuclei become enlarged, containing one or more acidophilic, acid-fast, inclusion bodies (Fig. 179). Some tubular cells appear to shrink; the cytoplasm and the nucleus become densely stained and these necrotic cells slough into the lumen as regenerating epithelial cells form beneath them. The inclusion bodies contain a lead-protein complex and may be seen in various tissues (Osheroff et al. 1982) and in sloughed renal epithelial cells in urine sediment (Colle et al. 1980; Zook et al. 1980a,b). The liver often contains enlarged hepatocyte nuclei with acid-fast inclusion bodies (Fig. 179). The number and size of hepatic inclusions tend to correlate positively with the concentration of lead in the liver. Other hepatic changes are less consistent but include fatty change (may be associated with anemia) and hemosiderosis. Nodular hyperplasia and portal fibrosis have been reported in a few primates (Hruban et al. 1986). Degenerative and atrophic changes also occur in the ovaries and testes of adults.

◀ **Fig. 173** (*above top*). Lead poisoning, juvenile macaque. Punctate basophilic stippling appears in one erythrocyte (*right*). Wright's-Giemsa stain, ×2000

Fig. 174 (*above bottom*). Adult rhesus, lead poisoning. Burtonian (lead) line is present in gums over incisors

Fig. 175 (*middle top*). Lead poisoning, juvenile macaque. Lead lines are present in metaphyses of the distal radius (*arrow*), ulna, and metacarpal bones. Radiograph

Fig. 176 (*middle bottom*). Metaphyseal sclerosis (*arrow*) corresponds to radiopaque line (Fig. 175) in distal radius of lead-poisoned zoo macaque. H and E, ×100

Fig. 177 (*below*). Brain of red-faced macaque (*Macaca arctoides*) affected with lead encephalopathy is swollen (flattened gyri and narrow sulci) and soft (spreading laterally)

The central nervous system of young primates is extremely sensitive to the harmful effects of lead (Levin et al. 1987; Lögdberg et al. 1988; Reuhl et al. 1989). Nonhuman primates poisoned by lead may develop apparent blindness, convulsions, or other signs of encephalopathy; death is frequently due to brain damage. Capillaries may become irregularly dilated, thrombosed, or undergo necrosis resulting in punctate or ring hemorrhages (Fig. 180). Hyaline droplets of PAS-positive material are characteristically found adjacent to injured capillaries (Fig. 181). Perivascular, and less commonly intravascular, deposits of exudate usually stain eosinophilic but may (apparently with time) become laminated and basophilic. Exudates containing fibrin may radiate from damaged vessels inducing a microglial response (Fig. 182). Capillary endothelial cells and astrocytes proliferate in damaged areas and often the acute hemorrhagic and necrotizing lesions are seen adjacent to areas of scarring (Figs. 183, 184). The vascular damage is most obvious in the white matter of the cerebrum and cerebellum and the resultant edema may cause lamina of spongy degeneration and necrosis in the cortex, especially in the pyriform and occipital lobes (Fig. 185). Focal areas of demyelination and necrosis surrounded by reactive glia may be found in the central white matter (Fig. 184).

Neurons in various locations degenerate; especially sensitive are Purkinje cells in the cerebellum and pyramidal cells in Rose's fields h1 and h2 of Ammon's horn. Hippocampal neurons frequently become eosinophilic, shrink, and undergo neuronophagia or lysis leaving a vacuolated neuropil (Fig. 186). Small acid-fast inclusions have been reported in the nuclei of rare astrocytes and in neurons of the substantia nigra (Osherhoff et al. 1982). The meninges may be thickened by a hyperplastic pia-arachnoid, but inflammatory cells are rare. Lead encephalopathy is most severe in young primates, but brain lesions also occur in older monkeys. Young primates can suffer subtle brain damage, as indicated on sensitive neuropsychiatric tests, in the apparent absence of detectable morphologic alterations. Segmental demyelination and axonal and Schwann cell degeneration of generally mild degree may be seen in peripheral nerves of adults (Purser et al. 1983).

Gross and microscopic lesions of lead poisoning vary primarily with the age of primate. Within age groups, lesions are reasonably consistent, no matter whether New or Old World species are

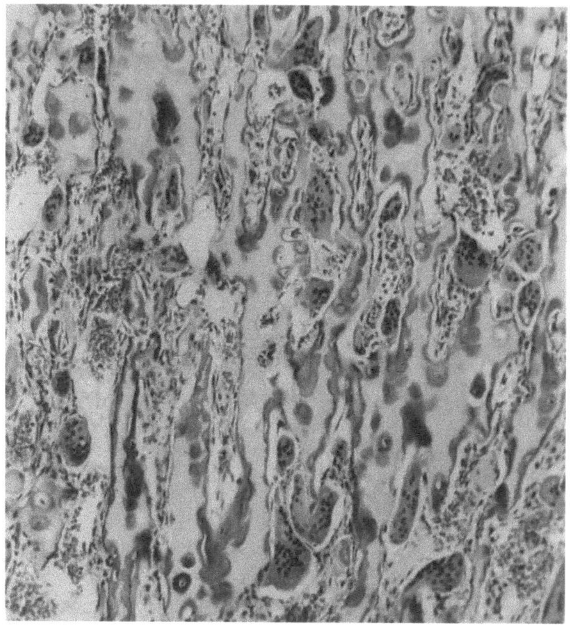

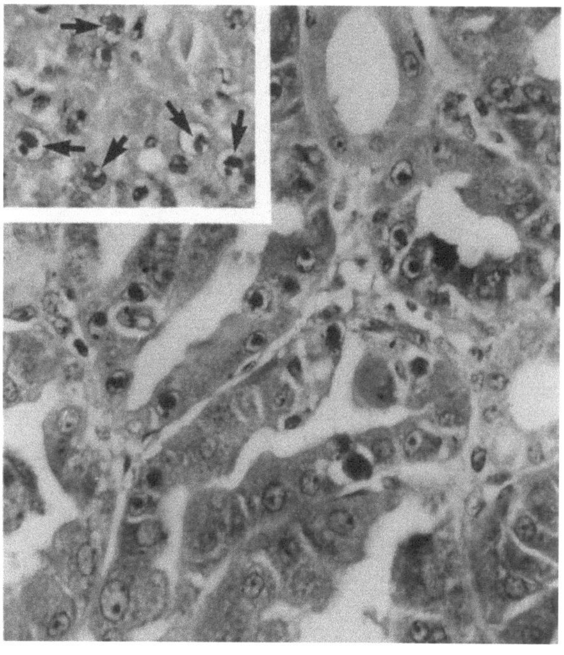

Fig. 178 (*above*). Lead poisoning, juvenile *Macaca actoides*. Metaphysis of radius contains multiple large osteoclasts between thick irregular trabeculae of mineralized cartilage sheathed with bone. H and E, ×100

Fig. 179 (*below*). Lead poisoning, macaque. Many renal proximal tubular epithelial cells are swollen or necrotic and sloughing. Many cells (*near center*) contain intranuclear inclusions. Acid-fast, ×400. *Inset*, Multiple, small, intranuclear inclusions (*arrows*) are present in enlarged hepatic nuclei. H and E, ×400

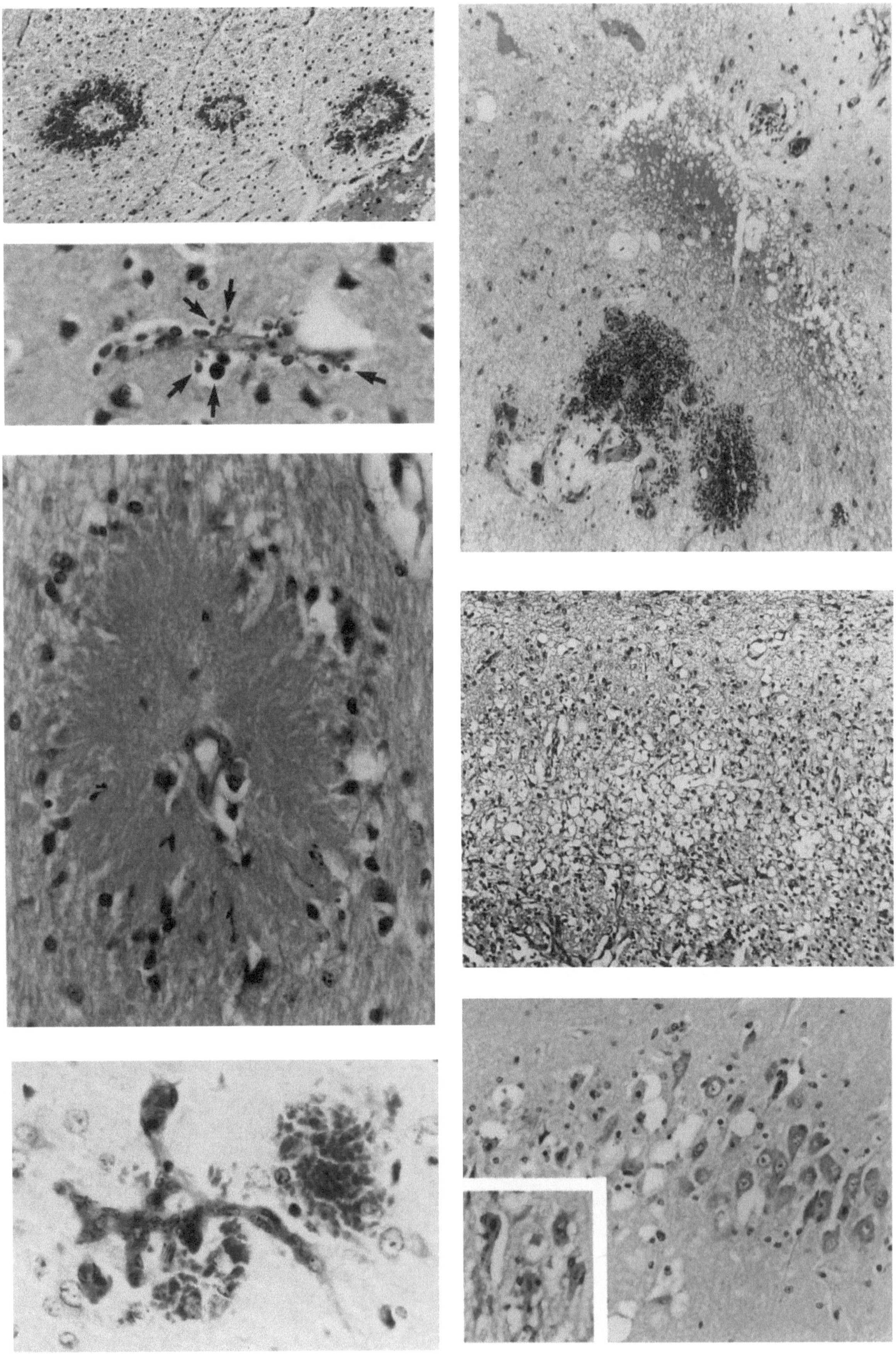

involved or whether they have been naturally or experimentally poisoned.

Ultrastructure

Transmission electron microscopy (TEM) of the lead-containing intranuclear inclusions reveals the bodies to have a characteristic dense amorphous center with an outer zone of fibrillar strands. Ultrastructural changes in the brain reveal endothelial cell injury and necrosis with open intercellular junctions and edema (Takeichi and Noda 1974). The PAS-positive globules seen by light microscopy are found by TEM to be located in perivascular astrocytes (Clasen et al. 1974a). Some of the vacuolated appearance of the brain seen by light microscopy is observed by TEM to be due to swelling of astrocytes.

Differential Diagnosis

Individual organ or tissue lesions associated with lead poisoning may resemble lesions caused by other processes. The Burtonian line can also be seen with some other heavy metal intoxications, notably bismuth poisoning. Basophilic stippling of erythrocytes has been reported in a variety of severe anemias associated with malignancies and other chronic or toxic diseases in humans. Radiodense bands in the metaphyses of growing long bones has been reported in phosphorus, bismuth, and vitamin D toxicosis. Nephrosis limited to the proximal tubules is seen with mercury and other heavy metal intoxications. Eosinophilic, acid-fast, intranuclear inclusions in renal and sometimes hepatic cells are virtually pathognomonic, as they are seen only in lead and bismuth poisoning (the latter being uncommon and associated only with its medicinal use). Peripheral neuropathy may be observed with thallium, arsenic, and other heavy metal poisonings. The lesions of lead encephalopathy, including the PAS-positive droplets found near leaking capillaries, once thought pathognomonic for lead encephalopathy (Clasen et al. 1974b), are very similar to the acute effects of ionizing radiation on the brain (Zook and Paasch 1978). The latter and other physical or chemical agents that result in cerebral capillary endothelial damage or cause marked breakdown of the blood-brain barrier could induce lesions resembling lead encephalopathy. A definitive diagnosis is confirmed by finding excess lead in blood, urine, or tissue in combination with typical signs and lesions.

◀ **Fig. 180** (*upper left top*). Lead encephalopathy, juvenile macaque. Ring hemorrhages are present in brain. H and E, ×100

Fig. 181 (*upper left bottom*). Lead encephalopathy, juvenile macaque. Multiple PAS-positive droplets (*arrows*) are adjacent to leaking cerebral capillary. PAS, ×400

Fig. 182 (*left middle*). Lead poisoning, brain, juvenile macaque. Microglia surround a rosette of fibrin-containing exudate radiating from injured cerebral capillary. H and E, ×400

Fig. 183 (*lower left*). Lead poisoning, brain, juvenile macaque. Capillary endothelial cells proliferate next to areas of capillary hemorrhage. H and E, ×400

Fig. 184 (*upper right*). Lead encephalopathy, juvenile macaque. Necrotic focus in central white matter contains lake of serous exudate and is partially bordered by excess glial cells. Irregularly dilated and leaking capillaries (*top left*) are present as well as hemorrhaging capillaries with proliferating endothelia (*lower center*). H and E, ×125

Fig. 185 (*right middle*). Lead poisoning, occipital cortex, juvenile macaque. Note layer of severe spongy degeneration; necrosis and gliosis are present. H and E, ×100

Fig. 186 (*lower right*). Lead poisoning, juvenile macaque. Pyramidal cell layer of Ammon's horn is undergoing pyknosis, shrinkage, and lysis, leaving vacuoles (*center* and *left*). H and E, ×200. *Inset*, Some pyramidal cells undergo neuronophagia. H and E, ×250

Biologic Features

In both children and nonhuman primates, signs of lead intoxication develop more commonly in summer and early autumn, possibly because vitamin D from sunlight assists in absorption of lead from the gut. Low levels of dietary calcium, and apparently protein and iron, also favor lead absorption, but phytic acid or monkey chow inhibit lead absorption. Absorption of lead also varies greatly with age; neonates and infant monkeys absorb about 60% of lead from ingested paint chips, whereas adults absorb only about 10%. The dose of lead required to poison primates experimentally varies with the source of lead, route of administration, diet and the age and perhaps species of primate. Oral doses of less than 1 mg lead oxide/kilogram body weight per day have caused encephalopathy and death in subadult squirrel monkeys, whereas 100 mg/kg per day in paint chips may be fed to subadult rhesus for months with little effect (Zook et al. 1976, 1980a,b). Cynomolgus monkeys (*Macaca fascicularis*) fed as little as 0.5 mg/kg per day from birth and maintaining only slightly elevated (15 μg/dl) blood lead levels reportedly develop behavioral deficits (Gilbert and Rice 1987). Permanent cognitive deficits have been measured in rhesus 8 years after sufficient lead was fed to induce peak blood lead levels of 250–300 μg/dl in infants followed by blood leads of 63–174 μg/dl during the first year of life (Levin et al. 1987). In addition, there are injurious effects of lead on the retina and vestibular system (Reuhl et al. 1989). The low dose lead-induced changes in young monkeys resemble those in the pediatric category of minimal brain dysfunction. Lead fed to pregnant monkeys causes placental lesions and abortion or congenital encephalopathy in neonates (Lögdberg et al. 1988; Tachon et al. 1983). Neural mitochondria and cell membranes are very sensitive to lead. Neurotransmission employing the transmitters dopamine, norepinephrine, acetylcholine, and γ-aminobutyric acid (GABA) has also been found to be altered by lead. In adult simian and human primates, the course of lead poisoning is generally chronic. In adult rhesus, few or no signs may be observed before a moribund state marked by anemia and nephropathy or sudden death.

Etiology and Frequency

The vast majority of cases of lead poisoning are caused by the ingestion of lead-containing paint by both nonhuman primates (Fujita et al. 1987) and children. In children, pica (the ingestion of nonfood items) and the accessibility to leaded paint in older poorly maintained dwellings has for many years resulted in a high incidence of poisoning. Blood lead levels sufficient to induce subclinical brain damage were found in over 4% of US children surveyed (Silbergelt 1985). Recently, levels of lead in blood and the incidence of lead poisoning in humans and dogs seems to be declining in concert with the decrease in lead content of gasoline and paint (Agency for Toxic Substances and Disease Registry 1988; Morgan et al. 1991). Among zoo primates, members of the subfamily Cercopithecinae are more commonly poisoned by lead, perhaps by virtue of their foraging and grooming habits which predispose them to contact with paint on or peeling from cage bars. Teething and behaviorally subordinant and socially stressed primates are also more likely to eat paint. The incidence of plumbism among primates at one large zoo was about 10%, with a mortality near 40% (Zook et al. 1980; Zook and Paasch 1980); after lead paint was removed from their environment the incidence dropped to zero.

Comparison with Other Species

Many species have been used successfully to study the effects of lead on biologic systems (Goyer and Rhyne 1973; Lampert et al. 1967; Goldstein et al. 1974; Pentschew 1965; Pentschew and Garro 1966). Nonhuman primates appear to be exceptionally good models for the study of lead poisoning in humans. The clinical signs, laboratory, and postmortem findings of lead intoxication in monkeys are nearly identical to those in humans (Akelaitis 1941; Allen et al. 1974; Clasen et al. 1974a,b; Purser et al. 1983; Hopkins and Dayan 1974; Zook et al. 1980a,b; Okazaki et al. 1963; Blackman 1936, 1937; Popoff et al. 1963; Whitfield et al. 1972). Lead poisoning of neonatal primates has, in the past 15 years, afforded insight into the neurochemical and behavioral effects of lead which appear to mimic those in children (Allen et al. 1974; Gilbert and Rice 1987; Levin et al. 1987; Rice and Gilbert 1990). Primates, by virtue of their large brain and intelligence, are especially well suited for study of the subtle neurologic deficits and behavioral changes induced by subclinical lead exposure.

References

Agency for Toxic Substances and Disease Registry (1988) The nature and extent of lead poisoning in children in the United States: a report to Congress. Public Health Service, US Department of Health and Human Services, Washington

Akelaitis AJ (1941) Lead encephalopathy in children and adults: a clinico-pathological study. J Nerv Ment Dis 93:313–332

Allen JR, McWey PJ, Suomi SJ (1974) Pathobiological and behavioral effects of lead intoxication in the infant rhesus monkey. Environ Health Perspect 7:239–246

Blackman SS (1936) Intranuclear inclusion bodies in the kidney and liver caused by lead poisoning. Bull Johns Hopkins Hosp 58:384–403

Blackman SS (1937) The lesions of lead encephalitis in children. Bull Johns Hopkins Hosp 61:1–61

Clasen RA, Hartmann JF, Starr AJ, Coogan PS, Pandolfi S, Laing I, Becker RA, Hass GM (1974a) Electron microscopic and chemical studies of the vascular

changes and edema of lead encephalopathy. Am J Pathol 74:215–240
Clasen RA, Hartmann JF, Coogan PS, Pandolfi S, Laing I, Becker RA (1974b) Experimental acute lead encephalopathy in the juvenile rhesus monkey. Environ Health Perspect 7:175–185
Colla A, Grimaud JA, Boucherat M, Manuel Y (1980) Lead poisoning in monkeys: functional and histopathological alterations of the kidneys. Toxicology 18:145–158
Fujita O, Nakazawa R, Ozasa Y, Ushijima I, Kondo M, Inagaki H, Osaka Y, Yamamoto H, Nigi H, Umeda M, Sugiyama M, Isoda M (1987) Lead poisoning in captive primates. Bull Nippon Vet Zootech Coll 36: 49–57
Gilbert SG, Rice DC (1987) Low-level lifetime lead exposure produces behavioral toxicity (spatial discrimination reversal) in adult monkeys. Toxicol Appl Pharmacol 91:484–490
Goldstein GW, Asbury AK, Diamond I (1974) Pathogenesis of lead encephalopathy: uptake of lead and reaction of brain capillaries. Arch Neurol 31:382–389
Goyer RA, Rhyne BC (1973) Pathological effects of lead. Int Rev Exp Pathol 12:1–77
Hausman R, Sturtevant RA, Wilson WJ (1961) Lead intoxication in primates. J Forensic Sci 6:180–196
Hopkins AP, Dayan AD (1974) The pathology of experimental lead encephalopathy in the baboon (Papio anubis). Br J Ind Med 31:128–133
Hruban Z, Meehan T, Wollmann RL, Glagov S, Wolff P, Pond J (1986) Diffuse nodular hyperplasia and fibrosis of the liver in lead-poisoned mandrills. J Med Primatol 15:245–258
Lampert P, Garro F, Pentschew A (1967) Lead encephalopathy in suckling rats. In: Klatzo I, Seitelberger F (eds) Symposium on brain edema, Vienna 1965. Springer, Berlin Heidelberg New York, pp 207–222
Levin ED, Bowman RE, Wegert S, Vuchetich J (1987) Psychopharmacological investigations of a lead-induced long-term cognitive deficit in monkeys. Psychopharmacology (Berl) 91:334–341
Lögdberg B, Brun A, Berlin M, Schutz A (1988) Congenital lead encephalopathy in monkeys. Acta Neuropathol 77:120–127
Morgan RV, Moore FM, Pearce LK, Ross IT (1991) Clinical and laboratory findings in small companion animals with lead poisoning: 347 cases (1977–1986). JAMA 199:93–97
Needleman HL, Gunnoe C, Leviton A, Reed R, Peresie H, Maher C, Barrett P (1979) Deficits in psychologic and classroom performance of children with elevated dentine lead levels. N Engl J Med 300:689–695
Okazaki H, Aronson SM, DiMaio DJ, Olivera JE (1963) Acute lead encephalopathy of childhood. Histological and chemical studies with particular reference to angiopathic aspects. Trans Am Neurol Assoc 88: 248–250
Osheroff MR, Uno H, Bowman RE (1982) Lead inclusion bodies in the anterior horn cells and neurons of the substantia nigra in the adult rhesus monkey. Toxicol Appl Pharmacol 64:570–576
Pentschew A (1965) Morphology and morphogenesis of lead encephalopathy. Acta Neuropathol 5:133–160
Pentschew A, Garro F (1966) Lead encephalomyelopathy of the suckling rat and its implications on the porphyrinopathic nervous diseases. With special reference to the permeability disorders of the nervous system's capillaries. Acta Neuropathol (Berl) 6:266–278
Popoff N, Weinberg S, Feigin I (1963) Pathologic observations in lead encephalopathy: with special reference to the vascular changes. Neurology 13: 101–112
Purser DA, Berrill KR, Majeed SK (1983) Effects of lead exposure on peripheral nerve in the cynomolgus monkey. Br J Ind Med 40:402–412
Reuhl KR, Rice DC, Gilbert SG, Mallett J (1989) Effects of chronic developmental lead exposure on monkey neuroanatomy: visual system. Toxicol Appl Pharmacol 99:501–509
Rice DC, Gilbert SG (1990) Sensitive periods for lead-induced behavioral impairment (nonspatial discrimination reversal) in monkeys. Toxicol Appl Pharmacol 102:101–109
Silbergelt EK (1985) Neurotoxicity of lead. In: Blum K, Manzo L (eds) Neurotoxicology. Dekker, New York, pp 299–322
Tachon P, Laschi A, Briffaux JP, Brain G, Chambon PC (1983) Lead poisoning in monkeys during pregnancy and lactation. Sci Total Environ 30:221–229
Takeichi M, Noda Y (1974) Electron microscopy of experimental lead encephalopathy – considerations on the development of mechanism of brain lesions. Folia Psychiatr Neurol Jpn 28:217–232
Whitfield CL, Chien LT, Whitehead JD (1972) Lead encephalopathy in adults. Am J Med 52:289–298
Zook BC, Paasch LH (1978) Lead poisoning in zoo primates: environmental sources and neuropathologic findings. In: Montali RJ, Migaki G (eds) The comparative pathology of zoo animals. Smithsonian Institution, Washington, pp 143–152
Zook BC, Sauer RM, Bush M, Gray CW (1973) Lead poisoning in zoo-dwelling primates. Am J Phys Anthropol 38:415–424
Zook BC, London WT, Sever JL, Sauer RM (1976) Experimental lead paint poisoning in nonhuman primates: I. Clinical signs and course. J Med Primatol 5:23–40
Zook BC, London WT, DiMaggio JF, Rothblat LA, Sauer RM, Sever JL (1980a) Experimental lead paint poisoning in nonhuman primates: II. Clinical pathologic findings and behavioral effects. J Med Primatol 9: 286–303
Zook BC, London WT, Wilpizeski CR, Sever JL (1980b) Experimental lead paint poisoning in nonhuman primates: III. Pathologic findings. J Med Primatol 9:343–360

Congenital Malformations in Nonhuman Primates

Andrew G. Hendrickx and Pamela E. Binkerd

Introduction

The nonhuman primate has been under close scrutiny for the characterization of spontaneous and experimentally induced malformations since the thalidomide catastrophe fueled a widespread search for a model of human teratogenicity. Early reports indicated spontaneous malformation rates of 0.44%–0.52% in a variety of nonhuman primates, mostly macaques (*Macaca* spp.) and baboons (*Papio* sp.) (Lapin and Yakovleva 1963; Hendrickx 1966; Wilson and Gavan 1967). More recent reviews have reported a somewhat higher incidence (1.02%–1.62% for Old World species; 6.0% in *Saimiri sciureus*) and a wide range of spontaneous defects affecting all major developmental systems in a variety of prosimians, New and Old World monkeys, and apes (Wilson 1978; Hendrickx and Prahalada 1986; Hoopes and Jerome 1987; Jerome 1987; Binkerd et al. 1988). More specifically, the cardiovascular and urinary systems represent the majority of cases studied in nine species (mostly macaques) in one survey (Hendrickx and Prahalada 1986). Jerome (1987) found that cardiovascular and central nervous system defects accounted for over half of the malformations occurring in a colony of *Macaca mulatta*, *Macaca fascicularis*, *Macaca arctoides*, and *Cercopithecus aethiops*. *Saimiri sciureus* appears to be susceptible to the spontaneous occurrence of cleft lip and/or palate; the fact that about one-half of these cases resulted from consanguineous matings suggests a genetic component for this particular defect (Hoopes and Jerome 1987). The reader is referred to these reviews (and references therein) for specific information on the reported defects, particularly incidence rates, discussed in the following summary. Due to the multifactorial nature of many malformations and different methods of determination, variable rates of naturally occurring defects have been reported in laboratory and domestic species and humans. While a broad spectrum of malformations is observed in most species, the frequency of individual malformations is species-specific (see Schardein 1985 and Wilson et al. 1978).

The experimental induction of malformations in primate teratology, as discussed here, has also been previously summarized (Wilson 1972, 1974; Hendrickx and Prahalada 1986). Only those reports which have since been published are specifically cited in the text that follows. Although only congenital malformations are discussed here, it is important to keep in mind that the other manifestations of developmental toxicity, i.e., embryolethality, growth retardation, and functional deficit, are equally important in the context of experimental studies, especially when extrapolating animal data to the process of risk assessment in humans (Hendrickx and Binkerd 1990). The occurrence of one or more of these toxic effects is related to the timing of treatment and dose. Further, the frequency and type of induced malformations depends on the susceptibility of specific organ systems at the time of exposure (Schardein 1985).

Skeletal System

A large number and variety of spontaneous congenital malformations in the skeletal system have been reported in nonhuman primates. Among the appendicular defects most commonly observed are various forms of oligodactyly, the total or partial absence of fingers or toes. These malformations have been collectively described as split hand and foot or lobster-claw deformities and monodactyly, depending on the severity of the condition. They have been reported in *Hylobates* sp., *Pan troglodytes*, *Mandrillus leucophaeus*, *Papio sphinx*, *Cercopithecus* sp., *Macaca mulatta*, and *Macaca fuscata*. The

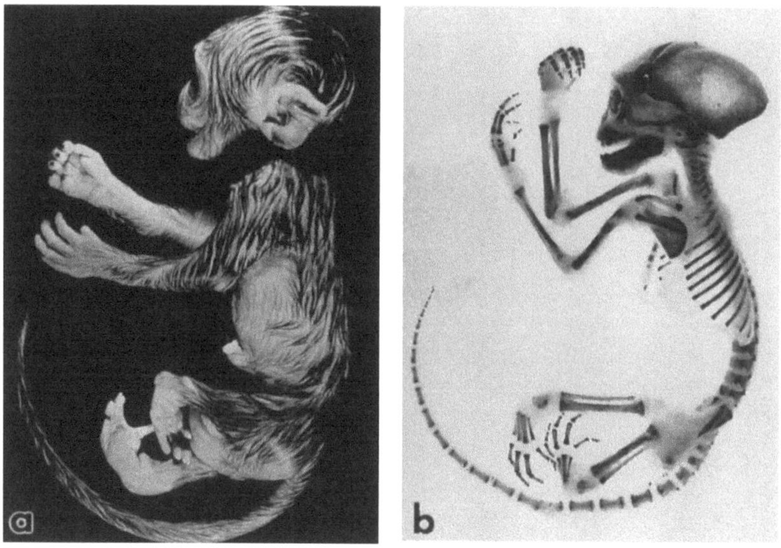

Fig. 187a,b. A 143-day *Macaca fascicularis* fetus treated with thalidomide (30 mg/kg, day 28) exhibiting bilateral meromelia. The femurs, first metatarsals, and phalanges are absent; the ankles are slightly flexed and rotated medially. The forelimbs are normal. **a** Gross specimen; **b** skeleton stained with Alizarin red S

incidence of these defects, in conjunction with syndactyly and more severe limb reduction defects (e.g., meromelia and amelia), is extremely high in *Macaca fuscata* (Morris 1971). It has been suggested that the prevalence of these malformations in several troops of these macaques is due to a common etiological factor which may be genetic, environmental (e.g., diet, teratogenic virus infection), or both. However, the role of genetic factors (dominant inheritance or recessive gene transmitted by intensive inbreeding) has not been substantiated experimentally and remains controversial (Wilson 1978).

Polydactyly, the occurrence of supernumerary digits, is common in *Hylobates* sp. This anomaly has also been reported in other species, including *Nasalis larvatus*, *Prebytis entellus*, *Pongo pygmaeus*, *Pan troglodytes*, and *Macaca* sp. Digital reduction, brachydactyly, has also been commonly reported in *Hylobates* sp. and less frequently in *Pan troglodytes* and *Mandrillus leucophaeus*. Syndactyly, the fusion of one or more digits, has been observed in *M. mulatta*, *Hylobates* sp., and *Saguinus nigricollis*. Other appendicular defects which have been sporadically documented include the following: clubfoot (talipes) in *Papio cynocephalus*, *Macaca mulatta*, and *Saimiri sciureus*; hallux anomaly in *Gorilla gorilla*; calcaneovalgus in *Macaca fascicularis*; wrist malformations with syndactyly in *Macaca mulatta*; deformed femurs with corkscrew tail in *Saimiri sciureus*; and arthrogryposis in *Macaca mulatta*. Thanatophoric dwarfism in combination with cardiovascular ventricular septal defect and genital (hermaphroditism) anomalies has also been reported in *Macaca mulatta* (Zeman and Baskin 1986). Interestingly, no ring constriction defects (digital or limb amputations), which are commonly seen in humans, have been reported in nonhuman primates.

Many appendicular defects have also been observed following exposure to developmental toxins. The most notable of these is the spectrum of limb reduction defects which comprises the thalidomide syndrome in eight species of nonhuman primates (*Macaca mulatta*, *radiata*, *arctoides*, *fuscata* and *fascicularis*, *Papio cynocephalus*, *Cercopithecus aethiops*, and *Callithrix jacchus*). Although species-specific difference exist in the severity of these anomalies and the relative frequency of fore- and hindlimb involvement, the major features of the syndrome, which have many similarities to those observed in humans, include limb reduction defects (e.g., amelia, meromelia, phocomelia), digital anomalies (triphalangism, polydactyly, oligodactyly, and syndactyly), malrotated wrist, and tail defects (hypoplasia, kinking) (Fig. 187).

Associated malformations, which also exhibit species specificity in severity and frequency of occurrence, involve the cardiovascular, uro-

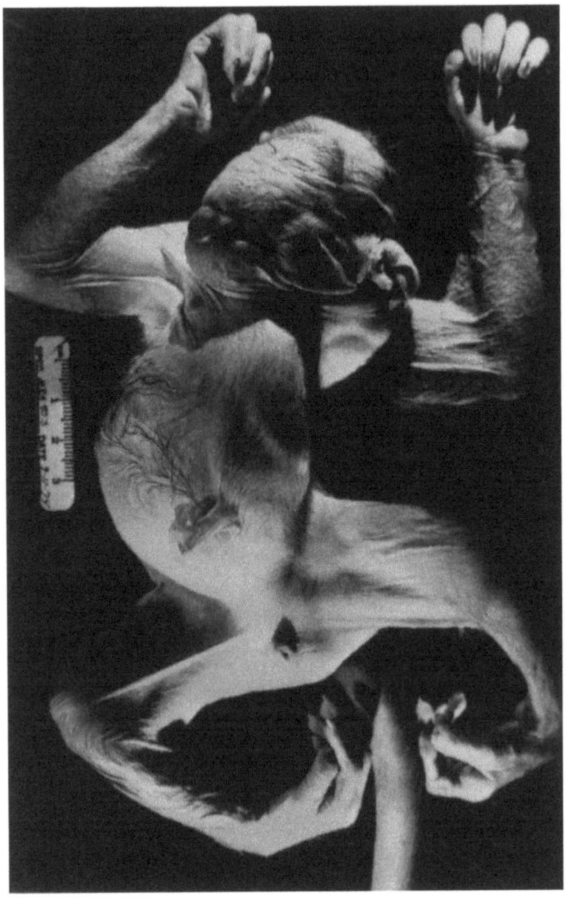

Fig. 188. Severely malformed *Macaca radiata* term fetus (185 days) following exposure to hyperthermia (day 26). External anomalies include: anencephaly, anophthalmia, arrhinia, scoliosis, and talipes

genital, gastrointestinal, and central nervous systems and the craniofacial skeleton. The similarity of defects in human and nonhuman primates following equivalent drug exposures strongly supports the use of these species as the most appropriate animal models for the study of limb reduction malformations and thalidomide teratogenesis (Hendrickx et al. 1983; Khera 1984; Merker et al. 1988).

Prenatal exposure to other agents causes less severe limb reduction and digital defects and various appendicular anomalies, often in conjunction with other manifestations of abnormal development. The teratogenic effects of the synthetic corticosteroid triamcinolone acetonide include syndactyly (*Macaca mulatta* and *radiata*), clubfoot (*Macaca radiata* and *Papio cynocephalus*), and leg hyperextension (*Macaca radiata*). All-*trans* retinoic acid treatment is associated with ectrodactyly (oligodactyly), syndactyly, and forelimb hypoplasia, with deformed radii in *Macaca nemestrina* and *Macaca mulatta* and minor clubfoot in *Macaca mulatta*. Administration of valproic acid to *Macaca mulatta* during organogenesis causes hypoplastic and/or aplastic phalanges and limb hypoplasia and abnormally curved long bones. Additionally, limb hypoplasia has been associated with X-ray irradiation in *Callithrix jacchus*; cyclophosphamide and vincristine sulfate cause syndactyly in *Macaca mulatta*; acetylsalicylic acid induces syndactyly and brachydactyly in *Macaca mulatta*; and clubfoot is a result of prenatal hyperthermia in *Macaca radiata* (Fig. 188). Lathyrogens induce abnormal curvature of the long bones and digital flexures in *Papio* sp. and arthrogryposis, in conjunction with craniofacial dysmorphia, results from all-*trans* retinoic acid treatment in *Macaca nemestrina*.

Spontaneous axial skeletal anomalies have been reported less frequently than appendicular defects. Scoliosis, with and without tail malformations, has been observed in *Lemur variegatus*, *Saimiri sciureus*, and *Macaca arctoides*; craniorachischisis has been noted in *Saimiri sciureus*; pectus excavatum has occurred in *Lemur variegatus*, *Macaca mulatta*, and *Pygathrix nemaeus*; and duplicate tails have been reported in *Macaca mulatta*. Additionally, spontaneous diaphragmatic hernia has been noted in several species – *Saimiri sciureus*, *Saguinus nigricollis*, *Leontophithecus rosalia*, *Macaca mulatta*, and *Papio* sp. In contrast, a number of axial defects have been induced under experimental conditions. Scoliosis has been observed following prenatal exposure to all-*trans* retinoic acid (*Macaca nemestrina*), hyperthermia (*Macaca radiata*, Fig. 188), and triamcinolone acetonide (*Macaca mulatta*, *Macaca radiata* and *Papio cynocephalus*). Kyphosis has been associated with all-*trans* retinoic acid exposure in *Macaca nemestrina*. Less severe vertebral anomalies (fusion, aplasia, and hemivertebrae), in combination with rib defects (fusion, hypoplasia), have been observed in *Macaca mulatta* following exposure to valproic acid and hydroxyurea. Other vertebral and rib changes (e.g., abnormal number) commonly occur in teratology studies using *Macaca* sp. and are generally considered developmental variations. Pectus excavatum resulted from prenatal exposure to all-*trans* retinoic acid in *Macaca mulatta*. Tail malformations (kinking) have been reported in *Macaca nemestrina* after all-*trans* retinoic acid treatment and in *Papio cynocephalus* after treatment with triamcinolone acetonide.

The disturbances in embryonic differentiation of the skeletal system, particularly in limbs, manifest themselves in many ways: reduction, absence, accessory, and fusion anomalies are observed spontaneously and after teratogenic exposure. While digital anomalies seem to commonly occur spontaneously, various types of limb reduction defects dominate the spectrum of experimentally induced anomalies (e.g., thalidomide studies). Although some animals with spontaneous anomalies exhibit a combination of skeletal defects, many are reported as isolated defects. In contrast, most of the anomalies resulting from teratogenic exposure are associated with perturbed development in other systems, especially cardiovascular, central nervous, and craniofacial.

The predominance of skeletal anomalies, particularly appendicular, in nonhuman primates, may be attributed to the relative ease of detection of these externally visible defects. Although specific incidences of spontaneous malformations are difficult to determine due to disparate methods of ascertainment, it has been suggested that certain species may be more susceptible to particular anomalies than humans (i.e., high incidence of digital anomalies in *Hylobates* sp. and reduction defects in *Macaca fuscata*).

Urogenital System

In nonhuman primates the kidneys are more often affected than other parts of the urogenital system. Unilateral and bilateral renal hypoplasia and aplasia have been reported at least once in *Lemur catta*, *Micro argentatus melanorus*, *Alouatta caraya*, *Saimiri sciureus*, and *Macaca radiata* and more frequently in *Macaca mulatta* and Papio sp. Similarly, cystic or polycystic kidney has been reported at least once in *Cheirogaleus major crossleyi*, *Microcebus coquerili*, *Saimiri sciureus*, *Macaca radiata*, and *Macaca nemestrina* and more frequently in *Macaca mulatta*. The spontaneous occurrences of supernumerary kidney (*Saimiri sciureus*), hydroureter (*Lemur variegatus*), and hydronephrosis with aplasia of the ureter (*Macaca mulatta*) have also been documented. Double ureter and retrocaval ureter have been observed infrequently in *Macaca mulatta* and *Papio ursinus*. Congenital fusion and ectopy of the kidney was noted in *Saimiri sciureus*. Enlarged kidneys (*Cercopithecus aethiops*, *Macaca fasciucularis*) and hydronephrosis accompanied by aplasia of the ureter (*Macaca mulatta*) have also been reported.

The urinary system, especially the kidneys, has also been perturbed by a number of exogenous agents. Exposure to thalidomide caused a variety of defects in *Macaca radiata* and, to a lesser extent, in *Macaca mulatta* and *Cercopithecus aethiops*. Hydroureters, enlarged renal pelvis, unilateral or bilateral renal aplasia, and pelvic kidney were most commonly observed. Cystic kidneys were noted occasionally after exposure to hyperthermia (*Macaca radiata*) and triamcinolone acetonide in association with central nervous and craniofacial defects (*Macaca radiata*, *Macaca mulatta*, and *Papio cynocephalus*). Retrocaval ureters, renal ectopia, and supernumerary kidneys with common ureters were observed after exposure to diphenylhydantoin hydroxyurea, and both unilateral renal and ureter agenesis and pelvic kidney were associated with exposure to the antineoplastic agent triacetyl-6-azauridine (*Macaca mulatta*).

The spontaneous occurrence of abnormal kidneys, especially cystic or polycystic kidney, was found more often than under experimental conditions, although the incidence for both appears to be low compared to anomalies in several other organ systems (i.e., skeletal, central nervous) for any given species. Similarly, other kidney disorders, including aplasia and/or hypoplasia, occurred infrequently in a number of different species and could only occasionally be induced.

Spontaneous malformations of the genital system are generally limited to the single occurrence of a specific defect and occur at a relatively low incidence, although most of the common defects reported in the human have also been observed in macaques and baboons. They include true hermaphroditism (*Macaca mulatta*), male pseudohermaphroditism (*Papio anubis*), female pseudohermaphroditism (*Papio anubis* and *Saimiri sciureus*), hypospadias, inguinal hernia and retention of Wolffian duct remnants (*Macaca mulatta*), retention of Müllerian duct remnants (*Cercopithecus talapoin*), paraovarian cysts (*Saimiri sciureus*), constricted vagina (*Macaca mulatta*), and masculinized female external genitalia (*Cercopithecus aethiops*). In addition, single cases of X-O ovarian dysgenesis (*Macaca mulatta*) and X-Y gonadal dysgenesis (*Papio ursinus*) have been reported. Although Schultz (1956) considered cryptorchidism to be more common in nonhuman primates than in humans,

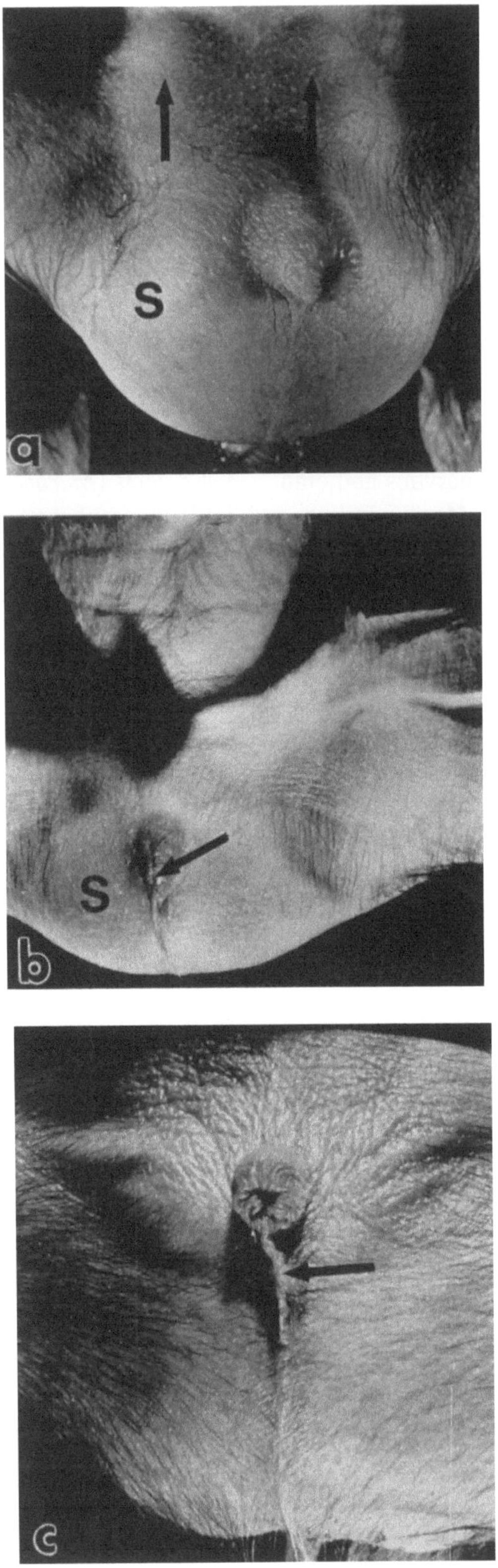

◀ **Fig. 189a–c.** External genitalia of *Macaca fascicularis* term fetus (day 152 ± 3). **a** Control male with prominent scrotum (*S*), testes in inguinal canal (*arrows*), and penis with central urethral opening. **b** Medroxy progesterone acetate-treated male (25 mg/kg, day 27 ± 2) with less conspicuous scrotal swelling (*S*) and shortened penis with ventral urethral opening (*arrow*). **c** Medroxy progesterone acetate-treated male (100 mg/kg, day 27 ± 2) with no scrotal swelling. The penis is short and directed caudally, and the median raphe (*arrow*) is prominent with a fold that extends over the ventral aspect to the tip of the penis

very few cases have been reported in the literature, presumably because of the minor nature of the defect and because of the variation in time of testicular descent associated with birth.

Only a relatively small number of malformations of the genital system have been experimentally induced and, with the exception of sex hormones, these anomalies have always been associated with more severe defective development of other systems. Of all the drugs and chemicals studied in nonhuman primates, excluding sex hormones, only two drugs, thalidomide and hydroxyurea, induced genital defects, namely unicornuate uterus, aplasia of the uterine body, and hydrocolpos in *Macaca mulatta* and *radiata*. By contrast, medroxyprogesterone acetate had paradoxical effects in both male and female fetuses. Exposure of *Macaca fascicularis* and *Papio cynocephalus* to medroxyprogesterone had a distinct masculinizing effect on female fetuses causing partial or complete labial fusion, prominent median raphe, clitoral hypertrophy, a distinct penile urethra, and increased anogenital distance. Male fetuses were feminized by exposure to this drug (Fig. 189). Diethylstilbestrol, a synthetic estrogen administered for varying periods of time during pregnancy, induced adenosis, metaplasia, and vaginal ridging, and/or cervical hooding in *Macaca mulatta* when examined between 3.5 and 12 years of age (Hendrickx et al. 1988b). Diethylstilbestrol also induced structural changes in the cervix, hypertrophy of the uterus, cervix, and vagina, vaginal and cervical adenosis and adenomatous, and clear cell hyperplasia of the endometrium in *Cebus apella*. Testosterone proprionate, dihydrotestosterone proprionate, and norethindrone, a progestin with androgenic properties, administered with or without ethinyl estradiol, induced masculinization of female fetuses (female pseudohermaphrodites) in *Macaca mulatta*. Male offspring exhibited some degree of cryptorchidism after exposure to either testosterone or norethindrone. The existing data suggest that induction of either masculinization of female or feminization of male offspring with target organ-specific sex hormones is more readily achieved experimentally than occurs spontaneously.

Central Nervous System and Eye

Both spontaneous and experimentally induced congenital malformations of the central nervous system have been reported in a number of different species. Spontaneous anencephaly is probably the most common malformation of the central nervous system and has been observed in *Macaca mulatta*, *Macaca fascicularis*, *Macaca arctoides*, *Saimiri sciureus*, and *Papio hamadryas*. Anencephaly, associated with other severe cranifacial, axial, and appendicular skeletal malformations, has only been observed once under experimental conditions and was induced by hyperthermia in *Macaca radiata* (Fig. 188). The spontaneous occurrence of hydrocephalus has been reported in five different species; it occurred most commonly in *Macaca mulatta* and *fascicularis* and has been reported as a single occurrence in *Saimiri sciureus*, *Cercopithecus aethiops*, *Macaca radiata*, and *Pan troglodytes*. Experimentally, this malformation of the ventricular system has been induced by maternal administration of hydroxyurea (*Macaca mulatta*) and triamcinolone (*Macaca mulatta*, *Macaca radiata*, *Papio cynocephalus*, Fig. 190). In addition, hydrocephalus has been induced by the intracerebral injection of the mumps and western equine encephalitis viruses in fetuses of *Macaca mulatta*. Acephaly and hydranencephaly have been reported in *Macaca mulatta* and *fascicularis*, respectively. The spontaneous occurrence and experimental induction of exencephaly, encephalocele, meningocele, and porencephaly have been reported infrequently. Of these malformations, only exencephaly and meningocele have been reported to occur spontaneously (*Lemur variegatus*, *Macaca mulatta*, *Macaca fascicularis*); experimentally, exencephaly was associated with ingestion of blighted potatoes in the common marmoset (*Callithrix jacchus*). Encephalocele has not been observed to occur spontaneously; however, both encephalocele and meningocele have been induced, in association with craniofacial effects, by triamcinolone in *Macaca mulatta*, *radiata*, and *Papio cynocephalus*. Encephalocele was also induced with vincristine sulphate, an antineoplastic agent, in *Macaca mulatta*. Porencephaly was also induced in *Macaca mulatta* by experimentally induced hypoxia.

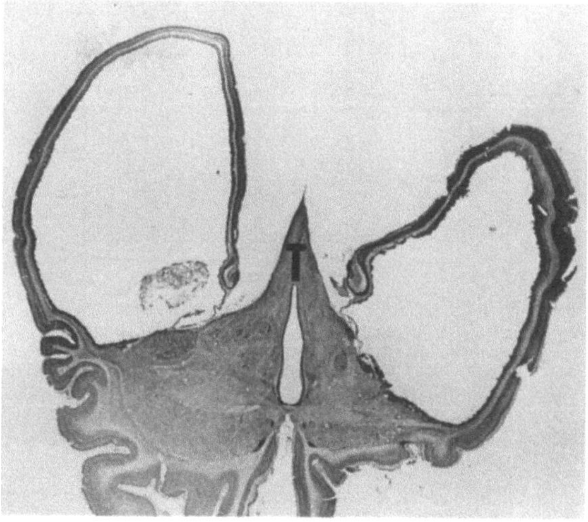

Fig. 190. Coronal section of the brain of a triamcinolone acetonide-treated (10 mg/kg, days 25, 27 and 29) *Macaca mulatta* term fetus (day 175) exhibiting hydrocephalus and midbrain beaking. The midbrain is severely distorted and the tectum (*T*) has a triangular shape. The ventricles are markedly dilated and there is compression of the cerebral cortex; ×4.8

Both cerebral and cerebellar hypoplasia, cerebral polygyria, microgyria and argyria, asymmetry of the brain, and mid-brain beaking and distortion have been induced experimentally but have not been reported as spontaneous malformations. Cerebral hypoplasia was induced by all-*trans* retinoic acid (*Macaca mulatta*), hydroxyurea (*Macaca mulatta*), lead exposure (*Saimiri sciureus*), and triamcinolone (*Macaca mulatta*, *Macaca radiata*, *Papio cynocephalus*). Cerebellar hypoplasia occurred after treatment with all-*trans* retinoic acid (*Macaca mulatta*), triamcinolone (*Macaca mulatta*, *Macaca radiata*, *Papio cynocephalus*), and after exposure to thalidomide (*Macaca radiata*). Interestingly, this defect was not reported for any other nonhuman primate species similarly treated with this antinauseant drug. Asymmetry of the brain was observed in *Papio cynocephalus* after exposure to thalidomide analogue (CG 3033) and in *Macaca mulatta* (tolbutamide, all-*trans* retinoic acid, and hydroxyurea). Cerebral polygyria/agyria occurred after exposure to all-*trans* retinoic acid (*Macaca mulatta*) and triamcinole (*Macaca mulatta*, *Macaca radiata*). Mid-brain beaking and distortion also developed

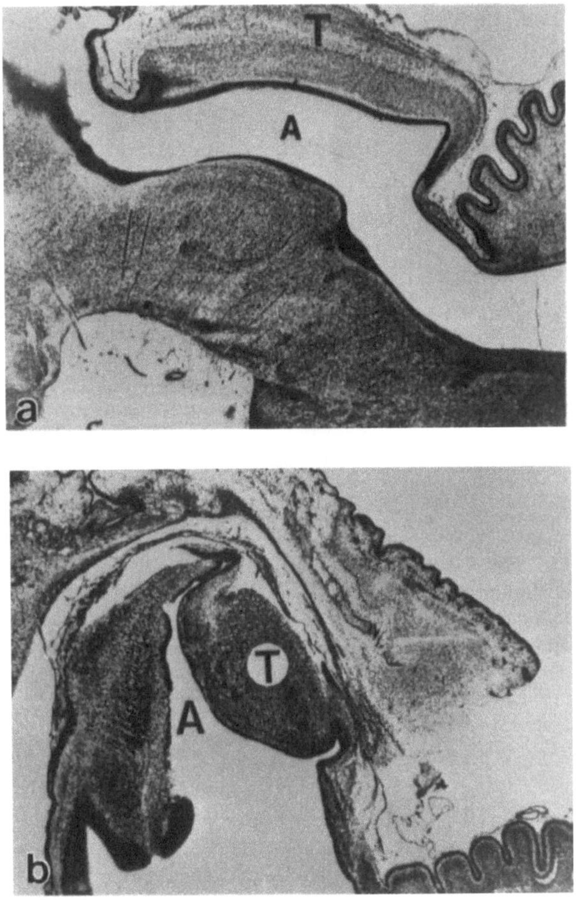

Fig. 191a,b. Sagittal sections of fetal brains from *Macaca mulatta* (day 70). **a** Control fetus with normal tectum (*T*) and aqueduct (*A*); ×23. **b** Triamcinolone acetonide-treated fetus exhibiting midbrain beaking. The tectum (*T*) has a wedge-shaped beaked appearance, and there is a marked dorsal-ventral elongation of the aqueduct (*A*) with extension to the apex; ×20

after triamcinole treatment in *Macaca mulatta* and *Macaca radiata* (Figs. 190, 191).

Spina bifida has been reported as a spontaneous malformation in two macaque species (*Macaca niger*, *Macaca mulatta*) and in *Papio hamadryas* in association with an Arnold-Chiari malformation; an Arnold-Chiari-like malformation was also observed as a single defect in the hamadryas baboon. A low incidence of spina bifida was induced with thalidomide in *Macaca fuscata* and with the lathyrogen β-aminoproprionitrile in *Papio cynocephalus*.

Spontaneous malformations of the eye have been documented in four different nonhuman primate species. Varying degrees of coloboma have occurred in *Macaca mulatta*, *Papio hamadryas*, and *Pan troglodytes*. Anophthalmia has been observed as a isolated defect in *Saimiri sciureus* and in *Macaca mulatta* in association with facial anomalies. Unspecified malformations of the eye, possibly microphthalmia, were associated with anencephaly in *Papio hamadryas*. Minor malformations of the eye, persistent pupillary membrane and arteriovenous malformation of the retina, and congenital blindness have been noted in *Macaca mulatta*. Anophthalmia was induced by prenatal exposure to hyperthermia (*Macaca radiata*, Fig. 188). In addition, microphthalmia was associated with exposure to hydroxyurea and exophthalmia was observed after valproic acid and all-*trans* retinoic acid treatment during early pregnancy (*Macaca mulatta*).

Both hydrocephaly and anencephaly seem to occur as spontaneous defects more often than under experimental conditions. For example, anencephaly was induced only once in *Macaca radiata* but has been reported as a spontaneous defect at least seven times in *Macaca mulatta* and at least once in four other species. Similarly, hydrocephalus was found nine times as a spontaneous defect in *Macaca mulatta* and at least once in five other species. In contrast, the induced incidence of hydrocephalus was relatively low except when associated with craniofacial malformations following exposure to triamcinole. No other obvious patterns emerge, except various degrees of cerebral and cerebellar hypoplasia were observed under experimental conditions but were not encountered in spontaneous cases.

Cardiovascular System

A large variety of cardiovascular anomalies affecting the ventricular and atrial septal complex, conotruncal and valve development, and the great vessels have been reported in many species of nonhuman primates. Spontaneous ventricular septal defects have occurred as isolated anomalies in *Papio cynocephalus*, *Macaca mulatta*, *Cercopithecus aethiops*, *Saimiri sciureus*, and *Gorilla gorilla* (Machado et al. 1989) and in combination with aortic arch and/or valvular defects in *Papio hamadryas*. Ventricular septal defect in association with other cardiac abnormalities, i.e., atrioseptal defect and cardiac hypertrophy, has also been reported in *Macaca fascicularis* and *Macaca mulatta*. Univentricular heart, the complete absence of the septum, has occurred in *Macaca mulatta* and a two-chambered heart has been observed in *Lemur variegatus*. Defects in atrial septal development have affected *Macaca mulatta*, *Macaca fuscata*, *Macaca arctoides*, *Pan satyrus*, and *Cercopithecus ascanius* and *aethiops*; patent foramen ovale has been observed in *Macaca mulatta*.

Dextrocardia and pulmonic stenosis have also been reported as spontaneous defects in *Macaca mulatta*. Observed abnormalities of the aortic arch complex include: patent ductus arteriosus in *Callicebus moloch*, *Cerocebus fulliginosus*, *Macaca radiata*, *Macaca mulatta*, and *Macaca arctoides*; aortic coarctation and stenosis in *Macaca mulatta*; double aortic arch in *Cercopithecus talapoin*; and anomalous origin of the right subclavian artery in *Saguinus nigricollis*. Patent ductus arteriosus in conjunction with other cardiac anomalies are common. Associated defects include: aortic hypoplasia and pulmonary valve defects (*Papio hamadryas*) and atrioseptae defects and aortic coarctation (*Cercopithecus. aethiops*, *Macaca mulatta*, and *fascicularis*). Anomalous superior vena cava has spontaneously occurred in *Pongo pygmaeus*; ectopia cordis has been observed in *Macaca fascicularis*; and acardia was reported in a *Macaca mulatta* twin. Multiple cardiac malformations (Ventrical and atrial septal defects, patent ductus arteriosus) accompanied skeletal anomalies in one case in *Pan troglodytes* with Down's syndrome. Nonimmune hydrops fetalis in *Macaca mulatta* was characterized by aortic and left atrioventricular valve atresia, aortic hypoplasia, and absence of the left ventricle (Cukierski et al. 1986; Fig. 192).

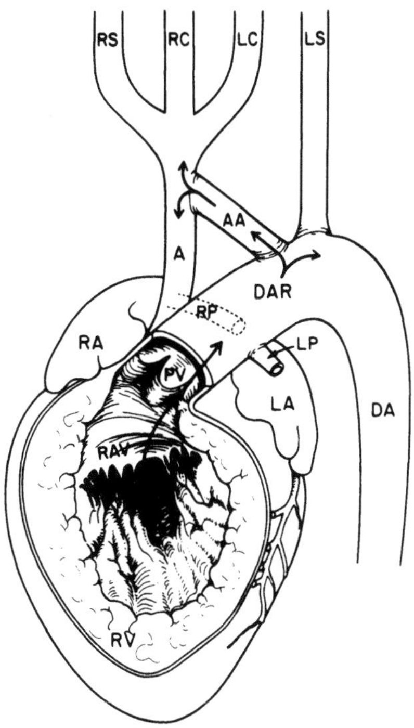

Fig. 192. Diagram of the abnormal heart and great arteries in a case of nonimmune hydrops fetalis in *Macaca mulatta* (day 68). The ascending aorta (*A*) and aortic arch (*AA*) were hypoplastic. The right subclavian (*RS*), right common carotid (*RC*), and left common carotid (*LC*) arteries were given off at a common branch point. The left subclavian (*LS*) artery branched at the junction of the hypoplastic aortic arch and the ductus arteriosus (*DAR*), which was large and continuous with the descending aorta (*DA*). The right and left pulmonary arteries (*RP* and *LP*), right and left atria (*RA* and *LA*), right atrioventricular valve (*RAV*), and pulmonary valve (*PV*) were all normal. The aortic and left atrioventricular valve were atretic and there was no connection of the aorta or left atrium with the single right ventricular chamber (*RV*). *Arrows* show the direction of blood flow

Cardiac malformations have also been reported in conjunction with other manifestations of developmental toxicity in experimental studies. The most common is ventricular septal defect which has been observed in *Macaca fascicularis* following exposure to 13-*cis* retinoic acid (Hummler et al. 1990), in *Macaca mulatta* after valproic acid and acetylsalicylic acid treatment, and in *Macaca radiata* in association with aortic stenosis following exposure to thalidomide. Delayed closure of the ventricular septum is a consistent feature of Bendectin treatment during organogenesis in *Macaca mulatta*, *Macaca fasciularis*, and *Papio cynocephalus* (Hendrickx et al. 1985). A combined atrioventricular septal defect was also observed following valproic acid treatment in *Macaca mulatta* (Hendrickx et al. 1988a). All-*trans* retinoic acid and diphenylhydantoin in combination with phenobarbital-induced dextrocardia in *Macaca mulatta* .

Transposition of the great vessels resulted from 13-*cis* and all-*trans* retinoic acid treatment in *Macaca fascicularis* and *nemestrina*, respectively. Hydroxyurea produced partial truncus communis with absent aortic arch in *Macaca mulatta* and exposure to hyperthermia during early pregnancy in *Macaca radiata* was associated with multiple congenital anomalies, including tetralogy of Fallot. The thalidomide syndrome in macaques also included ventricular hypertrophy with patent foramen ovale and valvular defects (*Macaca radiata*), abnormal branching of celiac and superior mesenteric arteries (*Macaca radiata* and *Macaca fascicularis*), and undefined cardiac anomalies in *Callithrix jacchus*.

It has been suggested that cardiovascular defects, particularly minor anomalies, are underreported in nonhuman primates since they are diagnosed only by thorough postmortem examination (Wilson 1978). Of those that have been documented, the majority of spontaneous anomalies are reported as single or multiple cardiovascular defects in single cases. The low level of associated anomalies are generally skeletal. In contrast, cardiac defects observed experimentally generally occur at a low frequency in combination with more predominant malformations which affect a variety of developmental systems, particularly the axial, appendicular, and craniofacial skeleton. The only agent which exclusively affects the heart is Bendectin, which retards, rather than perturbs, normal ventricular septal development. It has been proposed that isolated cardiac malformations which have been observed in birds, cattle, and humans have an evolutionary origin, i.e., they are attributable to genetic variants which cause organ-specific defects. In contrast, teratogens cause a more widespread insult to morphogenesis, and multiple anomalies often result depending on the time of exposure (Taussig 1988).

Gastrointestinal System

A few defects of the gastrointestinal tract have been reported in nonhuman primates. Spontaneous anomalies in *Macaca mulatta* have largely been limited to variations in the gallbladder. These include a low incidence of accessory and bilobed gallbladder in addition to diverticulosis. Pancreatic hypoplasia has also been observed as an inherent anomaly in these species. Additionally, single cases of anal atresia in combination with skeletal and craniofacial anomalies in *Saimiri sciureus*, persistent cloaca in *Macaca fascicularis*, and imperforate anus with an associated heart anomaly in *Cercopithecus aethiops* have been reported. The most prevalent anomaly induced experimentally in the gastrointestinal tract in *Macaca mulatta* is malrotation of the gut after exposure to methotrexate, acetylsalicylic acid, diphenylhydantoin in combination with phenobarbital, and thalidomide. Other drug-related malformations in this species include single occurrences of colonic atresia (trimethadione and hydroxyurea), intestinal diverticulum (hydroxyurea), esophageal stenosis (acetylsalicylic acid), and intestinal intussusception and bifid gallbladder (thalidomide). Although most of these are isolated defects, some have occurred in conjunction with experimentally induced skeletal anomalies. A few agents (e.g., thalidomide, antineoplastic drugs) have also produced accessory tissues, including liver and spleen; these are commonly observed (in addition to hemal nodes) in the authors' laboratory in teratology studies and are generally considered developmental variants.

Endocrine Glands

Accessory adrenal cortical tissue in the kidney has been reported in *Macaca mulatta* and in other abdominal sites in *Macaca mulatta*, *fascicularis*, and *Erythrocebus patas*. Spontaneous adrenal hypoplasia and aplasia have been observed in four macaque species: *Macaca mulatta*, *radiata*, *arctoides*, and *fascicularis* (Hendrickx and Prahalada 1986). This condition also occurred at a low rate after various drug and environmental exposures in *Macaca radiata* (triamcinolone, hyperthermia), *Macaca fascicularis* (Bendectin, medroxy progesterone acetate), and *Papio cynocephalus* (medroxy progesterone acetate). Accessory thymic tissue, both spontaneously occurring and drug-induced (retinoic acid), has also been noted in *Macaca mulatta*. Isolated instances of thymic hypoplasia have been reported after exposure to triamcinolone (*Macaca mulatta*, *Macaca radiata*, and *Papio cynocephalus*, Fig. 193) and the thalidomide analogue CG-3033 (*Papio cynocephalus*). Thyroid hyperplasia has also been observed in *Macaca fascicularis*. No unusual patterns of sensitivity to experimental agents were recognized for the endocrine tissues.

Nipples, Skin, and Hair

Polythelia, or supernumerary nipples, is considered to be the most commonly reported developmental variant in nonhuman primates (Wilson 1978), although it is not considered to be more common in other primates than in the human (i.e., approximately 1% incidence) (Schultz 1956) with the possible exception of one troop of *Macaca fuscata* in whom this variant was reported in about one-half of all females. It was also considered to be a common variant in *Macaca mulatta*, *fascicuilaris*, *Papio* sp., and in anthropoid apes (Wilson 1978). Anomalies of the skin, especially hemangiomas and nevi of the face, and port-wine nevus associated with arteriovenous malformation are infrequently reported in any species. Albinism, total or partial absence of skin and hair pigmentation, also appears to be rare; localized pigmentation of the distal extremities is relatively common in

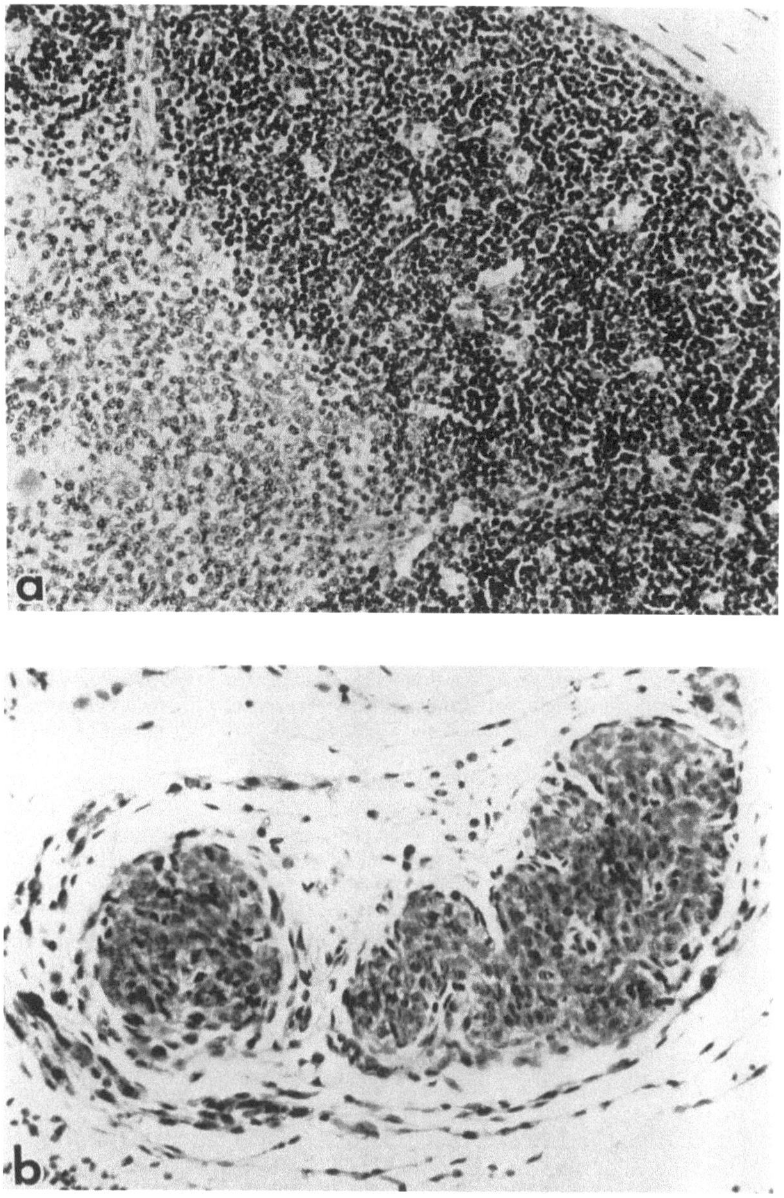

Fig. 193a,b. Sections of fetal thymus from *Macaca mulatta* (day 100). **a** Control thymus with well differentiated cortical and medullary regions; ×236. **b** A hypoplastic thymus following exposure to triamcinolone acetonide (13 mg/kg, days 60–63). The cortical and medullary regions are indistinct and the organ is depleted of lymphocytes; ×212

Macaca fuscata. None of these malformations have been associated with drug or chemical exposure during pregnancy.

References

Binkerd PE, Tarantal AF, Hendrickx AG (1988) Embryonic/fetal loss and spontaneous malformations in nonhuman primates. In: Neubert D, Merker H-J, Hendrickx AG (eds) Proceedings of the symposium on non-human primates – developmental biology and toxicology. Ueberreuter, Berlin, pp 115–127

Cukierski MA, Tarantal AF, Hendrickx AG (1986) A case of nonimmune hydrops fetalis with a rare cardiac anomaly in a rhesus monkey. J Med Primatol 15: 227–234

Hendrickx AG (1966) Teratogenicity findings in a baboon colony. In: Miller CD (ed) Proceedings of the conference on nonhuman primate toxicology, June 12–14. Government Printing Office, Washington, pp 120–123

Hendrickx AG, Binkerd PE (1990) Nonhuman primates and teratological research. J Med Primatol 19:81–108

Hendrickx AG, Prahalada S (1986) Teratology and embryogenesis. In: Dukelow R, Erwin J (eds) Reproduction and development. Comparative primate biology, vol 3. Liss, New York, pp 333–362

Hendrickx AG, Binkerd P, Rowland J (1983) Developmental toxicity and nonhuman primates: Interspecies comparisons. In: Kalter H (ed) Issues and reviews in teratology. Plenum, New York, pp 149–189

Hendrickx AG, Cukierski M, Prahalada S, Janos G, Rowland J (1985) Evaluation of bendectin embryotoxicity in nonhuman primates: I. Ventricular septal defects in prenatal macaques and baboon. Teratology 32:179–189
Hendrickx AG, Nau H, Binkerd P, Rowland JM, Rowland JR, Cukierski MJ, Cukierski MA (1988a) Valproic acid developmental toxicity and pharmacokinetics in the rhesus monkey: an interspecies comparison. Teratology 38:329–345
Hendrickx AG, Prahalada S, Binkerd PE (1988b) Long-term evaluation of the Diethylstilbestrol (DES) syndrome in adult female rhesus monkeys (Macaca mulatta). Reprod Toxicol 1:253–261
Hoopes CW, Jerome CP (1987) Oral-facial clefts and associated malformations in the squirrel monkey (Saimiri sciureus). J Med Primatol 16:203–209
Hummler H, Korte R, Hendrickx AG (1990) Induction of malformations in the cynomolgus monkeys with 13-cis retinoic acid. Teratology 42:263–272
Jerome CP (1987) Congenital malformations and twinning in a breeding colony of Old World Monkeys. Lab Anim Sci 37:624–630
Khera KS (1984) Adverse effects in humans and animals of prenatal exposure to selected therapeutic drugs and estimation of embryofetal sensitivity of animals for human risk assessment. In: Kalter H (ed) Issues and reviews in teratology. Plenum, New York, pp 399–507
Lapin BA, Yakovleva LA (1963) Development defects and monstrosities. In: Windle WF (ed) Comparative pathology in monkeys. Thomas, Springfield, pp 229–236
Machado C, Mihm FG, Noe C (1989) Diagnosis of a ventricular septal defect in a gorilla using in vivo oximetry. J Zoo Wildl Med 20:199–202
Merker H-J, Heger W, Sames K, Stürje H, Neurbert D (1988) Embryotoxic effects of thalidomide-derivatives in the non-human primate Callithrix jacchus: I. Effects of 3-(1,3-dihydro-1-oxo-2H-isoindol-2-yl)-2,6-dioxopeperidine (EM12) on skeletal development. Arch Toxicol 61:165–179
Morris LM (1971) Spontaneous congenital limb malformations in nonhuman primates: a review of the literature. Teratology 4:335–342
Schardein JL (1984) Chemically-induced birth defects. Dekker, New York, pp 1–46
Schultz AH (1956) The occurrence and frequency of pathological and teratological conditions and of twinning among non-human primates. Primates 1:965–1014
Taussig HB (1988) Evolutionary origin of cardiac malformations. J Am Coll Cardiol 12:1079–1086
Wilson JG (1972) Abnormalities of intrauterine development in non-human primates. Acta Endocrinol (Copenh) 166 [Suppl]:261–292
Wilson JG (1974) Teratologic causation in man and its evaluation in non-human primates. In: Motulsky AG, Lenz W (eds) Ebling FJG (co-ed) Birth defects. Proceedings of the 4th international conference, 1973, Vienna. Excerpta Medica, Amsterdam, pp 191–203
Wilson JG (1978) Developmental abnormalities: nonhuman primates In: Benirschke K, Garner FM, Jones TC (eds) Pathology of laboratory animals, vol II. Springer, Berlin Heidelberg New York, pp 1911–1917
Wilson JG, Gavan JA (1967) Congenital malformations in nonhuman primates: spontaneous and experimentally induced. Anat Rec 158:99–109
Wilson JR, Kalter H, Palmer AK, Hoar RM, Shenefelt RE, Beck F, Earl FL, Nelson NS, Berman E, Stara JF, Selby LA, Scott WJ Jr (1978) Developmental abnormalities. In: Bernirschke K, Garner FM, Jones TC (eds) Pathology of laboratory animals, vol II. Springer, Berlin Heidelberg New York, pp 1818–1910
Zeman DH, Baskin GB (1986) Lethal short-limbed dwarfism in a rhesus monkey. Vet Pathol 23:198–200

Craniofacial Defects Induced in Rhesus Monkeys by Prenatal Exposure to Triamcinolone Acetonide

Andrew G. Hendrickx and Pamela E. Binkerd

Gross Appearance

Craniofacial abnormalities have consistently been produced in the rhesus monkey (*Macaca mulatta*) by administration of the synthetic corticosteroid triaminolone acetonide (TAC) at critical times during early development (Hendrickx et al. 1980a). Craniofacial dysmorphia was characterized grossly in term fetuses by the following features: protrusion of the forehead, exophthalmia, low-set ears, perpendicular nasal angle, and shortened cranial length (Fig. 194). These were frequently accompanied by cleft or, more commonly, arched palate. Corresponding calvarial defects varied in severity according to the brain abnormalities. Cranium bifidum associated with meningocele and encephalocele was characterized by an underossification of the frontal, parietal, and occipital bones, with a thin membranous tissue (dense fibrous connective tissue) covering the unossified areas (Figs. 195, 196). The underossification was most apparent at the midline and involved the metopic, sagittal, coronal, and lamboid sutures. A less severe condition in some animals was manifested by cranium bifidum occultum associated with aplasia cutis congenita and absence of exterior gross lesions.

Some of the abnormalities were also present in younger specimens obtained for histological analysis in pathogenesis studies. At 35 gestational days (GD), TAC-treated embryos were unremarkable; however, by GD 43, occipital encephalocele, protrusion of the frontal region, low-set ears, and subcutaneous edema were apparent. Fetuses examined at GD 50 additionally exhibited both an abnormal nasal-frontal angle and shortened cranial length and arched and partial cleft palate. Exophthalmia was observed with this spectrum of malformation of 0-day fetuses. Although protrusion of the frontal region was not apparent, the facial features at GD 70 resembled those at GD 60.

Examination of the basicranium at birth revealed a distortion of the sphenoid bone which consisted of thinning of the lesser wings, hypoplasia or absence of the posterior clinoid process, a sella turcica that was abnormal in both shape and size, and a small and abnormally shaped foramen magnum. Direct cephalometric measurements revealed that the facial defects were associated with a severely abnormal sphenoid bone complex. The lesser wings were reduced to rudimentary structures, and the sella turcica and posterior clinoid process were absent. The transverse diameter of middle and posterior cranial fossae was also significantly reduced. Orbital volume and width were decreased but intraorbital distance and orbital height were increased.

Microscopic Features

In order to further define the craniofacial malformations induced by TAC in the rhesus monkey, histological features of embryos and

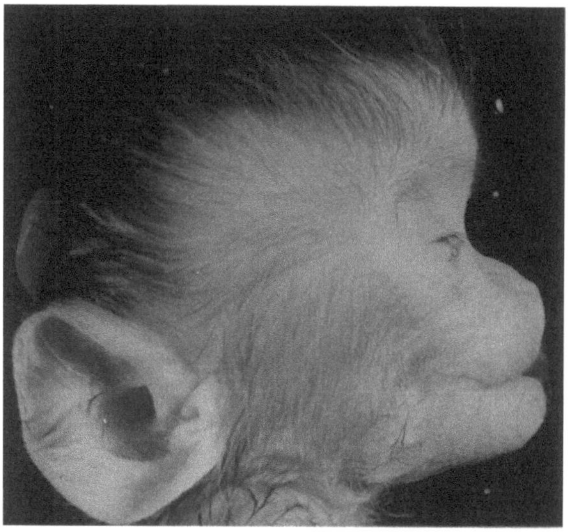

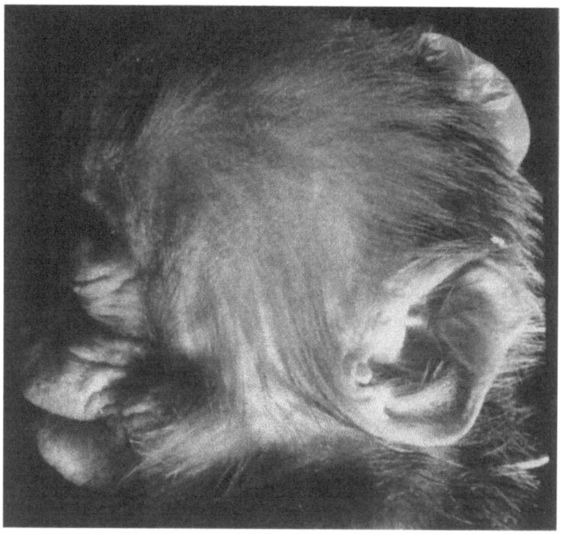

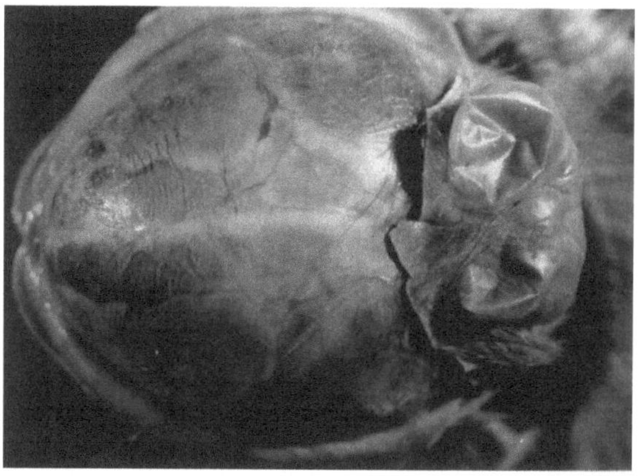

Fig. 194 (*above*). Lateral view of newborn rhesus monkey exhibiting craniofacial dysmorphia. The perpendicular nasal frontal angle, shortened cranial length, and building frontal area are well defined. Hydrocephalus is evident in the bulging frontal bones. Other defects include low-set ears and exophthalmia ▶

Fig. 195 (*middle*). Lateral view of newborn rhesus monkey with meningocele and hydrocephalus. Note the shortened cranial length, bulging frontal bones, and perpendicularity of the nasal frontal angle indicative of craniofacial dysmorphia

Fig. 196 (*below*). Dorsal view of newborn rhesus monkey skull (Fig. 189) demonstrating the position of the occipital meningocele and cranium bifidum. Note the perimeter of bone and the thin membraneous connective tissue covering the brain

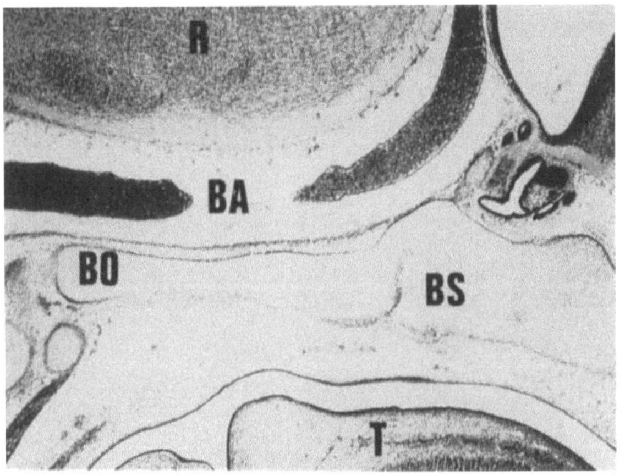

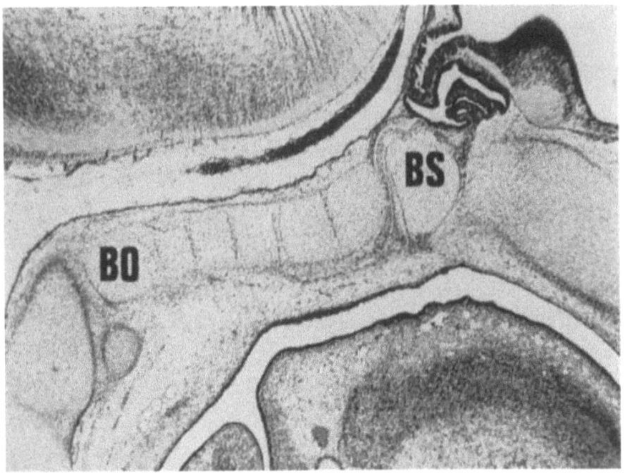

Fig. 197 (*above*). A median section through the notochord of a 42-day (stage 22) control rhesus monkey embryo. *BA*, basilar artery; *BO*, basioccipital cartilage; *BS*, basisphenoid cartilage; *R*, rhombencephalon; *T*, tongue; ×38

Fig. 198 (*below*). A median section through the notochord of a 42-day (stage 22) TAC-treated rhesus monkey embryo. *BO*, basioccipital cartilage; *BS*, basisphenoid cartilage; ×38

fetuses exposed to 10 mg TAC/kg on alternate days between GD 23 and 31 were compared to normal specimens on GD 35, 42, 50, 60, and 70 (Parker and Hendrickx 1982, 1983).

The 35-day embryos (stage 18) revealed a shortened, parachordal, mesenchymal condensation which was less advanced than controls and was characterized by less intercellular space, indicative of decreased extracellular matrix. Abnormal notochordal folding and a shortened cranial base were also apparent at this stage. The notochord configuration was altered from that of the control, showing sharp bends and anteroposterior compression in the shortened cranial base mesenchyme. The perinotochordal sheath appeared corrugated and the termination of the notochord in the basisphenoidal cartilage became corkscrew-shaped rather than a ventrally directed hook, as seen under normal conditions.

At GD 42 (stage 22), the parachordal cartilage was abnormal; the number of chondrocytes was reduced, the cartilage had irregular vacuolization, and the staining of the extracellular matrix was less dense. The perichondral surface was less regular and the cranial base angle was decreased from that of controls. The orbitosphenoid and alisphenoid cartilages were thin and underdeveloped and the anlage of the body of the sphenoid cartilage had a shallow pituitary fossa, a condition which persisted in all the fetuses examined. The basioccipital cartilage was short and thick and the exoccipital and supraoccipital cartilages were reduced. Intramembranous ossification was in progress in both the maxilla and the mandible. The frontal region consisted of a highly attenuated squama and a fibrous orbital plate and the distance between the orbital plate and eye was diminished. The mesoderm appeared compressed below the midbrain.

The notochord was noticeably thicker through the entire intracranial course, especially below the anterior portion of the basioccipital cartilage at this stage. The basioccipital cartilage was shortened and thicker in the dorsoventral direction and had a lordotic curvature (Figs. 197, 198). The terminal portion of the notochord remained convoluted and had penetrated the shallow hypophyseal fossa ending in the hypophyseal stem, whereas in the control the notochord terminated with a few fine branches and was not opposed to the adeno- or neurohypophysis.

In 50-day TAC-treated fetuses the frontal bone, maxilla, premaxilla, zygoma, palate, and mandible had abnormal ossification with thick trabeculae and decreased intertrabecular space. The frontal bone squama was primitive with a single layer of bone trabeculae. The orbital plate was thin and fenestrated and fewer osteoclasts were seen in the maxilla and mandible. Ossification was present in the arched palate but not in the cleft palate. The retarded development of the parachordal cartilage was more pronounced as development progressed. The orbitosphenoid was thin and shortened with an irregular flexion near the optic foramen. The cranial base angle remained decreased. The small exoccipital bone lacked the hypertrophic cartilagenous zone and the supraoccipital bone was thinner and shorter.

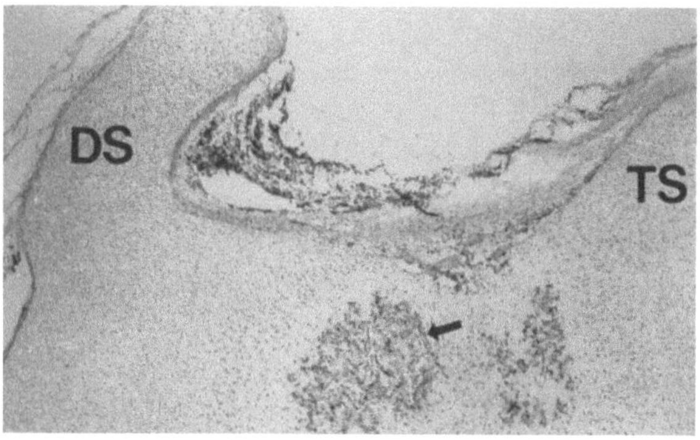

Fig. 199 (*upper left*). A section through the hypophyseal fossa of a 70-day control rhesus monkey fetus. *DS*, dorsum sella; *TS*, tuberculum sella; endochondral ossification (*arrow*); ×36

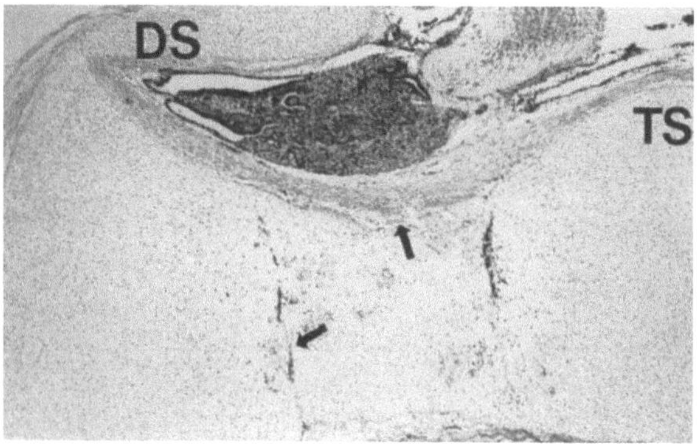

Fig. 200 (*lower left*). A section through the hypophyseal fossa of a 70-day TAC-treated rhesus monkey fetus showing the aberrant position of the dorsum sella (*DS*) resulting in an abnormal configuation of the hypophyseal fossa. *TS*, tuberculum sella; endochondral ossification (*arrow*); ×36

Fig. 201 (*upper right*). A section through the frontal bone of a 70-day control rhesus monkey fetus. *OP*, orbital plate; *SM*, supraorbital margin; *S*, squama; ×39

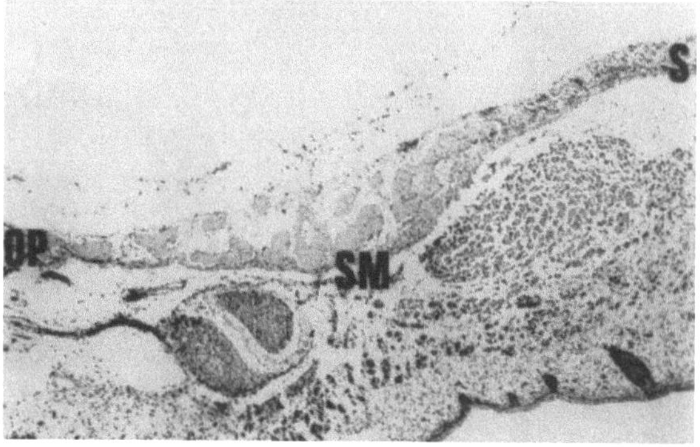

Fig. 202 (*lower right*). A section through the frontal bone of a 70-day TAC-treated rhesus monkey fetus. Note abnormal shape and decreased intramembranous bone formation. *OP*, orbital plate; *SM*, supraorbital margin; *S*, squama; ×39

The notochord of the TAC-treated 50-day fetus persisted throughout the entire intracranial course rather than degenerating as it did in the age-matched control fetus. It was markedly thickened in the short, thick, basioccipital cartilage and was convoluted in the area ventral to the anterior portion of the basioccipital cartilage. The perinotochordal sheath was not evident in this region and a space appeared around the notochord. The terminal portion of the notochord travelled dorsally between the anterior portion of the basioccipital cartilage and the basisphenoidal cartilage and emerged on the dorsal surface of the basisphenoid cartilage where it entered the shallow hypophyseal fossa and interdigitated with the adeno- and neurohypophysis.

The abnormal appearance of the cranial base continued at GD 60 with the cranial base angle still decreased. Growth retardation was evident in the basicranial region and the facial bones. The squama of the frontal bone was underossified, leaving a large metopic suture which had deficient connective tissue. The sagittal suture and fontanelles were wide, due to decreased ossification of the adjacent bones. Although the notochordal pattern in the TAC-treated 60-day fetus resembled that of the 5-day control, the portion of the notochord ventral to the anterior portion of the basioccipital cartilage was markedly thickened and the terminal portion ended within the hypophyseal fossa. The posterior basicranium was shortened and thickened.

The cranial base malformations increased in severity with growth, as evidenced in the GD 70 fetuses. The cartilaginous orbitosphenoid was thinner, shorter in anteroposterior and transverse diameters, and had a moderate lordotic flexion. Both the alisphenoid and the body of the sphenoid were underossified and the pituitary

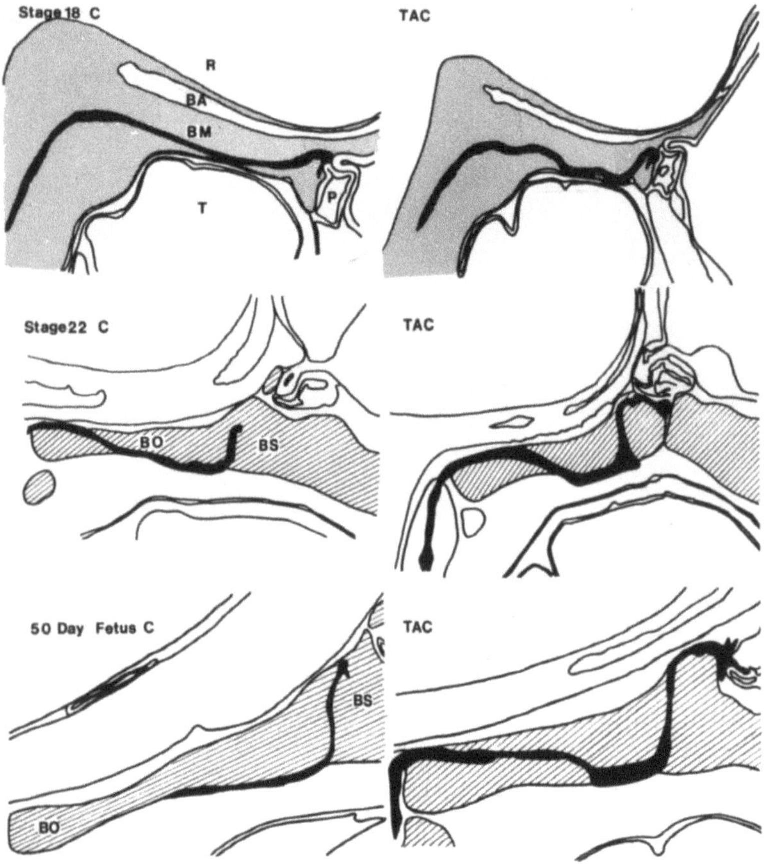

Fig. 203. The intracranial course of the notochord in control and TAC-treated rhesus monkey embryos and fetuses. *Top left*, stage 18 (35-day) control rhesus monkey embryo; *top right*, stage 18 TAC-treated rhesus monkey embryo; *middle left*, stage 22 (42-day) control rhesus monkey embryo; *middle right*, stage 22 TAC-treated rhesus monkey embryo; *bottom left*, 50-day control rhesus monkey fetus; *bottom right*, 50-day TAC-treated rhesus monkey fetus. *BA*, basilar artery; *BM*, basicranial mesenchyme; *BO*, basioccipital cartilage; *BS*, basisphenoid cartilage; *P*, pituitary; *R*, rhombencephalon; *T*, tongue

fossa remained shallow due to a reduction in the bones which comprise it (Figs. 199, 200). The basioccipital bone remained shorter and thicker than controls and was underossified. The exoccipital and supraoccipital were underossified and reduced in size. Ossification was diminished in the interparietal bone and was totally deficient in the region of the encephalocele.

The facial bones were similar to those noted in the 50- and 60-day fetuses although the frontal bone was less ossified and the thinned orbital plate did not extend under the orbitosphenoid bones. Both the supraorbital margin and frontal squama were underossified and underdeveloped compared to controls (Figs. 201, 202). The resulting metopic suture was still deficient in connective tissue. The calvarial sutures and fontanelles were characteristically widened and the connective tissue thinned, indicative of cranioschisis occulta. In the premaxilla, maxilla, and mandible, including the condyloid process, the trabeculae were thick and osteoclasts reduced in numbers. The palatine portion of the maxilla and the premaxilla was thinner. The squamae of the temporal and parietal bones were less ossified and the zygomatic process of the temporal bone was thinner and less trabeculated. The nasal, lacrimal, zygomatic, and vomer bones were reduced with characteristic thick trabeculae. No intracranial remnants of the notochord were found in the control 70-day fetus, whereas in the TAC-treated 70-day fetus remnants remained near the posterior portion of the basioccipital cartilage. The basioccipital cartilage was shortened and thickened and the endochondral ossification appeared delayed.

In summary, the basichondrocranium of the TAC-treated embryos and fetuses was shorter

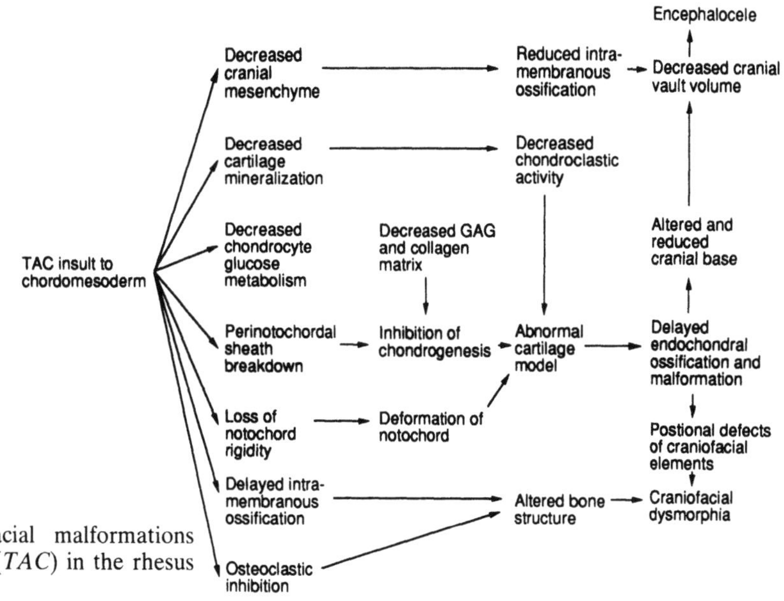

Fig. 204. Pathogenesis of craniofacial malformations induced by triamcinolone acetonide (*TAC*) in the rhesus monkey. *GAG*, glycosaminoglycans

than that of controls. The reduction of the longitudinal dimension of the cranial base was primarily caused by a shortened basioccipital cartilage. Reconstructions of this region showed the notochord folding in an accordian fashion, suggestive of a crowding of the notochord into a reduced space (Fig. 203).

Biologic Features

Pathogenesis

Using direct and indirect evidence, the following scheme for the pathogenesis of craniofacial malformations induced by TAC in the rhesus monkey is proposed (Fig. 204). The teratogenic insult of TAC causes: (1) breakdown of the perinotochordal sheath, (2) decreased induction of chondrogenesis, (3) decreased chondrocyte glucose metabolism and synthesis of glycosaminoglycans and collagen, and (4) decreased cartilage mineralization and chondroclastic activity. All these events alter the shape of the cartilaginous cranial base. Delayed and reduced intramembranous and endochondral ossification with decreased osteoclastic bone remodeling alter the bony morphology of the craniofacial skeleton, resulting in an altered and reduced cranial base and a decreased cranial vault volume, which, due to the expansion of the brain, leads to the formation of an encephalocele. The craniofacial dysmorphia is the result of positional alterations required by the facial skeleton to articulate with the abnormal cranial base and altered bone structure caused by delayed intramembranous ossification and altered remodeling due to osteoclastic inhibition (Parker and Hendrickx 1982). These studies corroborate the earlier findings of Marin-Padilla (1966, 1978, 1980) in both animals and humans, which indicated that cranium bifidum and related defects are the result of a primary mesenchyme insufficiency and retarded intramembranous ossification.

Related Abnormalities

In addition to these craniofacial abnormalities, prenatal exposure to TAC also produced adverse effects on the central nervous system. The major abnormalities include meningocele, encephalocele, and hydrocephalus; distortion and hypoplasia of the cerebellum, midbrain beaking, occipital lobe hypoplasia, agyria/polygyria, and thinned or absent corpus collosum were also observed. Growth retardation and minor malformation of the appendicular and axial skeleton and viscera were also observed. The most common defects in other organ systems were cutaneous syndactyly, hypoplastic thymus, and hypoplastic spleen. Terrell (1976) and Sawyer et al. (1977) extended observations of thymic hypoplasia in these experiments by studying the effects of 10–14 mg TAC/kg, administered for 4 consecutive days between GD 45 and 83. This treatment period coincides with differentiation of the thymus into cortical and medullary zones. Fetuses exhibited decreased viability, growth retardation, and hypoplasia of lymphoid organs, especially the thymus. Although some degree of lobulation was retained, the separation of the thymus into cortical and medullary zones was not evident. Some degree of interference of differentiation of lymphoid elements within the thymic rudiment (i.e., lymphocytopenia, preferential depletion of T lymphocytes, and diminished lymphocyte reactivity) was also observed.

Exposure

Malformations in rhesus monkeys resulted from oral administration of TAC (10–20 mg/kg) on single or multiple days during early pregnancy. This dose range is approximately ten times the human therapeutic dose. The incidence of TAC-induced defects was high – all 18 treated fetuses exhibited abnormalities of the craniofacial region; the sensitive period for these malformations was between GD 23 and 31. Lesions of similar distribution were induced with single- and multiple-day treatments, the principle difference being an increased severity with multiple doses (Hendrickx et al. 1980a). The effects of TAC (3–28 mg/kg) administered for 1–4 consecutive days during later periods of development (GD 37–48) included a high incidence of resorption and intrauterine death and a low level of malformations affecting the orofacial region (protruding forehead, widening of head, and expansion of the cranial sutures), thorax, hind limbs, thymus, adrenal, and kidney (Hendrickx et al. 1975). Further studies to investigate the sensitivity of primate species to corticosteroid-induced cleft palate involved administration of TAC during the period of secondary palate development. Treatment with 0.5–8 mg/kg between GD 35 and 43 resulted in craniofacial dysmorphia, cranium bifidum, meningocele, and thymic hypoplasia, but failed to induce cleft palate.

Comparison with Other Syndromes

Primates

Similar craniofacial syndromes have been experimentally induced in primates. McClure et al. (1979) treated rhesus monkeys with 5–10 mg cyclophosphamide/kg for 2–3 day periods during early and late organogenesis. Two distinct syndromes were observed in exposed offspring: (1) cleft lip and/or palate with ocular defects (open eyelids, hypertelorism, exophthalmus) following treatment between GD 27 and 29 and (2) craniofacial dysmorphia characterized by a flattened nasal bridge and arched palate and accompanied by either meningoencephalocele or persistent anterior fontanelle resulting from later treatment (GD 32–40). Related digital anomalies associated with these sydromes were poly- and ectrosyndactyly.

Retinoic acid (all-*trans*) also produces a consistent syndrome of craniofacial malformations in two primate species. Exposure to 20–40 mg/kg for 4–8 consecutive days during organogenesis (GD 17–45) produced a high incidence of rhesus monkey offspring with midfacial hypoplasia, arched palate, decreased cranial length, and abnormal ears. Associated skeletal and visceral defects which occurred at a low incidence were minor talipes, bilateral forearm hypoplasia with curved radius, kinked tail, dextroposition of the hear with unilateral lung hypoplasia, and athymia. The most critical sensitive period for these defects was between GD 24 nd 35 (Hendrickx et al. 1980b).

Pigtail monkeys (*M. nemestrina*) are also susceptible to the teratogenic effects of retinoic acid. The craniofacial anomalies most frequently observed following continuous treatment with 10 mg/kg between GD 20 and 44 were cleft palate, malformed ears, hypertelorism, exophthalmos, hypoplasia of the midface, mandibular abnormalities, and distortion of the cranium. Less frequent manifestations of the syndrome were fusion of the inferior coronal suture and/or midsphenoidal, and/or spheno-occipital synchondroses of the cranial base (Fantel et al. 1977; Newell-Morris et al. 1980). Associated musculoskeletal defects included syndactyly, missing phalanges (hand and foot), metacarpals, and metatarsals, severe muscular-joint contractures, tail vertebral defects, kyphosis, scoliosis, and malformed radii. Visceral anomalies were rare and consisted of transposition of the great vessels and genitourinary defects with anal atresia.

Other craniofacial defects have also been reported in primates following prenatal exposure to exogenous agents. Thalidomide induces otomandibular and temporozygomatic anomalies in marmosets (*C. jacchus*), green monkeys (*C.

aethiops), and long-tailed (*M. fascicularis*) and bonnet (*M. radiata*) macaques (Hendrickx and Prahalada 1986). Valproic acid is associated with craniofacial dysmorphia characterized by an abnormally shaped cranium, low-set ears, flattened nasal angle, hypoplasia of the maxilla and mandible, and zygomatic arch defects in the offspring of exposed rhesus monkeys (Hendrickx et al. 1988). Hyperthermia produces adverse effects on craniofacial development of marmosets (mandibular prognathism) and bonnet monkeys (mid-facial hypoplasia, arrhinia). Marmosets are additionally susceptible to the teratogenic effects (cleft palate, micrognathia, ear defects) of X-ray (Hendrickx and Prahalada 1986). Minor facial dysmorphia in pig-tailed macaques is associated with in utero exposure to alcohol (Sheller et al. 1988).

A low incidence of spontaneous craniofacial abnormalities, especially cleft lip and/or palate, in addition to a variety of calvarial and fontanellar defects have been reported in prosimians, New and Old World monkeys, and apes (see Schultz 1972; Hendrickx and Prahalada 1986; Hoopes and Jerome 1987).

Humans

The craniofacial malformations comprising the TAC syndrome as described herein have several features that are also commonly described in human infants, namely: mandibulofacial dysostosis (Treacher-Collins syndrome), cranioschisis, and cleft palate. Craniofacial malformations induced with retinoic acid in macaques also resemble, to some degree, the human condition of mandibulofacial dysostosis. Similarly, cyclophosphamide-induced defects of the craniofacial region present some features resembling those in the following human syndromes: acrocephalosyndactyly (Apert syndrome), oral facial digital syndrome, and Larsen's syndrome.

Effects of Corticosteroids in Other Species

Corticosteriods have been extensively studied as teratogenic agents for the last 40 years and have become a widely used tool for the study of cleft palate and other organ systems and developmental events in a number of laboratory animal species (Rowland and Hendrickx 1983). The demonstrated teratogenicity of corticosteroids in laboratory animals raises the question of possible human teratogenicity. The general conclusion is that corticosteroids probably do not have an adverse effect on human intrauterine development, although cleft palate and a number of other malformations have been reported in offspring exposed to corticosteroids during development. A number of deficiencies exist in the human studies, such as inadequate documentation of dose, time of treatment, and end points evaluated. In light of the equivocal results and inadequacies of the data base, the possibility that corticosteroids may be teratogenic cannot be eliminated. A syndrome closely resembling that described above for the rhesus monkey has been induced in both the bonnet monkey and baboon (*P. cynocephalus*), using the same protocol (Hendrickx et al. 1980a).

Mice have been the most extensively studied and many natural and synthetic corticosteroids have been used to induce cleft palate and other defects, such as spina bifida, shortening of the head and mandible, micromelia, abnormal digits, dilation of the marginal venous sinus, distal limb bud necrosis, subepidermal blisters, abnormal ovarian position, and cataracts. Different strains of inbred mice exhibit different degrees of susceptibility to corticosteroid-induced cleft palate. From a comparative standpoint, based on the dose required to induce a similar incidence of defects, TAC is the most potent teratogen, following by dexamethasone, prednisolone, cortisone, cortisol, and corticosterone.

Rats are considerably less susceptible to induction of cleft palate with corticosteroids, especially cortisol and cortisone; however, it has been accomplished by different routes of administration and increased dose. Cleft palate and other effects, namely umbilical hernias, undescended testes, decreased ossification of the skeleton, hypoplastic thymus, decreased fetal viability, and fetal growth retardation, are readily induced with TAC treatment.

Cleft palate is readily induced in the hamster fetus with the commonly used corticosteroids, except cortisone. Cranium bifidum is also commonly induced following exposure to TAC. Rabbits also are quite susceptible to induction of cleft palate with corticosteroids. Associated malformations included umbilical hernias and omphaloceles, ventricular and atrial septal defects, spinal bifida, exencephaly, and anencephaly. TAC and dexamethansone were the most effective in inducing cleft palate in fetuses on a dose basis. Only hydrocortisone has been studied in guinea pigs; cleft palate, umbilical hernia, edema, open eyelid, abdominal eventration, syringomyelomeningocele, syndactly, and limb torsion were induced following fetal exposure.

In summary, corticosteroids have been successfully used to induce abnormalities in several mammalian species, namely: cleft palate in mice, nonhuman primates, rabbits, hamsters, and guinea pigs; umbilical hernias in rats, guinea pigs, and rabbits; cranium bifidum with associated abnormalities in nonhuman primates and hamsters; genital defects in rats and mice; cardiovascular defects in rabbits; and limb anomalies and cataracts in mice. TAC is a very potent teratogen in all the species tested, inducing significant frequencies of abnormality at relatively low doses.

References

Fantel AG, Shapard TH, Newell-Morris LL, Moffett BC (1977) Teratogenic effects of retinoic acid in pigtail monkeys (Macaca nemestrina): I. General features. Teratology 15:65–72

Hendrickx AG, Prahalada S (1986) Teratology and embryogenesis. In: Dukelow R, Erwin J (eds) Reproduction and development. Comparative primate biology, vol 3. Liss, New York, pp 333–362

Hendrickx AG, Sawyer RH, Terrell TG, Osburn BI, Henrickson RV, Steffek AJ (1975) Teratogenic effects of triamcinolone on the skeletal and lymphoid systems in non-human primates. FASEB J 34:1661–1665

Hendrickx AG, Pellegrini M, Tarara R, Parker R, Silverman S, Steffek AJ (1980a) Craniofacial and central nervous system malformations induced by triamcinolone acetonide in non-human primates: I. General teratogenicity. Teratology 22:102–114

Hendrickx AG, Silverman S, Pellegrini M, Steffek AJ (1980b) Teratological and radiocephalometric analysis of craniofacial malformations induced with retinoic acid in rhesus monkeys (Macaca mulatta). Teratology 22:13–22

Hendrickx AG, Nau H, Binkerd P, Rowland JM, Rowland JR, Cukierski MJ, Cukierski MA (1988) Valproic acid developmental toxicity and pharmacokinetics in the rhesus monkey: an interspecies comparison. Teratology 38:329–345

Hoopes CW, Jerome CP (1987) Oral-facial clefts and associated malformations in the squirrel monkey (Saimiri sciureus). J Med Primatol 16/3:203–209

Marin-Padilla M (1966) Mesodermal alterations induced by vitamin A. J Embryol Exp Morphol 15:261–269

Marin-Padilla M (1978) Clinical and experimental rachischisis. In: Vinken PW, Bruyn GW (eds) Handbook of clinical neurology, vol 32. Elsevier/North-Holland, Amsterdam, pp 59–191

Marin-Padilla M (1980) Morphogenesis of experimental encephalocele (cranioschisis occulta). J Neurol Sci 46:83–99

McClure HM, Wilk AL, Horigan EA, Pratt RM (1979) Induction of craniofacial malformations in rhesus monkeys (Macaca mulatta) with cyclophosphamide. Cleft Palate J 16:248–256

Newell-Morris L, Sirianni JE, Shepard TH, Fantel AG, Moffett BC (1980) Teratogenic effects of retinoic acid in pigtail monkeys (Macaca nemestrina): II. Cranial features. Teratology 22:87–101

Parker RM, Hendrickx AG (1982) Triamcinolone-induced alteration of the notochordalbasicondrocranial relationship in the rhesus monkey. J Craniofac Genet Dev Biol 2:219–231

Parker RM, Hendrickx AG (1983) Craniofacial and central nervous system malformations induced by triamcinolone acetonide in nonhuman primates: II. Craniofacial pathogenesis. Teratology 28:35–44

Rowland JM, Hendrickx AG (1983) Corticosteroid teratogenecity. In: Cornelius CE, Simpson CF (eds) Advances in veterinary science and comparative medicine, vol 27. Academic, New York, pp 99–128

Sawyer R, Hendrickx AG, Osburn B, Terrell T, Anderson J (1977) Abnormal morphology of the fetal monkey (Macaca mulatta) thymus exposed to a corticosteroid. J Med Primatol 6:145–150

Schultz AH (1972) Developmental abnormalities. In: RN Fiennes (ed) Pathology of simian primates: I. General pathology. Karger, Basel, pp 158–189

Sheller B, Clarren SK, Astley SJ, Sampson PD (1988) Morphometric analysis of Macaca nemestrina exposed to ethanol during gestation. Teratology 38:411–417

Terrell TG (1976) Development of the lymphoid system in the rhesus monkey fetus: a comparison of normal development to teratogenic defects induced by the corticosteroid triamcinolone. Doctoral thesis, University of California, Davis

Congenital Anomalies, Tamarins

Laura V. Chalifoux

Synonyms. Achondroplastic dwarfism; chondrodystrophy, chondrodysplasia; polydactyly; supernumerary digits; syndactyly; bicornuate, bipartite, or septate uterus.

Gross Appearance

Two cases involving multiple congenital anomalies which were published previously (Chalifoux and Elliott 1986) are considered in this report.

Table 7. Body measurements of dwarf *Saguinous oedipus* compared to 16 normal neonates

	Dwarf	Normal
Weight (g)	37.8	37.8 ± 1.7
Crown-rump (cm)	8.8	8.7 ± 0.6
Crown-heel (cm)	11.6	13.1 ± 0.6
Hell-heel (cm)	5.7	8.8 ± 0.6
Heel-toe (cm)	2.2	2.7 ± 0.2
Finger-finger (cm)	8.9	12.6 ± 0.5
Head circumference (cm)	8.6	8.4 ± 0.4
Chest circumference (cm)	5.4	5.9 ± 0.4
Abdomen circumference (cm)	5.8	5.2 ± 0.6
Tail (cm)	9.6	11.3 ± 0.8

Case 1: A newborn male cotton-top tamarin (*Saguinus oedipus*) with achondroplastic dwarfism, polydactyly, and syndactyly. This new-born weighed 37.8g and had short extremities. His body measurements were compared with 16 normally proportioned infants (Table 7). These measurements were: from the crown to the rump: from the crown to the heel; from the heel on one foot to the rump and to the heel of the other foot (combined length of legs); from the heel to the tip of the longest toe; with arms outstretched, from the tip of the longest finger on one hand to the tip of the longest finger on the other; circumferences of head, chest, and abdomen; and the length of the tail. Radiographs were taken of this infant (Fig. 205) and a normal infant tamarin (Fig. 206). The measurements and radiograph indicated that the trunk (crownrump) was of normal length but the tail was short. The ribs were shortened, creating a small narrow thorax, and the abdomen was slightly distended. All measurements which included arm and leg lengths were shorter than normal. There was also a moderate curvature of both fibulas. A progressive narrowing of the interpedunculate distances of the lumbosacral vertebrae from anterior to posterior was seen. The palate was highly arched. The right hand had seven fingers and the left six. The first two digits of each hand were joined by a web of skin extending the length of the first joint. Separate metacarpals and phalanges were distinguishable for each digit in the radiograph (Fig. 205). There were seven toes on each foot (Fig. 207). Only six metatarsals and six of the first row of phalanges were apparent by radiograph. A web of skin joined the first two digits on the left foot for the entire length. On the right foot a small toe with no bone extended from the first joint of the second toe.

Case 2: A newborn female saddle-back tamarin (*Saguinus fuscicollis*) with scoliosis and bicornuate uterus. This infant weighed 31.2g. Radiographs were taken of the thoracolumbar scoliosis (Fig. 208). A hemivertebra was evident at T11 on the left side with corresponding abnormal spacing of the ribs. The lumbar vertebrae were increased in density. The presence of an additional hemivertebra or a defect of segmentation in the lumbar or anterior sacral region was unclear from the radiograph. A kink in the tail was also present. The uterus was bicornuate, having two uterine horns and two uterine corpora (Fig. 209).

Microscopic Features

A longitudinal section of the knee joint of the *S. oedipus* infant (case 1) was examined, which included both the distal femur and proximal tibia. Comparison with similar sections from normal neonates suggested no abnormality in the epiphyseal plate. Examination of other tissues revealed atelectatic lungs, incomplete glomerulogenesis in the peripheral renal cortex, and focal hepatic extramedullary hematopoiesis, all of which are common findings in tamarins which die during the neonatal period.

Examination of tissues from the *S. fuscicollis* (case 2) infant revealed focal hepatic necrosis for which no cause was determined, pulmonary atelectasis, and incomplete renal glomerulogenesis. Neither the skeleton nor the uterus was examined microscopically.

Ultrastructure

No ultrastructural study was done.

Differential Diagnosis

The gross diagnosis of congenital anomalies should present no difficulty, especially if normal specimens of similar age are available for comparison. There are numerous types of dwarfism recognized in humans, in some of which concomitant abnormalities occur, such as polydactyly and cleft or highly arched palate. The histologically normal appearance of the epiphyseal plate and the decreasing distances between pedicles of the lumbosacral vertebrae observed in the *S. oedipus* infant are consistent with find-

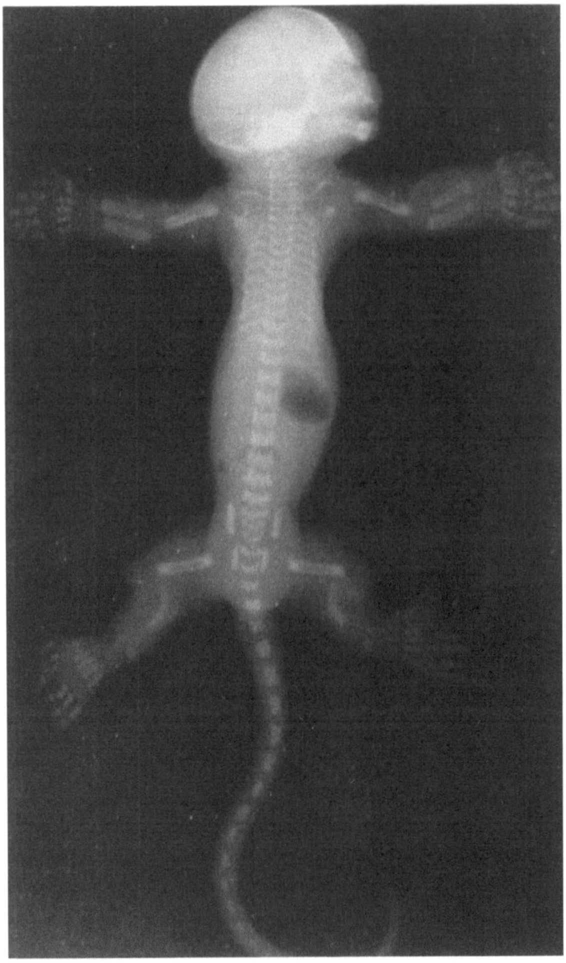

cm
1 2

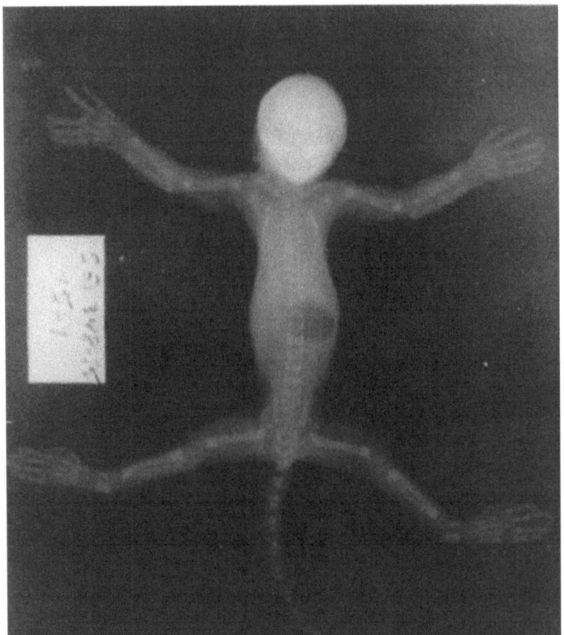

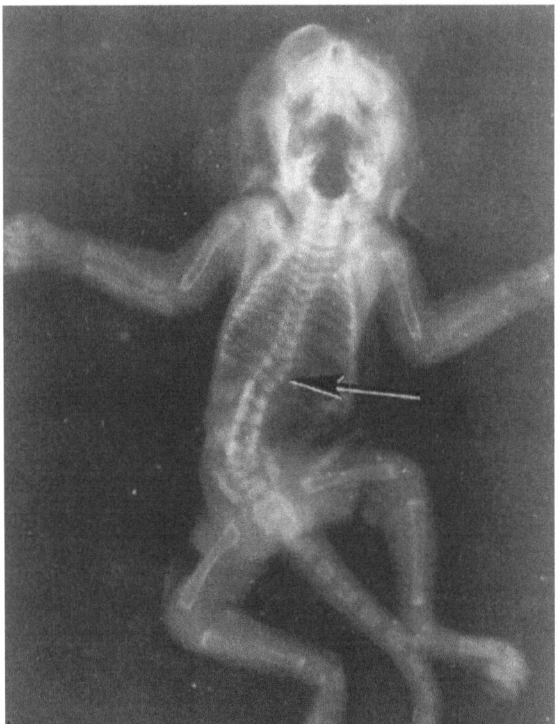

◀ **Fig. 205** (*upper left*). Radiograph of dwarf newborn *Saguinus oedipus*. Note short bones of appendages, small thorax, polydactyly, and syndactyly. (From Chalifoux and Elliott 1986)

Fig. 206 (*lower left*). Radiograph of normal neonatal *Saguinus oedipus*. (From Chalifoux and Elliott 1986)

Fig. 207 (*upper right*). Right feel of *Saguinus oedipus* neonates. A normal right foot is at *top*. *Below* is the right foot of the dwarf with shortened bones, polydactyly, and syndactyly. (From Chalifoux and Elliott 1986)

Fig. 208 (*lower right*). Radiograph of neonatal *Saguinus fuscicollis* with thoracolumbar scoliosis. Hemivertebra at T11 (*arrow*), increased density of lumbar vertebrae, and kink in tail (*arrow*). (From Chalifoux and Elliott 1986)

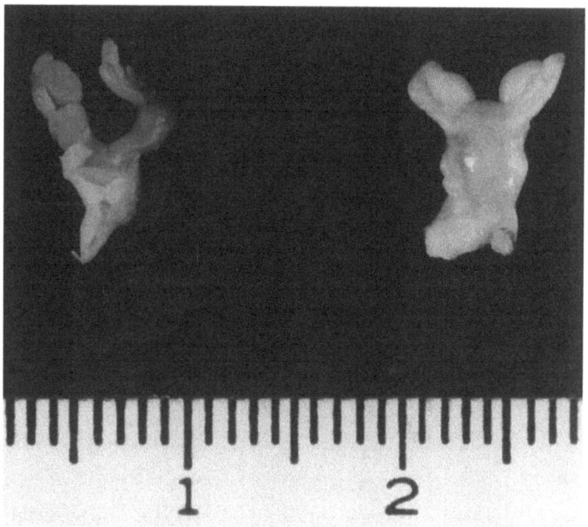

Fig. 209. Bicornuate uterus from *S. fuscicollis*. (*left*); normal tamarin uterus (*right*). (From Chalifoux and Elliott 1986)

ings reported in achondroplasia in humans (Kaufmann 1974; Rimoin 1974). Asphyxiating short rib syndromes, also associated with polydactyly, should be considered. The various syndromes have different modes of inheritance.

Other references available to aid in the diagnosis of congenital anomalies include: scoliosis (Moe et al. 1978), uterine development (King 1978; Patten 1964; Willis 1962; see also p. 170, this volume).

Biologic Features

Natural History

The infant male *S. oedipus* (case 1) was found dead in the cage and was presumed to be stillborn. His male twin was rejected by the parents but successfully raised in the infant nursery. The parents were laboratory-born twin siblings who were raised in a family group and who mated after the death of their parents.

The newborn *S. fuscicollis* female (case 2) was found hypothermic on the floor of the cage. She was brought to the infant nursery where she died at 2 days of age. A normal twin survived and was raised by the parents. The parents were a laboratory-born pair of twins which were inadvertently mated.

Behavioral studies have suggested that among the tamarin species the family group is organized to avoid incestuous relationships and that the presence of the mother inhibits breeding activity in her daughters (Epple 1978). In the two cases reported here, the incestuous breeding occurred between young twin siblings only after the death or separation of the parents.

Pathogenesis

The shortened extremities which characterize achondroplastic dwarfs reflect the failure of normal endochondral ossification (Teitelbaum 1985). Why this occurs is still unknown but may be related to rate of growth. There is debate between two schools of thought: one which believes the achondroplastic growth plate is histologically normal and another which believes it is not. A possible explanation of the controversy may be the inclusion of more than one type of dwarfism in the histologic studies.

Polydactyly, a common malformation in humans, follows the establishment of more than one center of organogenesis which may be the result of gene action. Syndactyly results from failure of parts to separate (Willis 1962).

The hemivertebrae causing scoliosis is a defect of formation resulting in absence of the lateral aspect of the vertebral body (Moe et al. 1978). The kink in the tail of this monkey (case 2) was also probably due to a defect in development of a caudal vertebra, although that was not clear from the radiograph.

Bicornuate uterus results from a defect in fusion of the caudal ends of the paramesonephric or Mullerian ducts (Patten 1964; King 1978; Willis 1962).

Etiology

Achondroplasia in humans and some animals is due to an autosomal dominant gene. In humans, 80% of cases are considered sporadic mutations (Teitelbaum 1985). Other syndromes of dwarfism may have different modes of inheritance. Most cases of polydactyly and many cases of syndactyly in humans are produced by autosomal dominant genes. However, the gene may be dominant in one case and recessive or sex-linked in others (Willis 1962).

The defect of vertebral body formation resulting in scoliosis is generally not considered to be inherited in humans. However, those with multiple anomalies are more likely hereditary than those with single anomalies, especially in consanguineous families (Moe et al. 1978). The consanguinity of the parents of these two tamarin neonates may have been a factor in the occurrence of these anomalies.

Frequency

There were 942 births of *S. oedipus* recorded from 1967 through 1988 at the New England Regional Primate Research Center. Except for the case reported here, no congenital anomaly was seen on necropsy or physical examination of these infants. In addition none has been seen in more than 800 necropsies on wild-caught tamarins over the years. The population of *S. fuscicollis* has been too small to determine significant incidence, with only 18 births recorded during 1976–1984. The intensity of interest and the availablity of newborn animals probably influence the reported frequency of anomalies in tamarins.

Comparison with Other Species

In other primate species, studies of congenital anomalies in large populations have been done in macaques (Furuya 1966; Iwamoto 1967; Koford et al. 1966; Rawlins and Kessler 1983). A few pertinent reports in New World monkeys include cases of cleft palate in squirrel monkeys, *Saimiri sciureus* (Baker et al. 1977); a common marmoset, *Callithrix jacchus* (Hill 1954); a *Cebus apella* (Schultz 1956); and a *S. fuscicollis* (Kraus and Garrett 1968). Skeletal defects resembling campomelic syndrome and other anomalies have been described in squirrel monkeys (Stills and Bullock 1981). A case of syndactyly has been reported in *S. nigricollis* (Hetherington et al. 1975) and a uterus didelphys has been described in an owl monkey, *Aotus trivirgatus* (King 1978). Malformations have been induced in common marmosets with blighted potato (Poswillow et al. 1972), but estimates of the spontaneous occurrence of congenital anomalies in monkeys have been made with caution (Hendrickx and Binkerd 1983, Wilson 1978). Scoliosis and kinked tail have been reported in *Lemur variegatus* infants sired by the same father (Benirshke et al. 1981).

Genes affecting skeletal development have been identified in the laboratory mouse (Jones and Hunt 1983). Chondrodysplasias, some of which are due to autosomal recessive genes in the homozygous state, occur in the Dexter and other breeds of cattle, rabbits and dogs (Jubb et al. 1985; Jones and Hunt 1983). Nonlethal achondroplasia is a breed characteristic of some dogs. Syndactyly occurs in sheep, cattle, and swine. Polydactyly has been observed in all species (Jubb et al. 1985) most commonly in cats, horses, dogs, and cattle.

The teratogenic affects of lupine consumption by cows during pregnancy causes scoliosis and other anomalies in calves. (Jones and Hunt 1983). Scoliosis also occurs as one of a spectrum of abnormalities in foals (Jubb et al. 1985).

Fusion of the Müllerian ducts progresses cephalad beyond the vagina to different degrees in different species (King 1978; Patten 1964; Willis 1962). In higher mammals fusion involves the caudal end of the uterus forming an unpaired cervix. In primates, fusion of the Müllerian ducts is complete, resulting in a pear-shaped uterus with a single lumen. However, many patterns of imperfect fusion may result in various anomalies in humans and monkeys. Anomalies in uterine development also occur in other species, most commonly in cattle and swine (Jones and Hunt 1983).

Acknowledgement. This work was supported by Public Health Service Award RR00168 from the National Institutes of Health.

References

Baker CA, Hendrickx AG, Cooper RW (1977) Spontaneous malformations in squirrel monkey (Saimiri sciureus) fetuses with emphasis on cleft lip and palate. J Med Primatol 6:13–22

Benirschke K, Kumatoto AT, Bogart MH (1981) Congenital anomalies in Lemur variegatus. J Med Primatol 10:38–45

Chalifoux LV, Elliott MW (1986) Congenital anomalies in two neonatal tamarins Saguinus oedipus and Saguinus fuscicollis. J Med Primatol 15:329–337

Epple G (1978) Reproductive and social behavior of marmosets with special reference to captive breeding. In: Gengozian N, Deinhardt F (eds) Marmosets in experimental medicine. Primates Med 10:50–62

Furuya Y (1966) On the malformations occurred in the Gagyusan troop of wild Japanese monkeys. Primates 7:488–492

Hendrickx AG, Binkerd PE (1983) Teratology and birth defects. In: Dukelow WR (ed) Nonhuman primate models for human diseases. CRC Press, Boca Raton, pp 131–158

Hetherington CM, Cooper JE, Dawson P (1975) A case of syndactyly in the white-lipped tamarin Saguinus nigricollis. Folia Primatol 24:24–28

Hill WCO (1954) Report of the Society's prosector for the year 1953. Proc Zool Soc (Lond) 124:303–311

Iwamoto M (1967) Morphological observations on the congenital malformation of limbs in the Japanese monkey. Primates 8:247–270

Jones TC, Hunt RD (1983) Veterinary pathology. Lea and Febiger, Philadelphia

Jubb KVF, Kennedy PC, Palmer N (1985) Pathology of domestic animals. Academic, Orlando

Kaufmann HJ (1974) New skeletal dysplasias in the newborn: new X-ray findings. In: Bergsma D (ed) Skeletal dysplasias. Excerpta Medica, Amsterdam, pp 1–9

King NW (1978) The reproductive tract. In: Benirschke K, Garner FM, Jones TC (eds) Pathology of laboratory animals. Springer, Berlin Heidelberg New York, pp 509–580

Koford CB, Farber PA, Windle WF (1969) Twins and teratisms in rhesus monkeys. Folia Primatol 4:221–226

Kraus BS, Garrett WS (1968) Cleft palate in a marmoset: report of a case. Cleft Palate J 5:340–345

Moe JH, Winter RB, Bradford DS, Lonstein JE (1978) Scoliosis and other spinal deformities. Saunders, Philadelphia

Patten BM (1964) Foundations of embryology. McGraw-Hill, New York

Poswillo DE, Sopher D, Mitchell S (1972) Experimental induction of foetal malformation with "blighted" potato: a preliminary report. Nature 239:462–464

Rawlins RG, Kessler MJ (1983) Congenital and hereditary anomalies in the rhesus monkeys (Macaca mulatta) of Cayo, Santiago Teratology 28:169–174

Rimoin DL (1974) Histopathology and ultrastructure of cartilage in the chondrodystrophies. In: Bergsma D (ed) Skeletal dysplasias. Stratton, New York, pp 1–18

Schultz AH (1956) The occurrence and frequency of pathological and teratological conditions and of twinning among non-human primates. Primatalogia I: 965–1014

Stills HF Jr, Bullock BC (1981) Congenital defects of squirrel monkeys (Saimiri sciureus). Vet Pathol 18: 29–36

Teitelbaum SL (1985) Metabolic and other nontumorous disorders of bone. In: Kissane JM (ed) Anderson's pathology, vol 2. Mosby, St Louis, pp 1705–1777

Willis RA (1962) The borderland of embryology and pathology. Butterworth, Washington

Wilson JG (1978) Developmental anomalies: nonhuman primates: In: Benirschke K, Garner FM, Jones TC (eds) Pathology of laboratory animals. Springer, Berlin Heidelberg New York, pp 1911–1917

OTHER PATHOLOGIC ENTITIES

Generalized Amyloidosis, Nonhuman Primates

James L. Blanchard

Synonyms. Reactive systemic amyloidosis; secondary amyloidosis; AA amyloidosis.

Gross Appearance

Many animals with significant systemic amyloidosis are debilitated and cachexic (Blanchard et al. 1986). There is an absence of subcutaneous fat and the hair coat is sparse and dry. Many animals have contracted hind limbs due to disuse atrophy or chronic arthritis. The most notable lesion, when present, is an enlarged liver (Fig. 210). Livers affected by mild to severe amyloid deposition are firm, mottled red to tan, very waxy, and have sharp margins. Severely affected livers may extend past the rib margin, almost to the pubis. The spleen is mild to moderately enlarged and firm. The only other organs that may have gross evidence of amyloid deposition are the small and large intestines. There may be a slightly thickened wall, particularly in the small intestine, if severe amyloidosis is present. Associated changes not directly related to the presence of amyloid, but common in animals with amyloidosis, are moderate to severe osteoarthritis, colitis (usually chronic ulcerative), colonic structures at the cecocolic junction, and a dilated fluid-filled colon with no gross evidence of inflammation (Blanchard et al. 1986; Slattum et al. 1989b). Other changes, such as generalized muscle atrophy and dehydration, and related to the debilitated condition of the animals.

Microscopic Features

Histologic evidence of systemic amyloidosis can be found in almost any organ (Blanchard et al. 1986; Slattum et al. 1989a). Amyloid deposits in the liver vary from mild, where pale eosinophilic material outlines hepatic cords and extends into sinusoidal spaces, to severe, where atrophic hepatocytes are embedded in lakes of amyloid (Fig. 211). The spleen has deposits of amyloid primarily in and at the margin of lymphoid follicles (Fig. 212). Initially amyloid is deposited at the periphery of follicles, but as the severity increases, the entire follicle is replaced by amyloid except for a few lymphocytes and stromal cells. In some areas it appears that amyloid is contained within intracellular packets. The small intestine is the most common location for amyloid deposition (Fig. 213). Villi are

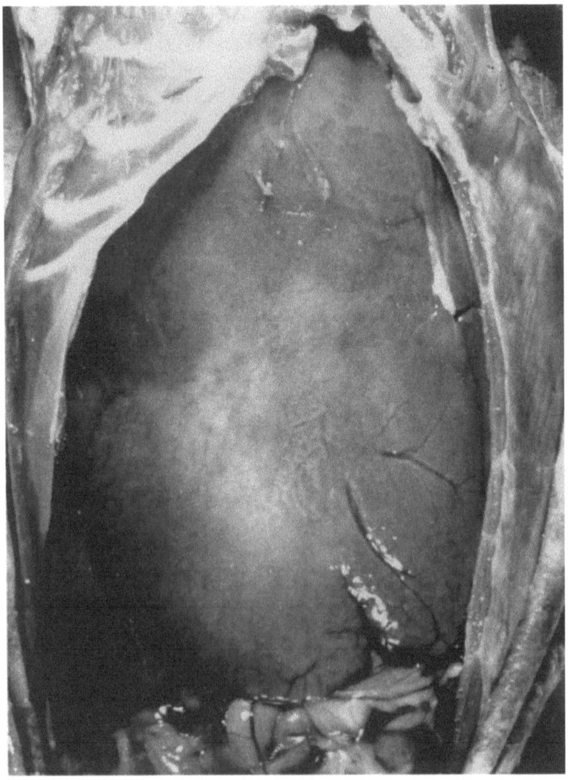

Fig. 210. Liver of adult rhesus monkey. Severe hepatomegaly due to amyloidosis

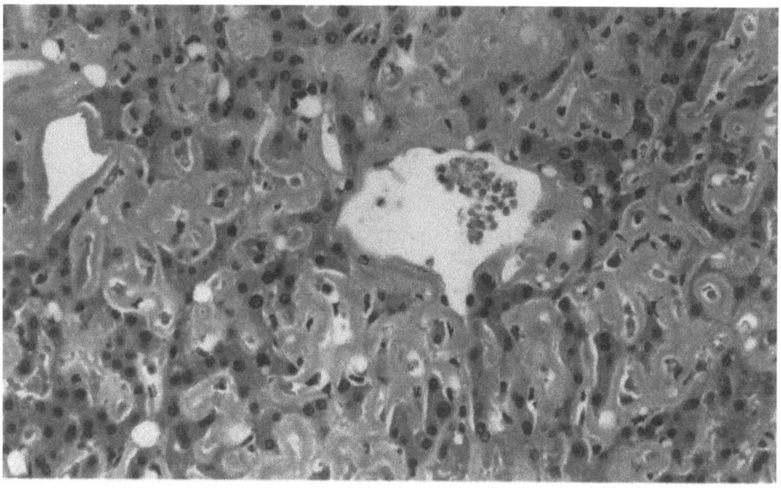

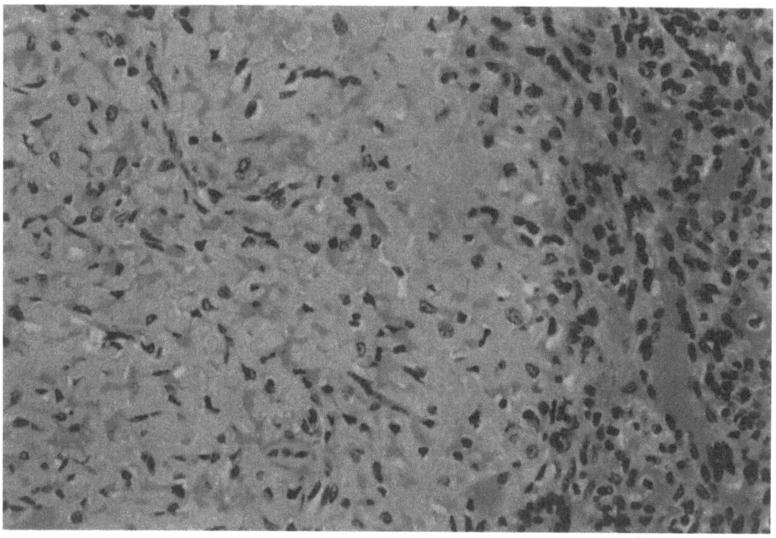

Fig. 211 (*above*). Central vein of liver from a rhesus monkey with severe amyloidosis. Note atrophy of hepatocytes and dilation of sinusoids. H and E, ×250

Fig. 212 (*below*). Splenic follicle from an adult rhesus monkey with depletion of lymphocytes and central amyloid deposits in packets. H and E, ×250

often blunt and widened, and mucosal cells are flattened and pyknotic. Amyloid is deposited in the lamina propria and in severe cases it separates crypts and stromal cells. Lacteals are often dilated. A mixed population of inflammatory cells, including lymphocytes and plasma cells, is also present. Amyloid is less frequently present in the large intestine and in smaller amounts than in the small intestine. Animals with systemic amyloidosis have a variety of associated changes in the large intestine. Some have severe chronic colitis with multiple ulcers and erosions characterized by infiltration of neutrophils, lymphocytes, and macrophages mixed with necrotic debris. Crypts are replaced by connective tissue and in some animals fibrous constriction of the cecocolic junction is seen. In these areas there is ulceration of the mucosa with severe thickening of the submucosa and muscular layers by edema, vascular dilation, and fibrosis. Older animals with dilated fluid-filled colons have little or no inflammatory response or may have slight thickening of the mucosa due to proliferation of the crypts. Amyloid deposits are less common in other organs. Lymph nodes have deposits usually in lymphoid follicles, but in severe cases the medulla may be involved. Lymph nodes are often depleted of mature lymphocytes and medullary areas have significant erythrophagocytosis. Amyloid is seen infrequently in other organs including the pancreas, adrenal glands, kidney (medullary interstitium), stomach, heart, lungs, brain, and gallbladder.

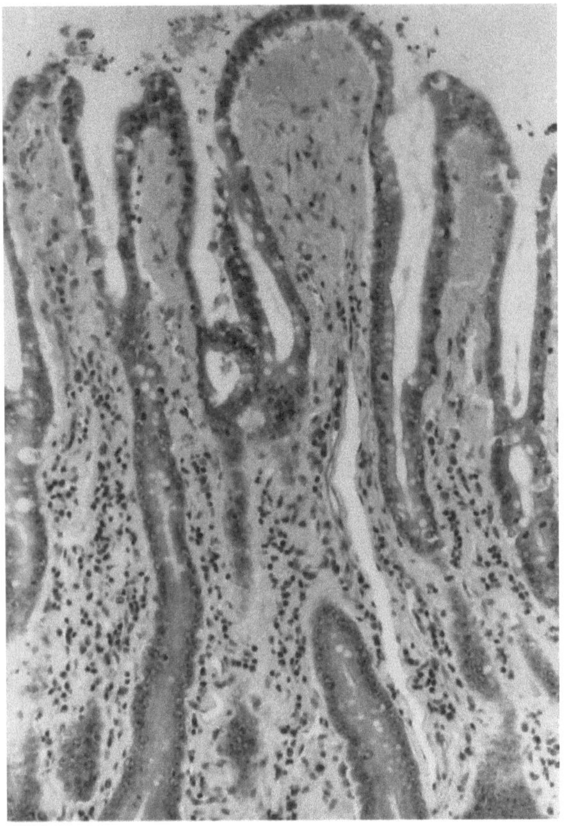

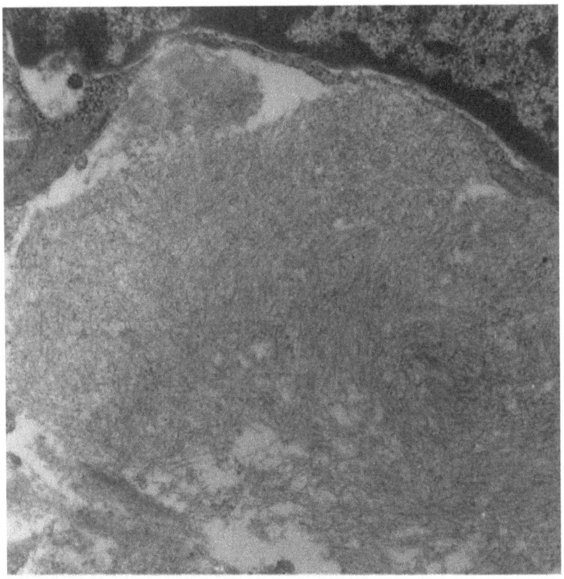

Fig. 213 (*above*). Small intestinal villi from a rhesus monkey with moderate amyloidosis. Note widened villus tips, deposition of amyloid within the lamina propria, and dilated lacteals. H and E, ×100

Fig. 214 (*below*). Mass of amyloid A fibrils in the liver from a rhesus monkey. TEM, ×11500

Ultrastructure

Systemic amyloidosis due to secondary or AA amyloid is characterized ultrastructurally by deposits of parallel arrays of fibers with a diameter of 70–100 A (Fig. 214). X-ray diffraction reveals the fibers to have a crossed-pleated sheet arrangement due to a large proportion of hydrophobic amino acids (Kisilevsky 1983).

Differential Diagnosis

Amyloid deposits in hematoxylin and eosin-stained sections appear as pale, eosinophilic, homogenous to fibrillar material. When amyloid is stained with Congo red and viewed under polarized light, it has a red-green birefringent appearance. To distinguish AA or secondary amyloid from AL or primary amyloid, sections are pretreated with potassium permanganate and then stained with Congo red. AA amyloid deposits will lose the typical red-green birefringence. Amyloid has a very typical ultrastructural appearance that most other protein deposits in tissues do not have.

Biologic Features

Amyloidosis is a naturally occurring condition in animals. It is most correctly used as a generic term, since the appearance of the pale staining material seen in histologic sections can be caused by many different related disease conditions and may be composed of vastly different types of protein fibrils (Kisilevsky 1983). Amyloid has classically been categorized as primary (AL) and secondary (AA), based on the origin of the protein. Primary amyloid is associated with plasma cell tumors and results in accumulations of AL proteins which are fragments of immunoglobulin light chains. AA amyloid is associated with numerous chronic inflammatory conditions and certain tumors and is deposited as a result of production of an acute reactive protein that is either overproduced or not degraded properly. Systemic amyloidosis in nonhuman primates is due to deposition of AA protein and has been associated with chronic diarrhea in several studies. The causes for development of systemic amyloidosis are not completely known. Current research indicates that AA protein begins as an acute reactive protein produced in the liver following an inflammatory condition. This protein, serum amyloid A (SAA), is then circulated throughout the body until it is degraded by reticuloendothelial cells. In normal individuals this process goes on without any detectable deposition of amyloid A. Alterations in the normal sequence of this portion of the inflammatory response resulting in deposition have been investigated. Overproduction of SAA could overwhelm the reticuloendothelial system, resulting in deposits of AA protein. This may occur due to an overstimulation of macrophages by antigens present in chronic infections which could, in turn, release an excess of amyloid-enhancing factor which could cause overproduction in the liver. Also, enzymatic deficiencies in the processing cells of the reticuloendothelial system would result in inadequate degradation of the SAA, ending in tissue deposition. Numerous case reports detail single and multiple cases of systemic amyloidosis in nonhuman primates (Blanchard et al. 1986; McClure 1984; Gribble 1972; Casey et al. 1972; Chapman and Crowell 1977). In most reports, amyloidosis is associated with chronic disease, especially colitis. In a recent study of 314 pigtailed macaques (*Macaca nemestrina*) with amyloidosis, a positive correlation with chronic diarrhea was observed (Slattum et al. 1989a). Amyloidosis can be induced experimentally in mice through injection of casein (McAdam and Sipe 1976). These studies have revealed much about the induction of amyoidosis and the potential for treatment with several agents. Thus far, no drug has been found that will reverse deposits of amyloid in tissues.

Comparison with Other Species

Amyloidosis is common in nonhuman primates, domestic animals, and humans. The amino acid sequence of AA protein is very well conserved between animal species with little variation between humans and mice (Kisilevsky 1983). The pattern and severity of deposition are variable between species and even differ between strains of mice. Amyloidosis in macaques appears to be a much more disseminated condition than in New World and African nonhuman primates. Renal amyloidosis is the most common form of amyloid in dogs (Slauson et al. 1970), but is rarely reported in nonhuman primates (Blanchard et al. 1986; Slattum et al. 1989a). A localized form of amyloidosis occurs in the islets of Langerhans of cynomologus monkeys, where it commonly causes diabetes mellitus (Sheldon and Geiser 1971).

References

Blanchard JL, Baskin GB, Watson EA (1986) Generalized amyloidosis in rhesus monkeys. Vet Pathol 23:425–430

Casey HW, Kirk JH, Splitter GA (1972) Generalized amyloidosis in a rhesus monkey. Lab Anim Sci 22: 587–593

Chapman WL, Crowell WA (1977) Amyloidosis in rhesus monkeys with rheumatoid arthritis and enterocolitis. J Am Vet Med Assoc 171:855–858

Gribble DH (1972) Granulomatous enteritis and intestinal amyloidosis in nonhuman primates. Vet Pathol 9:81–82

Kisilevsky R (1983) Amyloidosis: a familiar problem in the light of current pathogenic developments. Lab Invest 49:381–390

McAdam KPWJ, Sipe JD (1976) Murine model for human secondary amyloidosis: genetic variability of the acute phase serum protein SAA response to endotoxins and casein. J Exp Med 144:1121–1127

McClure HM (1984) Nonhuman primate models of human disease. Adv Vet Sci Comp Med 28:267–304

Sheldon WG, Gleiser CA (1971) Amyloidosis of the islets of Langerhans in a crab-eating monkey (Macaca fascicularis). Vet Pathol 8:16–18

Slattum MM, Rosenkranz SL, DiGiacomo RF, Tsai C-C, Giddens WE Jr (1989a) Amyloidosis in pigtailed macaques (Macaca nemestrina): epidemiologic aspects. Lab Anim Sci 39:560–566

Slattum MM, Tsai C-C, DiGiacomo RF, Giddens WE Jr (1989b) Amyloidosis in pigtailed macaques (Macaca nemestrina): pathologic aspects. Lab Anim Sci 39: 567–570

Slauson DO, Gribble DH, Russel SW (1970) A clinicopathologic study of renal amyloidosis in dogs. J Comp Pathol 80:335–343

Fatal Fasting Syndrome of Obese Macaques

John M. Gliatto and Roderick T. Bronson

Synonyms. Fatal fatty liver-kidney syndrome; obese macaque syndrome.

Gross Appearance

Despite the precipitous weight loss shortly before death in these obese macaques, there are still abundant deposits of body fat at necropsy (Bronson et al. 1982a; Laber-Laird et al. 1987). Areas of fat necrosis can be seen as myriads of white granules embedded in the normally yellow macaque fat (Bronson et al. 1982a). Fat necrosis can be focal to extensive, including most stores of body fat (Bronson et al. 1982a). The liver of these macaques is enlarged with rounded edges, pale, and friable (Bronson et al. 1982a; Laber-Laird et al. 1987). The kidneys are pale and soft. The pancreas may be pale and firm or red if acute pancreatitis is present (Bronson et al. 1982a).

Microscopic Features

There is severe diffuse fatty change in the liver (Fig. 215). The cytoplasm of hepatocytes is usually replaced by single large clear vacuoles. Vacuolation of proximal renal tubular epithelium is moderate to severe (Fig. 216). A few tubules may be necrotic and some contain granular or crystalline material. Vacuoles in both liver and kidney are fat, as demonstrated by positive Oil red O or Sudan IV staining (Bronson et al. 1982a; Laber-Laird et al. 1987). Pancreatic lesions of variable severity are present in the majority of affected macaques. Most areas of necrosis present are focal and small. Entire lobules or portions of lobules are ectatic (Fig. 217). Dilated acini are lined by cells with pale cytoplasm with little or no zymogen granules (Bronson et al. 1982a; Laber-Laird et al. 1987). Some dilated lumina contain floccular eosinophilic material that is PAS-positive, occasionally alcian blue positive (at pH 2.5), and mucicarmine negative (Bronson et al. 1982b). Colonic crypts may be dilated and filled with an eosinophilic material (Bronson et al. 1982a; Laber-Laird et al. 1987). Focal degeneration of myofibers can be seen in the heart (Bronson et al. 1982a; Laber-Laird 1987).

The white material detected grossly as small foci or larger masses in most stores of body fat are recognized microscopically as sharply delimited zones of altered adipose tissue. The affected fat cells are swollen, distorted, and take a grayish granular stain. Some of these cells are replaced by amorphous to roughly granular material, suggesting the formation of soaps. These groups of altered cells are surrounded by macrophages, foreign body giant cells, some leukocytes, and clefts of cholesterol (Fig. 218).

Ultrastructure

The only ultrastructural studies of these obese macaques have been of the pancreas. Ectatic pancreatic ducts are lined by acinar cells and centroacinar cells. Acinar cells have relatively large amounts of rough endoplastic reticulum, zymogen granules, autophagic vacuoles, and clumped nuclear chromatin. Centroacinar cells are electron-lucent and contain few cytoplasmic organelles and a few uniformly sized dark granules. Ectatic lumina contain loosely arranged filamentous material. This material is also present in the cytoplasm of macrophages, acinar cells, and centroacinar cells (Bronson et al. 1982b).

Differential Diagnosis

Hepatic lipidosis needs to be distinguished from fatty liver that can occur as a result of hypothyroidism, diabetes mellitus, or toxemias due to chemical or infectious agents. Some macaques with the fatal fasting syndrome have thyroids of normal size but with moderately to markedly dilated follicles filled with colloid and lined by flattened epithelial cells (Laber-Laird et al. 1987). Whether these histologic changes are associated with an actual hypothyroid state has not been evaluated. Lesions in islets of Langerhans reported in some macaques with this syndrome include hypertrophy of islet cells, deposits of amyloid, hyalinization, and hemor-

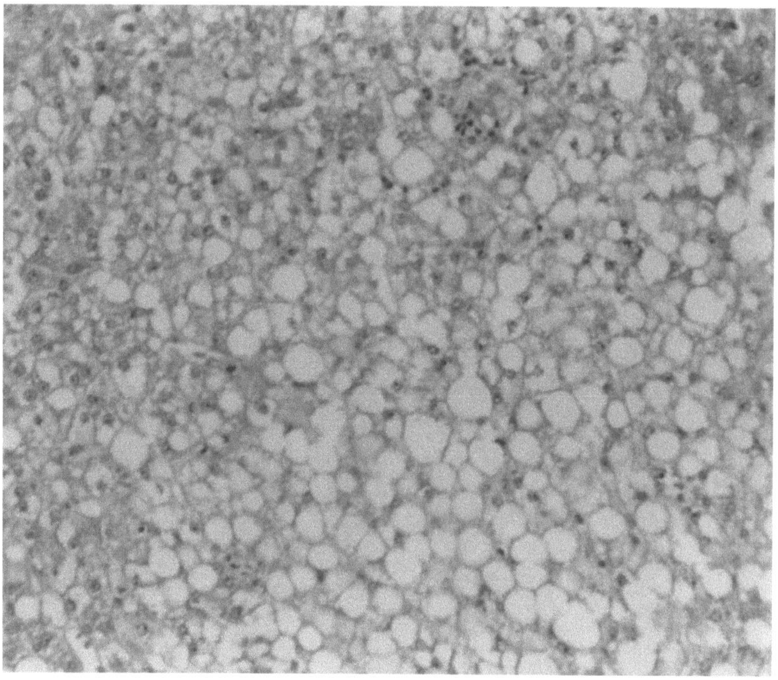

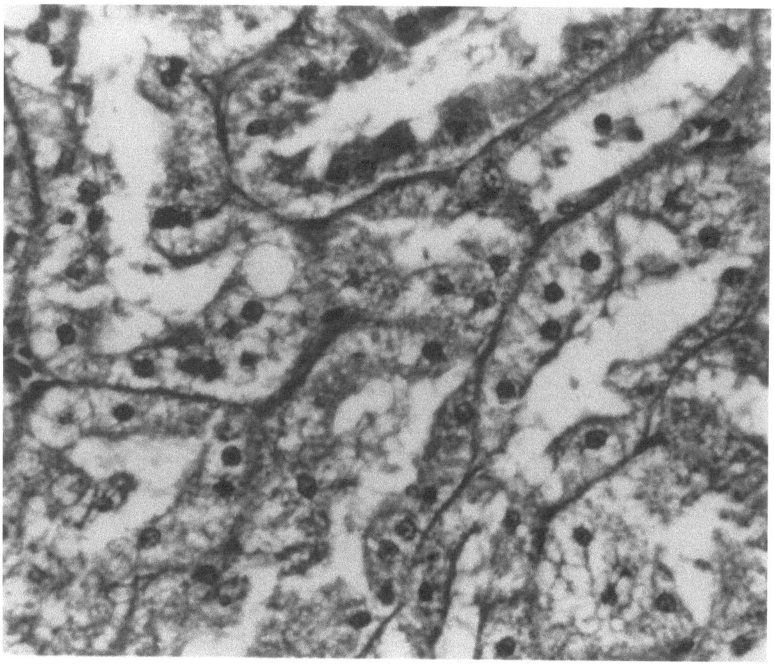

Fig. 215 (*above*). Fatal fasting syndrome, liver, macaque. Note severe, diffuse, fatty vacuolation of hepatocytes. H and E, ×100

Fig. 216 (*below*). Fatal fasting syndrome, kidney, macaque. Many epithelial cells lining proximal tubules contain prominent cytoplasmic fat vacuoles. H and E, ×400

rhage (Laber-Laird et al. 1987), but these lesions are not necessarily associated with the development of diabetes. Glucose intolerance has been seen in some macaques that become spontaneously obese, but glucose tolerance tests have not been reported for macaques that develop the fatal fasting syndrome (Kemnitz 1984; Laber-Laird et al. 1987). Since affected macaques are not seen clinically with signs of severe impairment of liver function, neither toxic nor infectious etiologies are likely (Laber-Laird et al. 1987). The pancreatic ectasia seen in many macaques

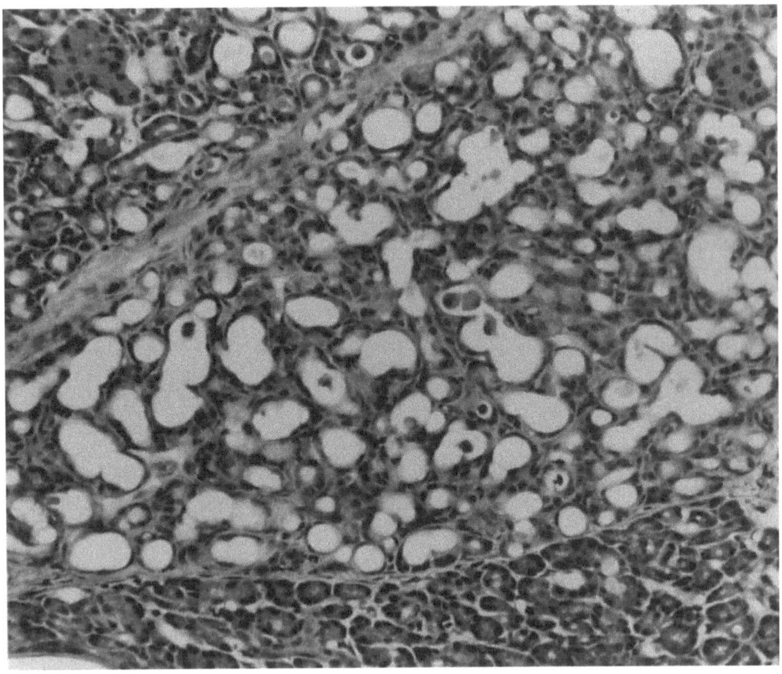

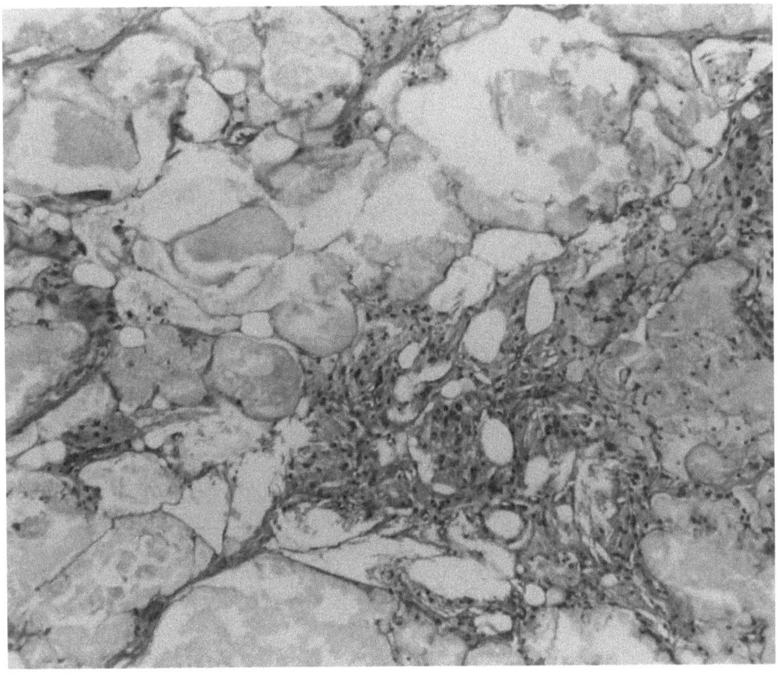

Fig. 217 (*above*). Pancreatic ectasia, pancreas, macaque. A pancreatic lobule contains numerous dilated acini lined by flattened cells. H and E, ×100

Fig. 218 (*below*). Fatal fasting syndrome, subcuticular fat, macaque. Saponified fat is infiltrated with sheets of macrophages. H and E, ×100

with fatal fasting syndrome has been compared to changes seen in cystic fibrosis. Pancreatic ducts in cystic fibrosis contain an inspissated material similar to that seen in pancreatic ectasia. However, in cystic fibrosis the intraluminal material forms concentric and crystalline concretions (Bronson et al. 1982b).

Biologic Features

The disease syndrome is defined as sudden death in obese macaques with no prior indication of illness or following brief periods of anorexia or weight loss of a few days to 2 weeks in length with as much as 20%–30% loss of body weight occurring during the terminal illness (Bronson et al. 1982a; Laber-Laird et al. 1987). In one colony of macaques 42 cases of the syndrome were described over a 10 year period involving four species of macaques (most of which were *M. fascicularis* or *M. mulatta*) and most of which were adult females (Bronson et al. 1982a). In another macaque colony 19 similar cases over an 8 year period were described, mostly involving adult female *M. fascicularis* and accounting for 11% of the deaths of this species in the colony (Laber-Laird et al. 1987). There is no apparent single husbandry practice or disease that accounts for the occurrence of this syndrome. A stress such as recaging is believed to be enough to precipitate rapid weight loss and the disease in some obese macaques (Bronson et al. 1982a; Laber-Laird et al. 1987). The pathogenesis is not understood, but the rapid weight loss implies a negative energy balance which can cause mobilization of free fatty acids from the adipose tissue stores with subsequent fatty change in a number of tissues (Laber-Laird et al. 1987). Elevated clinical parameters are more compatible with renal malfunction than with liver failure despite the diffuse hepatic lipidosis, as blood urea nitrogen and creatinine are consistently elevated. Blood triglyceride levels may also be elevated (Bronson et al. 1982a; Laber-Laird et al. 1987). The pancreatic ectasia is similar to that seen in uremic exocrine pancreopathy (Abu-Alfa et al. 1988) and can be observed in a number of conditions that disturb fluid balance and/or nutrition (Osborne and Butler 1988). The frequency of this obese macaque syndrome relates to the prevalence of obesity among macaques conventionally maintained under conditions of unlimited feeding. Macaques of both sexes and of several species can become obese spontaneously, but obesity rarely occurs until adulthood. In addition to lack of exercise and overfeeding, loss of social structure among macaques may contribute to obesity (Kemnitz 1984). Obesity is much less common among macaques in their natural environment. This may be the result of an undisturbed intact social structure, in which dominant individuals limit food consumption among a group, rather than the result of the more limited availability of food in the wild (Kemnitz 1984).

Comparison with Other Species

Some human diabetics can appear with fatty liver and kidneys, particularly following a serious metabolic disturbance such as ketoacidosis (Edmondson and Peters 1977; Warren et al. 1966). Sudden death may occur in humans with a diffusely fatty liver as the only postmortem lesion (Graham 1944; Randall 1980). Fatty change in other organs such as the kidney is not seen, most of these cases are associated with a history of alcoholism, and these individuals are not described as obese (Graham 1944; Randall 1980). Acute fatty change can be seen in pregnant women, but again this change is not associated with obesity (Breen et al. 1970). Hyperlipemia in ponies has many similarities to the macaque syndrome. Affected ponies are described as having been obese and stressed and have fatty change involving liver, kidney, adrenal gland, skeletal muscle, and myocardium in addition to marked elevation in blood lipids (Gay et al. 1978). Obese cats that become anorexic, perhaps because of stress, develop markedly fatty livers (Rogers and Cornelius 1985). In young chickens, acute fat accumulation in the liver, myocardium, and kidney is thought to be diet-related or may have a genetic basis (Riddell et al. 1971; Hemsley 1965). Pregnancy toxemia and fatty liver in sheep and guinea pigs often occur in obese animals. In guinea pigs acute extensive fatty change involving the liver, kidney, adrenal gland, and lung need not only be associated with pregnancy, since obese virgin females can develop identical lesions as a result of stress and inappetence (Ganaway and Allen 1971).

References

Abu-Alfa A, Ivanovich P, Mujais SK (1988) Uremic exocrine pancreopathy. Nephron 48:94–100

Breen KJ, Perkins KW, Mistilis SP, Shearman R (1970) Idiopathic acute fatty liver of pregnancy. Gut 11: 822–825

Bronson RT, O'Connell M, Klepper-Kilgore N, Chalifoux LV, Sehgal P (1982a) Fatal fasting syndrome of obese macaques. Lab Anim Sci 32:187–194

Bronson RT, Strauss W, Wheeler W (1982b) Pancreatic ectasia in uremic macaques. Am J Pathol 106:342–347

Edmondson HA, Peters RL (1977) Liver. In: Anderson WHD, Kissane JM (eds) Pathology. Mosby, St Louis, pp 1398–1399

Ganaway JR, Allen AM (1971) Obesity predisposes to pregnancy toxemia (ketosis) of guinea pigs. Lab Anim Sci 21:40–44

Gay CC, Sullivan ND, Wilkinson JS, McLean JD, Blood DC (1978) Hyperlipaemia in ponies. Aust Vet J 54: 459–462
Graham RL (1944) Sudden death in young adults in association with fatty liver. Bull Johns Hopkins Hosp 74:16–25
Hemsley LA (1965) The "fatty liver and kidney" syndrome of young chickens. Vet Rec 77:124–126
Kemnitz JW (1984) Obesity in macaques: spontaneous and induced. Adv Vet Sci Comp Med 28:81–114
Laber-Laird KE, Jokinen MP, Lehner NDM (1987) Fatal fatty liver-kidney syndrome in obese monkeys. Lab Anim Sci 37:205–209
Osborne BM, Butler JJ (1988) Pancreatic acinar ectasia and intraoperative needle biopsy. Hum Pathol 19: 480–483
Randall B (1980) Fatty liver and sudden death. Hum Pathol 11:147–153
Riddell C, Olsen GV, Grimson RE (1971) Fatty liver and kidney syndrome in a broiler flock. Avian Dis 15: 398–405
Rogers KS, Cornelius LM (1985) Feline icterus. Compend Contin Educ Pract Vet 7:391–398
Warren S, LeCompte PM, Legg MA (1966) The pathology of diabetes mellitus. Lea and Febiger, Philadelphia, pp 232–235

Vitamin C Deficiency, Old and New World Monkeys

E. Donald Roberts

Synonyms. Skeletal scurvy; "turban head;" nutritional osteodystrophy.

Gross Appearance

The response to vitamin C deficiency differs between the various primate species. The naturally occurring disease in squirrel monkeys (*Saimiri sciureus*) appears to involve mainly the head region, whereas in rhesus macaques (*Macaca mulatta*) the disease involves the long bones, teeth, and ribs. Petechiae to ecchymoses of gingiva, nasal cavity, subcutaneous tissue, and periosteum are basic lesions as vitamin C is essential to the production of collagen fibrils which are common to both vascular integrity and osteoid formation. In the squirrel monkey, the deficiency is manifest by pericranial hemorrhage forming large hematomas. The hematomas give the animal's head the appearance of a "turban" and stretch and distort the facial features (Fig. 219). The wall of the hematoma usually contains bone because the distorted pericranium produces large mantles of new bone (Fig. 220).

Naturally occurring vitamin C deficiency in rhesus macaques is manifest primarily as lameness. Epiphyseal fractures are common at most costochondral junctions, including the ribs (Fig. 221). The callus produced by the fracture gives a first impression of a calcium/phosphate-vitamin D problem; however, further evaluation reveals periosteal separation from the diaphyses of most long bones. The lameness period is usually preceded by gingival hemorrhage, loose teeth, and possibly anemia.

Microscopic Appearance

Table 8 lists the basic microscopic differences between squirrel monkeys and rhesus macaques with skeletal scurvy. The primary manifestation in squirrel monkeys is the pericranial hematomas

Table 8. Microscopic skeletal changes of skeletal scurvy

	Costochondral junction	Femur/tibia humerus	Cranium
Squirrel monkey	Reduced cellularity of chondral growth region; metaphyseal fibrosis	Closure of epiphyses; metaphyseal fibrosis	Multiple pericranial hematomas; pericranial exostoses
Rhesus macaques	Metaphyseal fracture of all costochondral junctions; periosteal hemorrhage	Metaphyseal fracture with periosteal hemorrhage	No change

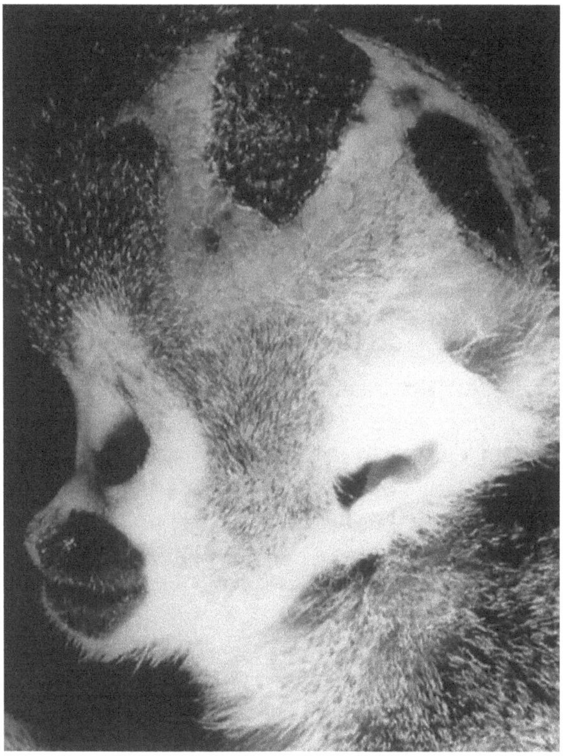

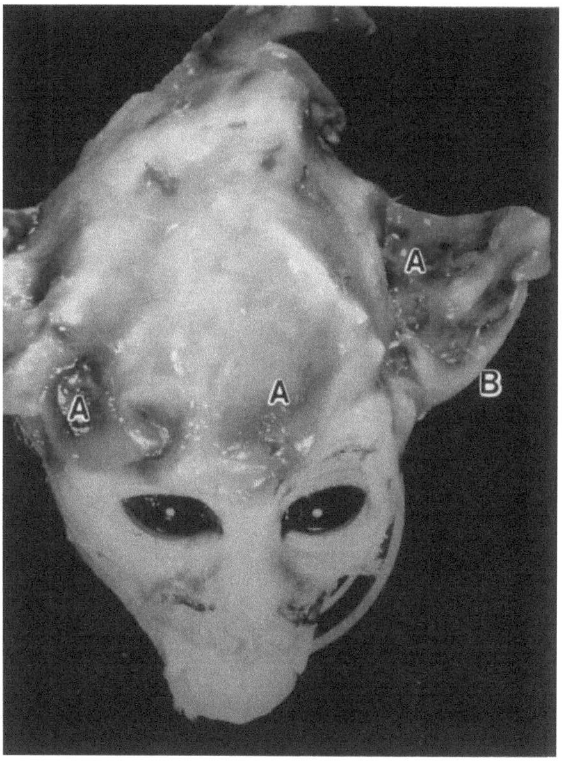

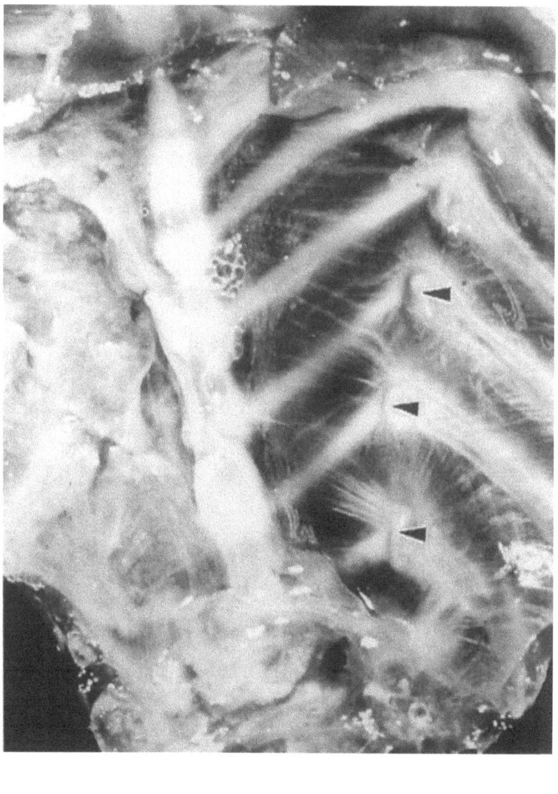

Fig. 219 (*upper left*). Squirrel monkey with naturally occurring vitamin C deficiency. Note the distended cranial region producing a distortion of the facial features. Focal, dark, cutaneous regions represent areas of previous hemorrhage

Fig. 220 (*upper right*). Squirrel monkey will incised hematoma (*A*). Note the areas of exostoses (*B*)

Fig. 221 (*lower right*). Costochondral junction of a rhesus affected with naturally occurring vitamin C deficiency. Metaphyseal fractures (*arrow*) are evident at each costochondral junction

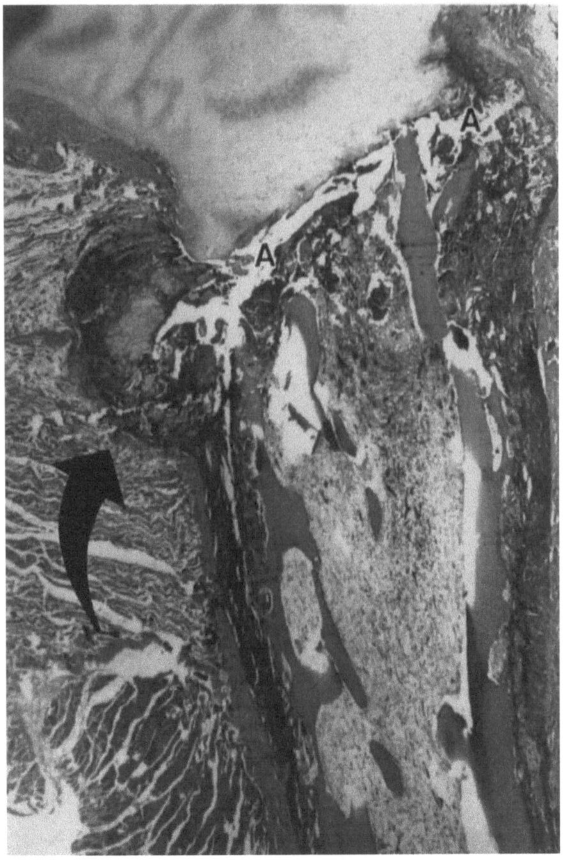

Fig. 222. Costochondral junction from a rhesus with vitamin C deficiency. The gross lesion is shown in Fig. 215. Metaphyseal fracture (*A*) with a callus (*arrow*) is shown. H and E, ×10

varying from 1 mm to 8 cm in diameter. The surrounding tissue changes consist of very loosely arranged fibrillar connective tissue with bone formation in the displaced periosteum. The changes of reduced cartilage cellularity and premature closure are most probably related to inanition associated with the deficiency.

The primary manifestation in rhesus macaques is the metaphyseal fracture (Fig. 222), with periosteal separation affecting most long bones. Changes of the affected periosteum vary from a loosely arranged fibrillar fibroblast to hyperplasia of an immature type fibroblast. Arterial structures are disorganized and the intercellular fibrillar matrix is poorly developed.

Ultrastructure

Diagnostic features of this deficiency in primates have not been reported.

Differential Diagnosis

Any osteodystrophy of rhesus macaques affecting growth plates, a primary calcium/phosphate-vitamin D imbalance, should be considered in a differential diagnosis. Normally such imbalances result in an excess of osteoid formation and, when incised, the osseous tissues are soft. Varying degrees of fibrosis will be present. Vitamin C deficiency does not produce soft pliable bones nor is excess osteoid produced. Epiphyseal plate change will be manifest as a hypocellular poorly organized plate with epiphyseal closure and blunting of the metaphyseal mineralized matrix spicules (Fig. 223). If metaphyseal fractures are part of the lesion, the fracture callus will mobilize cellular resources of the animal to form the callus. Periosteal hemorrhage will be prominent. The oral changes of gingival hemorrhage, fibrous gingival hyperplasia, and loose teeth should not be confused with other nutritional problems (Alvares and Siegel 1981; De Klerk et al. 1973). Hematomas of the head in squirrel monkeys are diagnostic and should not be confused with other nutritional osteodystrophies.

Biologic Features

Vitamin C deficiency is a naturally occurring disease of humans, nonhuman primates, (Blackwell et al. 1974; Demaray et al. 1978; Ratterree et al. 1989), and guinea pigs (Richmond and Stokstad 1969; Machlin et al. 1979; Manning et al. 1984). The deficiency is characterized by hemorrhagic diathesis, defective wound healing, defective bone and teeth formation (Alvarez and Siegel 1981), and anemia (Machlin et al. 1979; Robbins et al. 1984). These signs result from a failure of osteoid, dentine, and intracellular cement formation (Wolback 1937) since each substance contains collagen as an essential component. Those species requiring vitamin C lack cellular L-glulono-γ-lactone oxidase which converts L-glulono-α-lactone to L-ascorbic acid (Nishikimi and Udenfriend 1976). The hemorrhage that is prominent in this disease is related to the structural defects of the vasculature (Friederici et al. 1966).

The naturally occurring vitamin C deficiency in nonhuman primates is the direct result of inadequate vitamin C in the ration. This may result from inadequate storage of the feed or faulty manufacturing of the feed. One outbreak resulted

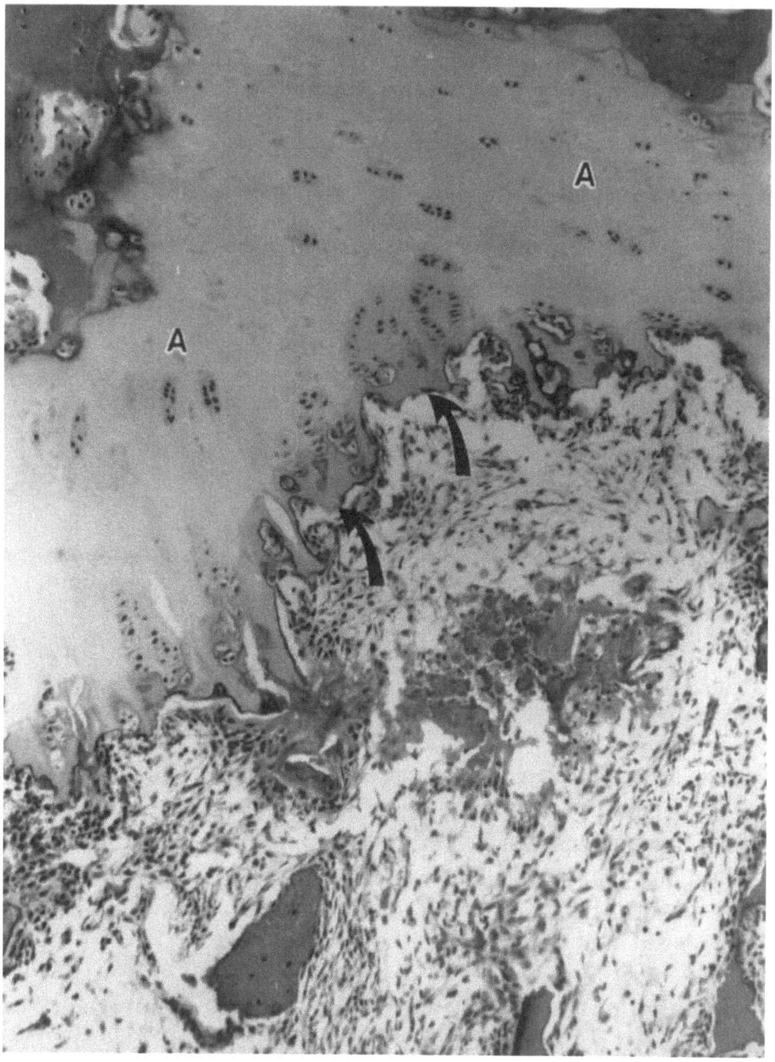

Fig. 223. Epiphyseal plate of a squirrel monkey with typical vitamin C deficiency. Note the chondral hypocellularity (*A*) and blunting of the metaphyseal mineralized spicules (*arrow*). H and E, ×100

from various shipments of feed which contained minimal to no vitamin C (Ratterree et al. 1989). Supplemented fruits provided added protection against the deficiency disease.

Skeletal manifestations of vitamin C deficiency in squirrel monkeys have been well documented (Blackwell et al. 1981; Demaray et al. 1978; Lehner et al. 1968). Cephalohematomas and cranial hyperostosis were observed. In rhesus macaques, experimentally produced scurvy resulted in gingival hemorrhage, oral ulceration, loose teeth, swollen joints, anemia, and inability to move (Ostergaard and Löe 1975; Machlin et al. 1979). The deficiency in vervet monkeys (*Cercopithecus aethiops*) is manifest as fibrous gingival hyperplasia (De Klerk et al. 1973).

References

Alvares O, Siegel I (1981) Permeability of gingival sulcular epithelium in the development of scorbutic gingivitis. J Oral Pathol 10:40–48

Blackwell CA, Manning PJ, Hutchinson TC, Fisk SK (1974) Cranial hyperostosis of squirrel monkeys (Saimiri sciureus). Lab Anim Sci 24:541–544

De Klerk WA, du Plessis JP, Van der Watt JJ, Jager A, Laubscher NF (1973) Vitamin C requirements of the vervet monkey (Cercopithecus aethiops) under experimental conditions. S Afr Med J 47:705–708

Demaray SY, Altman HH, Ferrel TL (1978) Suspected ascorbic acid deficiency in a colony of squirrel monkeys (Saimiri sciureus). Lab Anim Sci 28:457–460

Friederici HHR, Taylor H, Rose R, Pirani CL (1966) The fine structure of capillaries in experimental scurvy. Lab Invest 15:1442–1458

Lehner NDM, Bullock BC, Clarkson TB (1968) Ascorbic acid deficiency in the squirrel monkey. Proc Soc Exp Biol Med 128:512–514
Machlin LJ, Garcia F, Kuenzig W, Brin M (1979) Antiacorbutic activity of ascorbic acid phosphate in the rhesus monkey and the guinea pig. Am J Clin Nutr 32:325–331
Manning PJ, Wagner JE, Harkness JE (1984) Biology and diseases of guinea pigs. In: Fox JG, Cohen BJ, Lowe FM (eds) Laboratory animal medicine. Academic, Orlando, pp 149–181
Nishikimi M, Udenfriend S (1976) Immunologic evidence that the gene for L-gulono-gama-lactone oxidase is not expressed in animals subject to scurvy. Proc Natl Acad Sci USA 73:2066–2068
Ostergaard E, Löe H (1975) The collagen content of skin and gingival tissues in ascorbic acid-deficient monkeys. J Periodont Res 10:103–114
Ratterree MS, Didier PJ, Blanchard JL, Clarke MR, Schaeffer D (1989) Vitamin C deficiency in captive nonhuman primates fed commercial primate diet. Lab Anim Sci 39:101–106
Richmond V, Stokstad ELR (1969) Effect of ascorbic acid on guinea pig skin collagen synthesis: I. Total collagen. J Dent Res 48/5:863–871
Robbins SL, Cotran RS, Kumar V (eds) (1989) Pathologic basis of disease, 4th edn. Saunders, Philadelphia, pp 356–359
Wolbach SB (1937) The pathologic changes resulting from vitamin deficiency. JAMA 108:7–13

Filariasis, New World Primates

Laura V. Chalifoux

Gross Appearance

Filariae are nematodes of the superfamily Filarioidea. Infection by these parasites, filariasis, is a very common finding in necropsies of imported, wild-caught, South American monkeys. However, monkeys born and raised in the laboratory environment are free of these parasites due to the lack of arthropod vectors. Adults of *Dipetalonema* sp. are long and slender filariae found, often in large numbers, in the peritoneal cavity, particularly in squirrel monkeys, spider monkeys, and cebus monkeys. Occasionally a mild chronic peritonitis with fibrous adhesions may be associated with the filariae. The length of the adults may vary depending on the host species. For example, female worms of the species *Dipetalonema gracile* are reported to vary from 16 to 21 cm in a *Cebus* monkey to 50 cm in a black spider monkey (McCoy 1936). The male is considerably smaller than the female and distinguishable by his corkscrew-shaped tail. Adults of other filarids, such as *Tetrapetalonema marmosetae*, may be found in the subscapular connective tissue of tamarins and other species. The adults responsible for many of the microfilariae we have identified in blood smears have not been found, in spite of careful search. Filariasis in macaques, raised in the wild or laboratory, is a rare occurrence in our experience.

Some helpful review articles (Orihel and Seibold 1972) include: South American monkeys (Dunn and Lambrecht 1963) and (Dunn 1968), Southeast Asian monkeys (Dunn and Ramachandran 1968), the gorilla (van den Berghe et al. 1964), and chimpanzee (Orihel 1970). Webber (1955a,b) reviewed the literature on most of the common filarial parasites found in primates in the 1950s. A recent comprehensive review (Toft 1986) covers the filarial and other parasites of primates.

Microscopic Features

The adult female filaria is viviparous and produces larvae, microfilariae, which may be seen in the circulating blood or sequestered in tissues. Large numbers of microfilariae are routinely seen in blood smears of infected New World monkeys. These microfilariae may be distinguished by histochemical demonstration of acid phosphatase (Chalifoux et al. 1973). The naphthol AS-TR method of Barka (Chalifoux and Hunt 1971) precipitates the red azo dye at the site of acid phosphatase activity which is unique for each species of microfilaria.

Eleven distinct types of microfilariae were identified in seven species of New World monkeys from 282 blood smears stained for acid phosphatase. Each type of microfilaria was assigned a letter designation until the larva was identified and the species name determined. These data are summarized in Fig. 224. Micrometer measurements were done on formalin-fixed wet mounts and the morphology of the larvae was studied in thick smears prepared by the method of Fulleborn

Name	Length (μm)	Mean width (μm)	Sheath	Distribution of acid phosphatase activity	
A *Tetrapetalonema marmosetae*	294	3.1	Not present	2 bands area excretory and anal-pores.	
B	121	5.0	Present	3 distinct spots. Posterior close to tip tail	
C *Dipetalonema gracile*	132	4.5	Present	2 spots, band area exc. pore, small spot anal pore. Spots not distinct, body reddish between. Inner body conspicuous.	
D	239	4.3	Not present	Body uniformly red. Anal pore and excretory pore more dense.	
E *Parlitomosa*[a] *zakii*	299	4.4	Not present	Distinct bands anal and excretory pores with swellings. Head red. Pink spot near end tail.	
F *Dipetalonema*[a] *caudispina*	143	5.5	Present	2 distinct spots. Body sometimes reddish between.	
G *Tetrapetalonema*[a] *atelensis*	211	3.9	Not present	Body uniformly dark pink. Dark red spot post, area anal pore.	
H *Tetrapetalonema*[a] *panamensis*	247	5.0	Not presnt	3 distinct spots, anal and excretory pores and head. Body red anterior to excretory pore. Inner body seen.	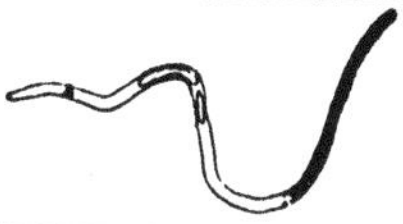
I *Dipetalonema obtusa*	135	3.9	Not present	3 distinct spots, anal and excretory pore and head.	
J *Tetrapetalonema tamarinae*	380	1.9	Not present	2 spots area excretory and anal pores.	
K	217	3.3	Not present	2 distinct spots excretory and anal pores. Head pink.	

Fig. 224. Morphologic and acid phosphatase characteristics of eleven species of microfilaria. [a] Tentative identification

(1929). Four species of microfilariae were specifically identified, and tentative identification was made of four other species. Three additional microfilariae were distinguishable, but their characteristics did not conform with known species and their rarity prevented adequate examination. However, the acid phosphatase staining characteristics of these microfilariae were distinctive. The distinctive identifying characteristics of these microfilariae are:

1. Microfilaria A (Fig. 225) is long and thin, measuring 294 μm long × 3.1 μm wide. It is sheathless and striations are visible. The precipitated red azo dye indicating acid phosphatase activity is confined to two bands

near the excretory pore and the anal pore. From these characteristics, this microfilaria can be identified as *Tetrapetalonema marmosetae*. It has been found in *Saimiri sciureus* (squirrel monkey), *Saguinus oedipus* (cotton top tamarin), and *Saguinus nigricollis* (white-lipped tamarin).

2. Microfilaria B (Fig. 226) measures 121 μm in length × 5.0 μm in width. It has a sheath and both ends are blunt. Acid phosphatase activity is evident in three spots, the posterior spot being close to the end of the tail. This microfilaria was found in only four blood samples obtained from *Saimiri sciureus* and one from *Saguinus oedipus* and has not been identified further.
3. Microfilaria C (Fig. 227) is sheathed and measures 132 μm long × 4.5 μm wide. The tail is sharply pointed with nuclei to the tip. Acid phosphatase activity is seen as a band anteriorly and a small spot posteriorly. The spots are often diffuse and the body quite red between the spots. The inner body is usually very conspicuous with this stain. This microfilaria is found frequently and in large numbers in *Saimiri sciureus* and occasionally in *Ateles geoffroyi*. It has been identified as *Dipetalonema gracile*.
4. Microfilaria D (Fig. 228) has no sheath. It measures 239 μm long × 4.3 μm wide. It stains uniformly bright red with the acid phosphatase substrate. It occurs occasionally in *Saguinus oedipus* and in *Cebus apella*. This microfilaria is unidentified.
5. Microfilaria E (Fig. 229) is 299 μm long × 4.4 μm wide and unsheathed. Distinct spots of acid phosphatase activity occur at the anal pore, excretory pore, and head; a light pink spot is present in the tail. At the points of activity the body appears swollen. It is seen rarely in each species studied except *Cebus* sp. This microfilaria is tentatively classified as *Parlitomosa zakii*.
6. Microfilaria F (Fig. 230) measures 143 μm long × 5.5 μm wide. A sheath is present and often extends a greater distance beyond the tail than in microfilariae B and C. Two distinct spots of acid phosphatase activity are demonstrable. This microfilaria appears similar to B, but the tail is pointed and the posterior spot of enzyme activity is not as distally located. It is found occasionally in *Samiri sciureus*, *Ateles geoffroyi*, *Saguinus nigricollis*, and *Cebus albifrons*. Tentatively it is identified as *Dipetalonema caudispina*.
7. Microfilaria G (Fig. 231) is unsheathed and has obvious striations. It is 211 μm long × 3.9 μm wide. Nuclei continue to the tip of the tail, and the last nucleus is larger than those preceding it. Acid phosphatase activity is uniformly distributed throughout the body. Although uniformly dark pink, this microfilaria is not as intensely stained as microfilaria D. One red spot is seen near the anal pore. This microfilaria is only found in *Ateles geoffroyi*. For this reason, and because of its characteristics, it is identified as *Tetrapetalonema atelensis*.
8. Microfilaria H (Fig. 232) measures 247 μm long × 5.0 μm wide and is unsheathed. Acid phosphatase activity is seen at two distinct spots near the excretory and anal pores, and the head is bright red with a bright pink zone extending posteriorly to the excretory pore. This color seldom extends farther posteriorly with any intensity. The inner body is easily seen with this stain. This microfilaria is identified very frequently in *Saimiri sciureus* and often in *Cebus* sp., but always in very small numbers. It is occasionally seen in *Saguinus oedipus* and *nigricollis* and in *Aotus trivirgatus*. It is tentatively identified as *Tetrapetalonema panemensis*.
9. Microfilaria I (Fig. 233) measures 135 μm long × 3.9 μm wide. It has no sheath and is usually found flexed in a characteristic obtuse angle, especially in wet formalin-fixed preparations. The ends are rounded and the nuclei closely packed. Three distinct spots are evident with acid phosphatase stain: at the head, the excretory pore, and the anal pore. It is seen only in *Cebus albifrons* and is identified as *Dipetalonema obtusa*.
10. Microfilaria J (Fig. 234) is sheathless and similar in appearance to microfilaria A. It averages 380 μm long × 1.9 μm wide. Acid phosphatase activity is seen in two spots near the anal and the excretory pores. It was identified in one *Saguinus nigricollis* as *Tetrapetalonema tamarinae*.
11. Microfilaria K (Fig. 235) measures 217 μm × 3.3 μm. It is unsheathed. Two spots of acid phosphatase activity are seen, near the anal and the excretory pores. Slight activity is evident in the head. This microfilaria was seen in three *Saguinus nigricollis* and has not been identified.

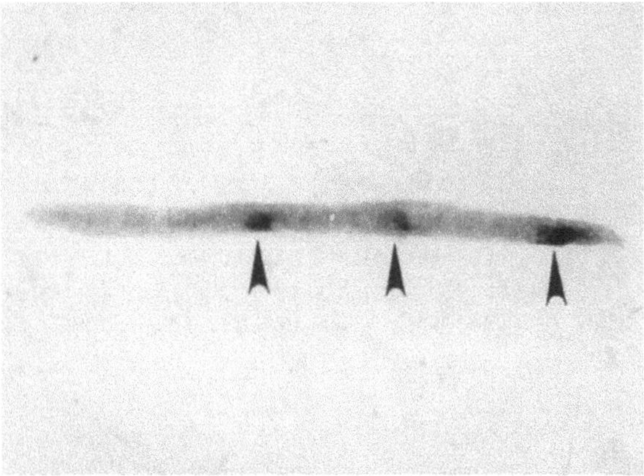

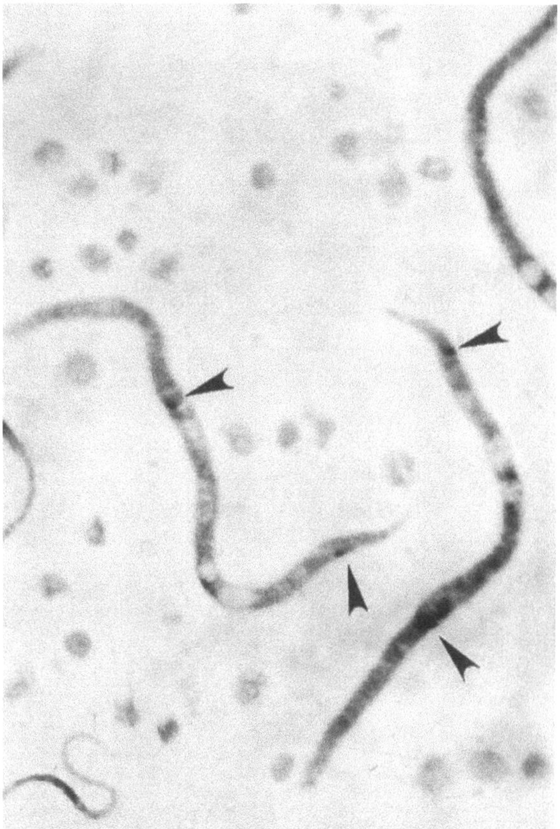

Fig. 225 (*above*). Microfilaria A, *Tetrapetalonema marmosetae*. Acid phosphatase staining at anal pore and excretory pore (*arrowheads*)

Fig. 226 (*lower left*). Microfilaria B, unidentified. Three distinct spots of acid phosphatase staining (*arrowheads*)

Fig. 227 (*lower right*). Microfilaria C, *Dipetalonema gracile*. Areas of enzyme staining (*arrowheads*) vary between specimens

Ultrastructure

No ultrastructural study was done.

Differential Diagnosis

Standard techniques identify microfilariae by measurements and morphology. However, measurements vary with methods of fixation and staining and there is evidence that variations in size occur depending on the host (McCoy 1936). Acid phosphatase staining characteristics are an aid to diagnosis of microfilariae in blood smears. Table 9 provides a key using these characteristics to identify 11 different microfilariae. The differential staining is particularly helpful in distinguishing microfilariae A, D, E, G, and H as they are similar in size and appearance in the usual preparations. Microfilariae E, I, and K all stain at the head, excretory pore, and anal pore. However, their difference in length and general appearance distinguishes them. Similarly, although *T. marmosetae* (A) and *T. tamarinae* (J) stain alike, the greater length of *T. tamarinae* provides a distinguishing characteristic. Microfilariae B,

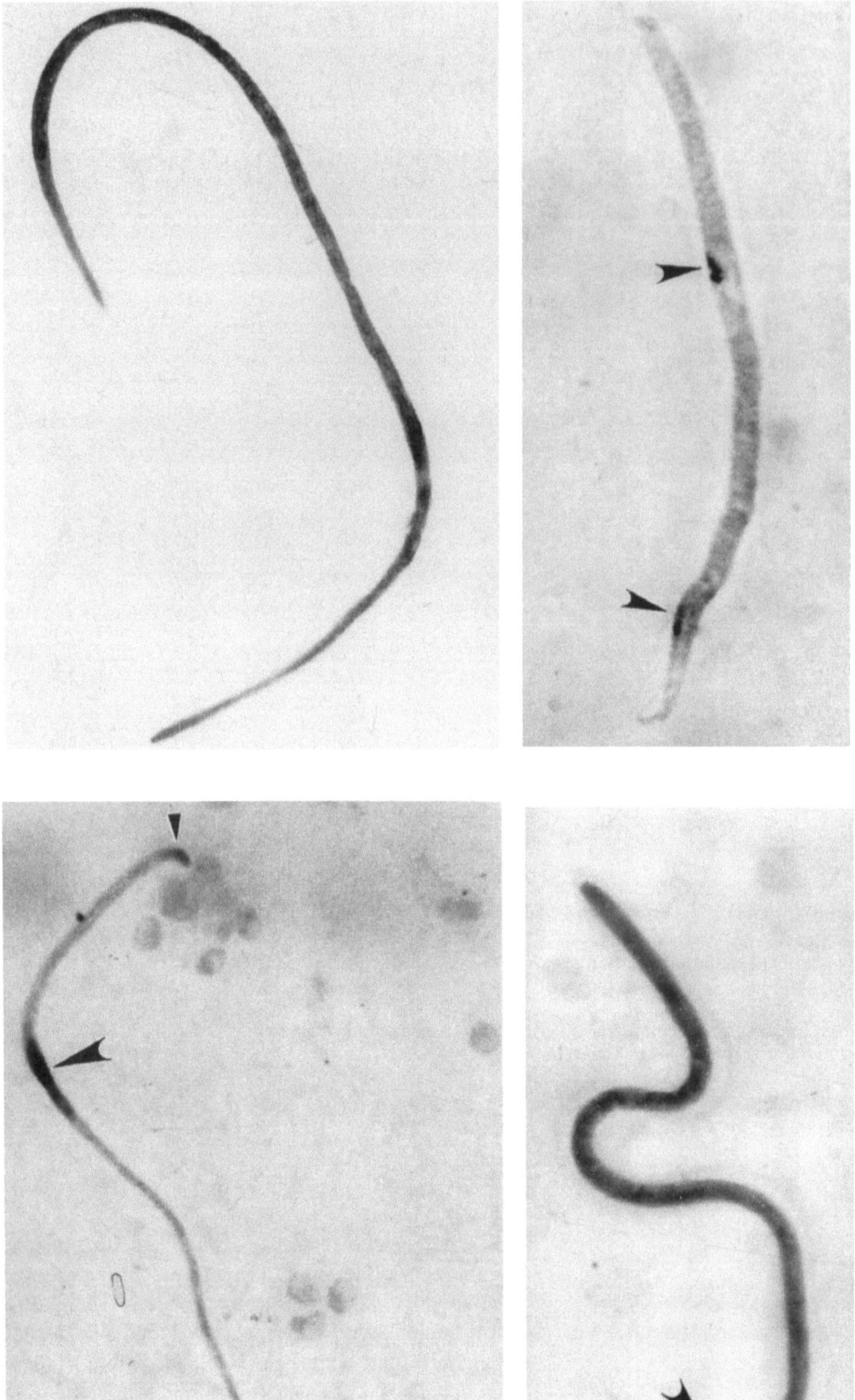

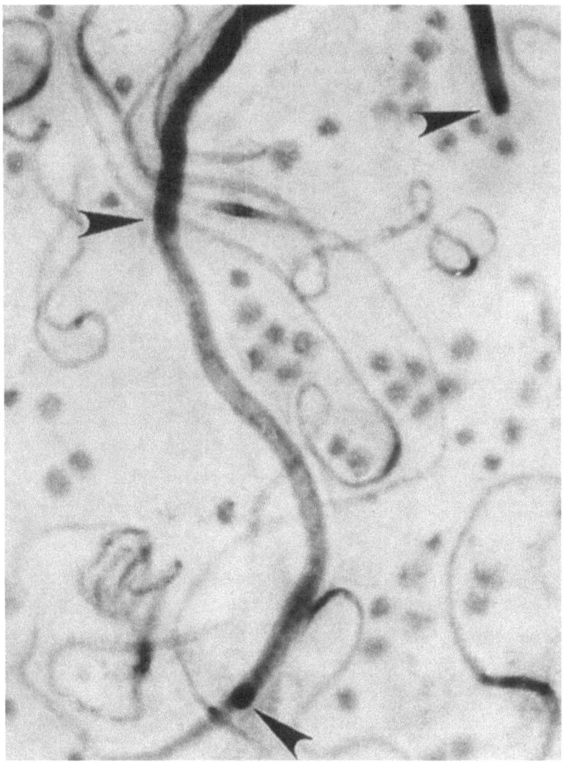

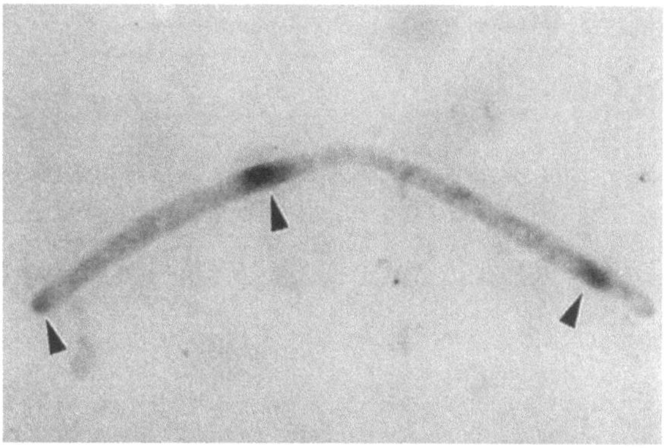

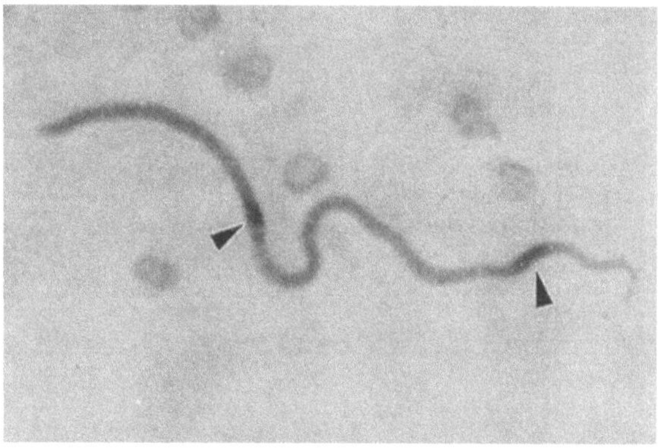

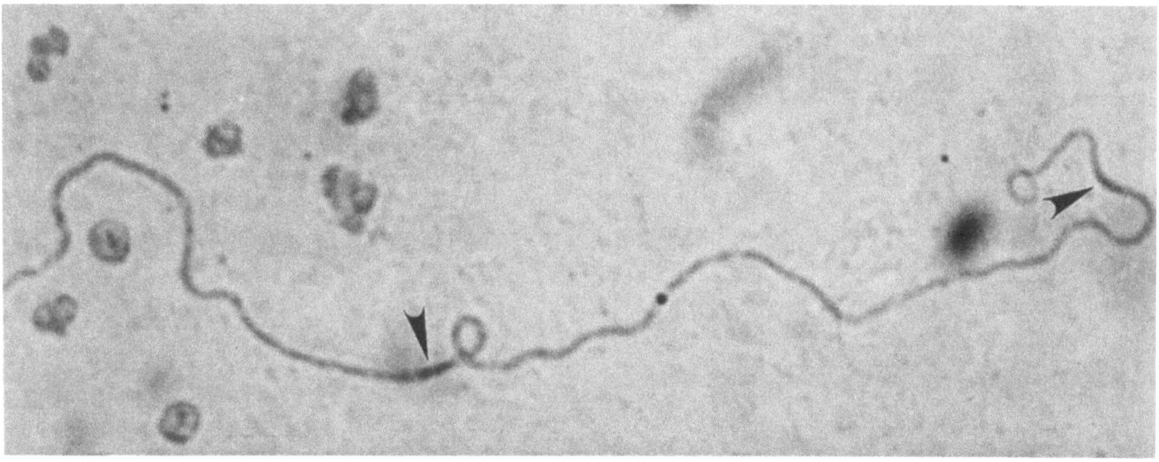

◀ **Fig. 228** (*upper left*). Microfilaria D, unidentified. Body is uniformly stained

Fig. 229 (*lower left*). Microfilaria E, tentatively identified as *Parlitomosa zakii*. Acid phosphatase staining at anal and excretory pores (*large arrowheads*), lighter areas at head (*top*) and tail (*small arrowheads*)

Fig. 230 (*upper right*). Microfilaria F, tentatively identified as *Dipetalonema caudispina*. Two distinct spots of enzyme staining (*arrowheads*)

Fig. 231 (*lower right*). Microfilaria G, *Tetrapetalonema atelensis*. Uniform staining throughout body, but lighter than microfilaria D, darker staining at anal pore (*arrowhead*)

Fig. 232 (*upper left*). Microfilaria H, tentatively identified as *Tetrapetalonema panamensis*. Distinct areas of staining at head, excretory and anal pores (*arrowheads*), and between head and excretory pore. *T. marmosetae* is in background

Fig. 233 (*upper right top*). Microfilaria I, *Dipetalonema obtusa*, in characteristic obtuse angle with staining at head (*left*), excretory and anal pores (*arrowheads*)

Fig. 234 (*below*). Microfilaria J, *Tetrapetalonema tamarinae*. Enzyme staining at excretory and anal pores (*arrowheads*)

Fig. 235 (*upper right bottom*). Microfilaria K, unidentified. Acid phosphatase staining excretory and anal pores (*arrowheads*). Head (*left*) is stained much lighter

Table 9. Key to identification of microfilariae

Sheathed	
Three well defined areas of acid phosphatase activity; posterior area close to end of tail	Microfilaria B
Two areas of acid phosphatase activity; posterior area not close to end of tail	
Approximately 130 µm; spots of acid phosphatase activity not sharply defined; body usually red between spots	*Dipetalonema gracile*
Approximately 145 µm; spots of acid phosphatase activity more sharply defined; body less likely to stain red between spots	Microfilaria F (tentatively *Dipetalonema caudispina*)
Nonsheathed	
Acid phosphatase activity in distinct limited areas	
Acid phosphatase activity at excretory pore and anal pore	
Length around 300 µm (248–330)	*Tetrapetalonema marmosetae*
Length around 400 µm (361–402)	*Tetrapetalonema tamarinae*
Acid phosphatase activity at head, excretory pore, and anal pore	
Length around 135 µm (116–147); bent in obtuse angle	*Dipetalonema obtusa*
Length around 217 µm (208–232)	Microfilaria K
Length around 299 µm (287–323); pink spot tail	Microfilaria E (tentatively *Parlitomosa zakii*)
Acid phosphatase activity more diffuse throughout body	
Body uniformly red	Microfilaria D
Body uniformly dark pink or rose; dark red spot at anal pore	*Tetrapetalonema atelensis*
Body anterior to excretory pore red; dark red spots at excretory and anal pores and head	Microfilaria H (tentatively *Tetrapetalonema panamensis*)

C, and F are the only sheathed types and are therefore distinguishable from the others. There may be some overlap in the staining characteristics of these types. Individuals which have lost their sheaths may stain somewhat differently than those retaining them in the preparation. Some types of microfilariae occur in small numbers, sometimes in mixed infections. Multiple infections occurred in up to 62% of the blood samples examined from each species. Some may also be periodic, found in circulating blood only at certain times, usually nocturnally.

Biologic Features

Natural History

An intermediate arthropod vector is necessary in the transmission of filariae. Species of mosquitoes belonging to five genera are capable of transmitting filariasis in humans (Beaver et al. 1984), and black flies, *Simulium* spp., are involved in transmission of onchocerciasis. Gnats, midges, fleas, and lice may also serve as vectors. The adult female filaria produces a first-stage larva, the microfilaria. In some species the egg membrane is retained and becomes the sheath, while in others it ruptures and the larva is unsheathed or naked. The circulating microfilariae are ingested by a blood-sucking arthropod in which the second and third larval stages evolve. The third stage larva is the infective stage which is then transmitted to the vertebrate host by the arthropod. Many species of microfilariae are periodic; the circulating larvae are present in greater numbers during the time of day or night when the arthropod vector is most likely to feed.

Pathogenesis

Although filariasis may be a serious disease problem in humans and other species, the parasites and their monkey hosts seem to have evolved a relatively compatible relationship. Considering the large numbers of circulating microfilariae found in some monkeys, in our experience they are surprisingly benign. A mild chronic peritonitis or pleuritis may occur as a result of the presence of adult filariae and an occasional granuloma with foreign body giant cells occurs in response to a dead adult parasite.

Etiology

Filariasis is an infection by nematode worms of the superfamily Filarioidea after transmission by an arthropod intermediate host vector.

Table 10. Occurrence of microfilarial infection

Host species	Samples (n)	Positive		Microfilarial type										
		(n)	(%)	A	B	C	D	E	F	G	H	I	J	K
Saimiri sciureus	95	84	88	50	3	48		1	8		50			
Saguinus oedipus	43	25	58	49			7	2			5			
Saguinus nigricollis	11	5	46	9	9				9		18		9	27
Ateles geoffroyi	14	8	57			29		7	14	36				
Aotus trivirgatus	50	4	8			2		6			4			
Cebus albifrons	58	12	21			2	3		2		7	10		
Cebus apella	11	9	82				73				55			

Frequency

Filarial infections were found in each of the seven species of New World monkeys examined. These data are summarized in Table 10.

Comparison with Other Species

An estimated 240 million people are infected with lymphatic filariae in endemic tropical areas and about 40 million Africans have onchocerciasis or river blindness (von Lichtenberg 1989). The presence of adult worms in the lymphatics may produce lymphedema of the scrotum, leg, or other parts and in extreme chronic cases causes the changes known as elephantiasis. Two species are found in the lymphatics: *Wuchereria bancrofti*, found worldwide in tropical areas, and *Brugia malayi*, limited to Southeast Asia. Cats and monkeys can serve as reservoirs of strains of *B. malayi*. *B. malayi* and the morphologically similar species *B. pahangi*, which infects animals, can be differentiated by the acid phosphatase method (Montgomery et al. 1974). Occult filariasis, in which microfilariae are not found, may cause eosinophilia and asthma-like respiratory symptoms. Various species of mosquitos serve as vectors. *Onchocerca volvulus* is transmitted by black flies and occurs in Africa and South America. The adults may produce skin nodules but the microfilariae, deposited into the subcutaneous tissue, may produce dermatitis or blindness. Other filarids may cause dermatitis, eosinophilia, and fever but seldom cause serious disease problems. Zoonotic filariae, such as the dog heartworm *Dirofilaria immitis*, rarely produce lesions in humans.

There are many filarial parasites of animal species (Jones and Hunt 1983), the best known of which is *Dirofilaria immitis*, the dog heartworm. The adults are found in the right ventricle and may cause right heart failure or pulmonary embolism. The circulating microfilariae of *D. immitis* may also be differentiated from the innocuous *Dipetalonema reconditum* by acid phosphatase staining (Chalifoux and Hunt 1971). Diethylcarbamazine or, more recently, ivermectin is efficacious in preventing the development of larvae in the dog. Ivermectin has proved efficacious in the treatment of the larval agent in river blindness and some promise has been shown in the treatment of lymphatic filariae (Ottesen et al. 1990), although there is controversy as to whether the adults are affected.

Acknowledgement. This work was supported by Public Health Service Award RR00168 from the National Institutes of Health.

References

Beaver CB, Jung RC, Cupp EW (1984) Clinical parasitology. Lea and Febiger, Philadelphia

Chalifoux LV, Hunt RD (1971) Histochemical differentiation of Dirofilaria immitis and Dipetalonema reconditum. J Am Vet Med Assoc 158:601–605

Chalifoux LV, Hunt RD, Garcia FG, Sehgal PK, Comiskey JR (1973) Filariasis in New World monkeys: histochemical differentiation of circulating microfilariae. Lab Anim Sci 23:211–220

Dunn FL (1968) The parasites of Saimiri in the context of platyrrhine parasitism. In: Rosenblum LA, Cooper RW (eds) The squirrel monkey. Academic, New York, pp 31–68

Dunn FL, Lambrecht FL (1963) On some filarial parasites of South American primates, with a description of Tetrapetalonema tamarinae n. sp. from the Peruvian tamarin marmoset, Tamarinus nigricollis (Spix, 1823). J Helminthol 37:261–286

Dunn FL, Ramachandran CP (1968) South East Asian filaroids with special reference to those normally parasitic in vertebrates other than man. Proceedings of the seminar on filariasis and immunology of parasitic infections and laboratory meeting, Singapore

Fulleborn F (1929) Filariosen des Menschen. In: Kolle W, von Wasserman A (eds) Handbuch der pathogenen Mikroorganismen 3rd edn, vol 6. pp 1043–1224

Jones TC, Hunt RD (1983) Veterinary pathology. Lea and Febiger, Philadelphia

McCoy OR (1936) Filarial parasites of the monkeys of Panama. Am J Trop Med 16:383–403

Montgomery CA, Redington BC, Jervis HR, Hockmeyer WT (1974) Histochemical differentiation of Brugia malayi and Brugia pahangi. Fed Proc 33:592

Orihel TC (1970) Filariasis in chimpanzees. In: Bourne GH (ed) The chimpanzee, vol 3. University Park Press, Baltimore, pp 56–70

Orihel TC, Seibold HR (1972) Nematodes of the bowel and tissues. In: Fiennes RNT-W (ed) Pathology of simian primates: II. Infectious and parasitic diseases. Karger, Basel, pp 76–103

Ottesen EA, Vijayasekaran V, Kumaraswami V, Perumal Pillai SV, Sadanandam A, Phil M, Frederick S, Prabhakar R, Tripathy SP (1990) A controlled trial of ivermectin and diethylcarbamazine in lymphatic filariasis. N Engl J Med 322:1113–1117

Toft JD II (1986) The pathoparasitology of nonhuman primates: a review. In: Benirschke (ed) Primates. The road to self-sustaining populations. Springer, Berlin Heidelberg New York, pp 571–679

van den Berghe L, Chardome M, Peel E (1964) The filarial parasites of the eastern gorilla in the Congo. J Helminthol 38:349–368

von Lichtenberg F (1989) Infectious disease. In: Cotran RS, Kumar V, Robbins SL (eds) Robbins pathologic basis of disease. Saunders, Philadelphia, pp 307–433

Webber WAF (1955a) The filarial parasites of primates: a review: I. Dirofilaria and Dipetalonema. Ann Trop Med Parasitol 49:123–141

Webber WAF (1955b) The filarial parasites of primates: a review: II. Loa, Protofilaria and Parlitomosa, with notes on incompletely identified adult and larval forms. Ann Trop Med Parasitol 49:235–249

Subject Index*

* Page numbers in **boldface** indicate the principal discussion; Figures are designated by the letter "f" following the page number; Tables are found on page numbers followed by the letter "t."

MIX
Papier aus verantwortungsvollen Quellen
Paper from responsible sources
FSC® C105338

If you have any concerns about our products,
you can contact us on
ProductSafety@springernature.com

In case Publisher is established outside the EU,
the EU authorized representative is:
Springer Nature Customer Service Center GmbH
Europaplatz 3, 69115 Heidelberg, Germany

Printed by Libri Plureos GmbH
in Hamburg, Germany